DISPUTES AND INTERNATIONAL PROJECTS

Disputes and International Projects

DAVID G. CARMICHAEL

The University of New South Wales, Sydney, Australia

A.A. BALKEMA PUBLISHERS / LISSE / ABINGTON / EXTON (PA) / TOKYO

Library of Congress Cataloging-in-Publication Data

Applied for

Cover design: Studio Jan de Boer, Amsterdam, The Netherlands.
Typesetting: Grafische Vormgeving Kanters, Sliedrecht, The Netherlands.
Printed by: Krips, Meppel, The Netherlands.

© 2002 Swets & Zeitlinger B.V., Lisse

ISBN 90 5809 326 3 (hardbound)

Contents

About the author

David G. Carmichael is a graduate of The University of Sydney (B.E., M.Eng.Sc.) and The University of Canterbury (Ph.D.) and is a Fellow of The Institution of Engineers, Australia, a Member of the American Society of Civil Engineers, a Graded Arbitrator with The Institute of Arbitrators and Mediators, Australia and a trained mediator. He is currently a Consulting Engineer and Professor of Civil Engineering, and former Head of the Department of Engineering Construction and Management at The University of New South Wales.

He has acted as a consultant, teacher and researcher in a wide range of engineering and management fields, with current strong interests in all phases of project management, construction management and dispute resolution. Major consultancies have included the structural design and analysis of civil and building structures; the planning and programming of engineering projects; the administration and control of infrastructure projects and contracts; and various mining, construction and building related work.

He is the author and editor of sixteen books and over sixty-five papers in structural and construction engineering, and construction and project management.

CHAPTER 1

Introduction

1.1 OVERVIEW

The nature of major projects is such that there will always be disputes between contracting parties. The source of these disputes variously might be personalities, different opinions, values, desires, needs and habits, performance, insufficient attention to contract documentation, unexpected eventualities, and so on. The parties might be regarded as being in competition. Also, the English language is not precise, and this lends itself to alternative interpretations of the contract. Because of the uniqueness of projects and because project personnel are only human, and when both egos and money are at stake, it is perhaps surprising that disputes are not more prevalent than they are.

Disputes have the potential to convert an otherwise successful project into an unsuccessful one.

Disputes are anticipated with the inclusion of dispute resolution clauses in most standard contracts. And this is sound thinking from an efficiency viewpoint. However, the existence of such clauses might be taken by some as a bad signal; as well, in the euphoria of starting a new project, the anticipation of future disputes may be the last thing considered and not considered well. Cultural differences and changing international circumstances make the need greater for disputes clauses in international contracts.

Disputes are something project personnel will have to face many times during the life of a project and may continue long after a project has ostensibly finished. The Pareto (or the 80:20) rule would suggest that the 'Pareto' disputes and their causes, the 20% having the most impact on the project outcomes, are the ones to which most attention should be paid, though all disputes need to be addressed in the interests of the harmony of a project.

Outline

This book focuses on a number of issues to do with contractual disputes within projects and provides this in an international context:
- It would seem reasonable that a first attempt in managing projects would be aimed at the contracting parties trying to work together and avoiding disputes wherever possible.

Chapter 2 - Dispute Avoidance Practices and Non-Adversarial Projects
Chapter 3 - Trust, Goodwill and Cooperation
Chapter 4 - Formal Cooperation.
• Nevertheless there are situations where disputes will arise in spite of the best of intentions of both parties. A first attempt at resolving such disputes is commonly through negotiation.
Chapter 5 - Negotiation
Chapter 6 - Negotiation — Case Studies I
Chapter 7 - Negotiation — Case Studies II.
The case studies in Chapter 6 only involve one culture. The case studies in Chapter 7 contain more than one culture, and this may have influenced the courses taken by the negotiations.
• Where negotiation fails, other means are sought to resolve the disputes.
Chapter 8 - Methods of Dispute Resolution
Chapter 9 - Disputes and Verbal Contracts
Chapter 10 - Disputes — Case Studies I
Chapter 11 - Disputes — Case Studies II.
The case studies in Chapter 10 only involve one culture. The case studies in Chapter 11 contain more than one culture, and this may have influenced the courses taken by the disputes.
Case studies are used to illustrate the application of the ideas in practice.

This book is a companion book to 'Contracts and International Project Management', by the same author and published by A.A. Balkema, Rotterdam, 208 pp., 2000, ISBN 90 5809 324 7 / 333 6.

Conflict or dispute

... is 'conflict' synonymous with 'dispute'? Although in ordinary parlance the two words are used interchangeably, they are not synonymous. Conflict is in fact the precursor to a dispute. Conflict means an inter-reaction between people who are pursuing incompatible or competing claims. ...

When does a conflict develop into a dispute? ... A dispute starts with conflict, with competing interests — it represents a crisis in the parties' relationship. Conflict in the commercial context is usually preceded by a transaction. A transaction occurs when two or more parties get together and deal — they bargain, they sell, they lease, they buy. For a dispute to arise, the deal has to be perceived to have failed by one party to the transaction. ...

... there is a subjective element to dispute development. Before disputes can arise the injured parties must first perceive that they have been injured ... and secondly they must know that the injury can be remedied. ...

In a study on the process by which a conflict (or in their terms a 'grievance') becomes a dispute, Felstiner, Abel and Sarat called the process 'transformation'. The first step which they identified in a transformation, namely that of 'saying to oneself that a particular experience has been injurious', they call 'naming'. The second step, that of 'attribut[ing] an injury to the fault of another individual or social entity', they call 'blaming'. The third step, that of voicing the grievance to the person or entity believed to be responsible and asking for a remedy, they call 'claiming'. In this transformation process a claim is only finally formed into a dispute when the party to whom it is directed rejects the claim.

(Fulton, 1989)

Alice began to feel very uneasy: to be sure, she had not as yet had any dispute with the Queen, but she knew that it might happen any minute, 'and then,' thought she, 'what would become of me? They're dreadfully fond of beheading people here: the great wonder is, that there's any one left alive!'

Lewis Carroll, Alice in Wonderland, Ch. 8

A *contractual dispute* arises when one party claims something, and the other party rejects the claim, or disagrees over liability either expressly or by conduct. Legally there might also be requirements for communicating the claim and rejection, and disagreeing with the rejection. A claim by itself does not constitute a dispute. Whether or not there is a legal remedy available is another issue.

A *difference* is where two parties fail to agree.

Terminology uses

Terms used interchangeably by writers include:
• Owner, client, principal, employer, developer, proprietor, purchaser.
• Contractor, builder.
• Owner's representative, superintendent, architect, engineer.
Generally, the first term in each of these lists is the one adopted in this book.

Owner–contractor

The book is written largely in terms of an owner–contractor relationship. The notes generally apply to the contractor–subcontractor relationship, and dealings with consultants and suppliers as well. Only where there is a specific need to refer to consultants or subcontractors, rather than the more general term contractor, is such specific terminology used.

Party

A contract is assumed to be between two entities called parties, typically here an owner and a contractor. Any dispute that develops is assumed between these two parties. Only two-party disputes are considered, but there is an obvious extension to multi-party disputes.

Gender comment

Commonly references use the masculine 'he', 'him', 'his' or 'man' when referring to project personnel possibly because the majority of project personnel have historically been male. However such references should be read as non-gender specific. Project management is not an exclusive male domain.

Aphorisms

Many of the aphorisms given in this book are from: A. Bloch 'Murphy's Law Complete', Methuen, 1986; and J. Peers '1001 Logical Laws', Hamlyn, 1981. Aphorisms are given within a double-lined border.

Acknowledgment

The book contains numerous case studies contributed by as many people. Their contribution is gratefully acknowledged.

1.2 WHY DISPUTES?

It has been said axiomatically that conflict cannot be excluded from social life [Max Weber], nor can it be excluded from the substratum of commercial life. ... an underlying assumption of the market economy is that people act out of self-interest. There must therefore be incessant conflict between competing individuals in the market place and disputes both actual and potential are virtually assured in these conditions. ...

'[R. Hellard] In today's industrial and commercial climate few would disagree that good management is the preventative medicine of dispute. But in any relationship brought about by contract, particularly in areas at the forefront of technology such as computers and microchip applications, or in complex operations such as large construction projects, disputes will inevitably arise. There will always be misunderstandings and conflicts of interest between those involved — if only because of the general problems of communication across technical and commercial disciplines.'

A further factor is that business persons when transacting in the market place pay more attention to describing the performance of the object of the transaction, and setting an exchange price, than they do to planning for contingencies or defective performances or to taking steps to ensure the legal enforceability of their bargain. Thus there is always a potential for commercial conflict in the everyday transactions of the market.

(Fulton, 1989)

Two men enter. One man leaves.
Fighting leads to killing and killing gets to warring.
Bust a deal. Face the wheel.
Justice is only a roll of the dice, a flip of the coin, a turn of the wheel ...

"Aunty," Ironbar yelled up to her. "Two men in dispute."
"These our witness, Aunty," Master shouted, waving a hand at the crowd. "Us suffer bad. Want justice. Want Thunderdome."
Entity put her hands on her hips. "You know the law." She shouted the ritual warning. "Two men enter. One man leaves ..."
"Two men enter ..." the crowd around Max murmured, like a sigh. "One man leaves." *(page 66)*

"Listen on ... listen on ..." Dr Dealgood shouted. The crowd's racket began to fall away. "This is the truth of it," he cried, repeating the words that he spoke before every contest, fervent preacher and impassioned teacher and cynical showman rolled into one. "Fighting leads to killing and killing gets to warring. And that was damn near the death of us all ... Look at us now, busted up and everyone talkin' about hard rain. But we've learned. By the dust of 'em all, Bartertown's learned ..." He paused for breath, for effect, lowering his upraised arms. "Now when men get to fighting it happens here. And it finishes here. Two men enter. One man leaves."

The crowd repeated his words like a credo: "Two men enter. One man leaves." *(pages 70, 71)*

And when Max refuses to kill Blaster:
[Entity] flung out her hand, sent an accusing finger at Max like an arrow, "This man has broken the law. Right or wrong — we had a deal. And the law says: Bust a deal — face the wheel." (page 82)

"But aint it the truth," Dr Dealgood cried, "you take your chances with the law. Justice is only a roll of the dice, a flip of the coin, a turn of the wheel ..."

Tweedledummer pointed eloquently toward the roof of Thunderdome. An enormous wheel of fortune hung suspended there, a nearly perfect imitation of the kind once used on television game shows. Civilization might be dead on earth, but its memory lingered on ... The wheel began to drop slowly into the arena. Below it hung a sign: WHEEL OF JUSTICE. *(pages 83, 84)*

Max stood woodenly, staring down at the wheel with dull eyes. A Guard grabbed his wrist, slapping his hand against the edge of the wheel to set it spinning. His choice of fates clattered past him in an unreadable blur; the wheel began to slow again, as the rubber flipper at its edge batted against one raised spoke after another.

The words in their pie-sliced sections grew readable again, passing more and more slowly, as the Wheel settled toward its judgment: LIFE IMPRISONMENT, SPIN AGAIN, FORFEIT GOODS, AUNTY'S CHOICE, UNDERWORLD ...

... Barely moving now, the Wheel quivered as the flipper toyed with ACQUITTAL

... The flipper trembled against the final peg ... clicked once more into [the] next section: GULAG. *(page 84)*

J.D. Vinge 'Mad Max, Beyond Thunderdome',
QB Book/Progress Publications, Sydney, 1985

Former times

| A man's word is only as good as his handshake. |

In days gone by, this was an often quoted saying. Boys were encouraged to develop a strong handshake. Even today, people comment on the handshake of others. The handshake has to be up to a required but undefined 'standard' — for example, 'He has the handshake of a dead fish'. Today, though, someone's good character cannot be gauged any more by the handshake, but rather formal contracts are used to confirm character. The nature of project industries is that contracts are at the very core of business.

In the past, it is suggested, people would engage each other to perform services, and the only evidence of a contract may have been a single piece of paper. In a large number of cases that piece of paper didn't even exist, or for that matter didn't need to exist. The contract would be verbal and confirmed by a simple handshake. This is not the case today, however. Even riding on a train or bus has been formalised contractually. And there are penalties if the person fails to buy a ticket and gets caught. (Such contracts have always existed in law of course, but people never thought of them or treated them as such.)

Today, people enter into contracts for most activities. Yet they are uneasy about the relationships. A complicated document produced for the protection of the parties doesn't remove people's anxiety. This uneasiness is not limited to personal transactions such as buying a house or a car, but occurs in commercial dealings as well.

Modern technology and society allows dealings with a vast range of people, many of whom have never met and spoken face-to-face. This is compared with former village type relationships. The world has become smaller, and dealings are with people, not just around the corner, but on the other side of the world. Personal rapport has disappeared. Some people suggest that this is a main reason why people have found the need to protect themselves through formal contracts, not only at a commercial level, but also at the personal level. In the commercial world, liquidated damages clauses are frequent and bonus clauses are rare.

The increasing scale of projects, beyond the capabilities of a single organisation, where there is a need to outsource components of the project, and specialisation have also contributed. Fragmentation within a project and boundaries between people have increased. Fragmentation and a sectionalised approach to projects have tended to create adversarial attitudes amongst project parties.

Today, litigation has also become an unfortunate but acceptable norm, and litigation and lawyers are all too easy to access. Lawyers look to protect their owner clients through the legally enforceable agreements that are contracts. As well, contractors have improved their professional strength and have become more active in pursuing their rights.

People are still buying the same things, more or less, as previous generations did, but now with increased complication and heartache.

Contractual relationships should not be fundamentally adversarial, but rather cohesive and bind the parties together. Experience has shown that in practice, however, contracting parties take adversarial positions which lead to the erosion of the working relationship and industry ill-feeling. Projects may degrade to a process of fault finding and defensiveness.

Project versus product contracts

In simple contracts for the sale of products, an item is made available for sale, an offer is made for its purchase and, upon acceptance of the offer, a contract is formed, ownership of the item passes, and that is usually the end of the matter. Project contracts differ, principally due to the intangibility of what is being exchanged.

Products that are the subject of day-to-day contracts are available for inspection ('off-the-shelf'), or an almost identical item is available for trial or inspection, or the item is well known to all concerned. For example, goods in a department store are available for inspection, or a service such as the provision of a daily newspaper, is relatively well understood. Even the sale of real estate involves a contract that, whilst the commercial structure of the contract can become complex, the property itself needs only a relatively simple description because what is being sold is available for thorough scrutiny.

Contracts within project industries differ from such product contracts, because what is being contracted for is considerably more difficult to define with sufficient precision for there to be a complete understanding by both parties as to the full extent of what is being contracted. Every project is a one-off prototype, with both foreseeable and unforeseeable problems.

Team thinking

Successful projects would seem to follow from both the owner and contractor working together to produce a single goal — namely the owner's project at the right price, on schedule and with the requisite quality. The success depends very much on the people involved. However too often in projects, the parties take a 'them and us' attitude; the parties become adversaries.

The proposition that successful projects have the parties working together to produce a single goal, is reasonable until the impact of the commercial circumstances of the parties is examined more closely.

Both the owner and the contractor are obliged to look after their own interests. The project is merely a means whereby each party seeks to better its interests, the owner having the benefit of an end-product for whatever purpose, the contractor making a profit as a result of doing the work. It is inevitable that the owner will be keen for completion of the work to the required standard for the least cost. The contractor, on the other hand, is keen for the work to be completed at the greatest profit.

A simplistic view would hold that the two interests have only the remotest chance of being jointly served. A more realistic view would be that the contractor would balance an 'immediate profit' approach with one of 'ongoing profit from future work'. From an owner's perspective, too, there is more to the situation than merely driving for the least cost in terms of what is paid to the contractor. A sensible owner will balance its drives to have the project completed at least payment to the contractor with assessment of the consequences of that pursuit on the quality and timeliness of the work, and subsequent commercial matters such as further projects.

Relationships

> One of the conclusions of the NSW Royal Commission into Productivity in the Build-
> ing Industry research was that project outcomes were heavily influenced by the
> relationships between the parties involved in the industry.
>
> In general, the Commission found that relationships had severely degraded into
> a destructive and costly adversarial approach characterised by mistrust, lack of
> respect and the ever-present threat of litigation.
>
> This adversarial approach has evolved over many years. It often has its roots in
> the earliest stages of project inception with each party developing [its] own manage-
> ment team. Each team then independently formulates its own goals and decision
> frameworks for the project without regard for the other parties' interests or expecta-
> tions.
>
> As a consequence, communication is limited and often non-existent — with the
> result that conflicts are inevitable as paths diverge and expectations are not met.
>
> An adversarial management style then takes over and the goals each party had
> for the project get lost in a proliferation of paperwork and posturing, with the sole
> purpose of guarding each [party's] position during the execution of the project. The
> stage is then set for conflict and, often, litigation.
>
> The bottom line is clear. These current adversarial management relationships
> jeopardise the ability ... of the parties to realise their expectations. The result has
> been increased costs for the Government and declining profit margins for the private
> sector elements of the industry.
>
> This adversarial climate has severely reduced the productivity of the construc-
> tion industry and consequently its ability to achieve the primary goal of [quality
> end-products] on time and within budget.
>
> NSW Government, 1993

Some adversarial attitude is baggage from previous projects and experiences, some is personality driven, but there are other contributing factors. People who got their 'fingers burnt' on one project may take forward suspicion and caution to the next project.

It is believed that even at the starting point of a contractual relationship, be it prequali-fication or invitation to tender, the problems have already begun. The contract represents a fixed term relationship, where the owner 'gets' some product, and the contractor 'gets' paid. The project may be one-off and the owner and contractor may never cross paths again. The success of the relationship is measured by whether both parties 'got' what they wanted, and in time. Long-term relationships are not allowed to develop, and there is potentially no future to hold any intrigue. Public sector owners may be driven (via competitive tendering) to one-off projects for probity reasons. The owner and contrac-tor, knowing that the work is only for a fixed period, see no gain in putting effort into developing a relationship. In fact, such relationships may be seen as getting in the way, because they promote an annoying thing called 'conscience', which interferes with true commercial decision making.

Long-term relationships are seen as beneficial, yet owners commonly are 'one-off' purchasers. The contractor knows that there is not going to be any further business. If there is a future project, then the contractor has to compete with others to win the work, and winning is uncertain.

Much of the onus for initial relationship building falls on the owner, yet there are a number of issues working against putting in this effort, but rather remaining detached from the contractors:

- Only one tenderer will be awarded the work, and the remaining tenderers will have spent money only to be disappointed.
- The owner may have to negotiate with one or more of the preferred tenderers before agreement is reached.
- Probity (at least in the public sector), where it has to be seen or perceived that all tenderers are treated equally and in a transparent fashion.
- Projects contain unknowns, and owners like to transfer risks to contractors.

One manifestation of such detachment leads to, for example, the contractor-annoying practice of charging contractors for the tender documentation. Owner-transferred risks lead contractors to inflate their prices, which annoys the owner, and in turn leads to more risk transfer to the contractors; the contractors are perceived as being unscrupulous.

Some relationship-building however is possible starting with prequalification or an invitation to tender, dealing with inquiries, clarifications, revisions etc. As a counter example, repeatedly asking for new tenders based on updated information, and extending the tendering period alienates the tenderers. This indicates a lack of preparation and consideration on behalf of the owner, and creates a bad relationship between the contractor and owner right from the start. It starts cycles of blame and accusation, and this prevents any meaningful relationship being developed.

There is a challenge for people to treat others how they themselves would like to be treated. This implies openness, trust, mutual respect and relationships that outlive individual projects.

Asked to sum up his teaching, Confucius gave what is known in the West as a form of the 'Golden Rule':
... 'Is there a single word such that one could practise it throughout life?' The Master replied: 'Reciprocity, perhaps? Do not inflict on others what you yourself would not wish done to you.'

M. Thompson, Eastern Philosophy, T. Y. Books, 1999, p. 139

Costing

In project industries, every cent of the owner's dollar is expected to be accounted for in 'bricks and mortar'. Money spent on non-project work, such as developing relationships, is regarded as dead money. Should the contractor offer more than 'bricks and mortar', its price will be higher than its competitors and more than likely will not win the work, or its price will stay the same and it will be regarded as a skimper on quality, as unscrupulous, or silly.

Profit

Contractors make their living by doing work defined by a continuing series of contracts. A contractor's reputation and ability to win future work depends on successfully completing each contracted work parcel. Part of successful completion is that the contractor makes a profit, thereby ensuring the contractor's continued existence. Each project could represent a threat to the continued existence of the contractor, because a significant loss on one project may mean the end of the contractor's business. The situation may be further exacerbated during a business growth phase where the contractor's capabilities and business are expanding. Similar commercial pressures may be on the owner.

In such a climate, it is only natural that people will protect their livelihood, and this may lead to an adversarial attitude.

The competitive nature of project industries influences profit levels and, in turn, the way a contractor goes about its work and looks to claims for extra work, and disputes. Competitive tendering forces low margins.

History

Some project industries have a long history of disputes being common. Projects are rarely completed without some problems occurring. There, employers look for employees with contracts administration and previous battle experience. Adversarial attitudes exist from day one of a project.

Disputes become an accepted way of doing business. There have developed sizeable claims consultant, legal consultant and dispute resolution consultant practices based on disputes. Many people earn a living from disputes.

Example. The construction industry
In the Australian construction industry, not much seems to have changed since at least the 1960s and 1970s. See for example Antill (1975, 1979).

Over the second half of the twentieth century, there has been a movement by public sector owners from using day labour to having work done under contract, for numerous reasons (Carmichael, 2000). This transfer of work has often been strongly resisted by unions and in some cases management. Those supervising contract work, such as engineers and foremen, often had a vested interest in making life difficult for contractors. They were often eager to prove that contractors were 'shifty' and only profit oriented with no regard for providing quality. A view is that most supervisors of contract work in the 1960s and 1970s had an attitude that was anti-contractor and pro the concept of day labour. A view is that there were a reasonable number of unscrupulous contractors operating in that period as well. Contracting was expanding and it drew a lot of newcomers into the industry, some with little financial resources and not much business experience.

In the 1960s and 1970s, contractors were often 'lorded over' by the owner's representative. The general conditions of contract operating at the time were biased against the contractor, and often contractors were not aware of their contractual rights.

As the contracting industry became more sophisticated and more powerful and united, it was able to bargain for more equitable conditions for contractors.

In the 1980s, as day labour forces were increasingly reduced, the balance of power swung to contractors, and the construction industry became very dependent on them.

Over the years, the industry has been characterised by:
- Contractors bidding competitively to win work and then fighting for variation claims in order to make a profit; loophole engineering.
- Unfair documentation biased against contractors; risk transfer to the contractors.
- The use of non-standard documentation.
- Disputes.
- Collusive practices (Gyles Royal Commission into Productivity in the Building Industry in New South Wales, 1992).
- A growth in people making a living from disputes.

It is acknowledged by all that the industry has to be improved; the culture of disputes has to be eradicated (NPWC/NBCC, 1990).

Size

The relative size of the owner's and contractor's organisations can influence behaviour. An owner may be wary of a large contractor, yet treat a small contractor poorly, perhaps even refusing to pay for work done. A smaller and weaker contractor can easily be put out of business through the actions of an owner.

Uncertainty

Many projects are characterised by high uncertainty of events occurring, in turn influencing the success or failure of the projects. No two projects are alike. The inherent unpredictability and risks influence both owner and contractor behaviour in which bears what responsibility, and act as a source for disputes.

Estimating a project's cost and duration is not a precise activity. Similar projects can differ markedly in cost and duration.

Establishing when design or construction is deficient as opposed to being acceptable is not clear cut.

And when something happens on a project, it can be difficult establishing a direct causal relationship. For example the link between management and performance, what caused the delays etc.

Prevention v cure

Contracting and contract documents make little, if any, mention of prevention. Instead they concentrate on prescribing cures. The cure mentality is a serious problem facing project industries. It is a cause of the way people behave in administering contracts.

Unchangeable?

Many of these practices and much of this thinking are integral to society and may be unchangeable. It may be difficult to change contracting parties into showing mutual respect, and practising in an ethical fashion.

1.3 SOME COMMON DISPUTE ISSUES

> *If you dig deeply into any problem, you will get people.*
>   W.J. Wilson, The Growth of a Company, 1966

Some example dispute issues follow; this is in no way a complete list.

Documentation

Anecdotal evidence suggests that the majority of disputes relate to design and documentation issues:
*   *Drawings*
    Late issue of drawings; this may be a result of poor time management on the part of the owner and its consultants. Inadequate information on, or errors in drawings; this may be a quality issue. Lack of information. (Frequent) re-issuing of drawings; this may be in response to the owner's ideas consolidating as the project progresses. Even though the re-issuing of drawings may not delay the project, the contractor may use this excuse to gain extra time and/or money. Frequent amendment of drawings. Design-as-you-go. Poor design.
*   *Documents generally*
    Poorly written documents; this includes insufficient detail and technical requirements. Lack of information. Inadequate documentation. Interpretation of the specification. Lack of a written contract. Precedence of documents. Lack of coordination. Late supply. Errors, omissions, inconsistencies, ambiguities.
*   *Errors in survey/set out information, data provided*

On site

*   *Possession*
    Late or inadequate possession of the site; restricted access may be unintentional or a result of an owner's optimistic program. Lack of owner-supplied facilities.
*   *Latent conditions*
    Site conditions different to that anticipated.
*   *Neighbours*
    Disputes with neighbours; the project impacting on neighbours may lead to injunctions and disruptions.

- *Workmanship and materials*
  Alleged poor workmanship or materials; there may be two opinions as to whether the workmanship and materials satisfy the contractual requirements. Problems with designated materials and sources. Delay in the supply of materials. Defects. Rectification.
- *Subcontractors*
  Nominated subcontractors. Late nomination. Default. Alleged poor workmanship or materials.
- *Inefficiencies*
  Through uncertainties, delays, disruption, cost overruns, poor quality work, ...

Contract scope

- *Variations, extras*
  The question may arise as to whether a change is a variation, and to its payment. Valuing the variations. Large scale variations. Variations may be caused by the owner, designer or other professionals, or promoted by the contractor.
- *Quantities*
  Measurement of quantities. Large quantity changes. Adjustment to quantities and rates.
- *Rise and fall*
  Interpretation.

General administration

- *Communication*
  Communications may not be written or clear. Late claims. Poorly documented and supported claims. Responses to claims.
- *Record keeping*
  When it comes to arguing for or against a claim, records may be found to have not been systematically kept or filed. Records include diaries, technical reports, letters, programs, receipts and discussions.
- *Payment*
  Payment may be late, withheld or insufficient, perhaps in retaliation for perceived poor workmanship. Retention. Variations payment.
- *Progress*
  Suspension of the work. Disruption of the work. Completion. Differences of opinion over the amount of liquidated damages. Management's inability to maintain control of the project; poor management practices. Program changes; changes in order and sequence. Method changes. Delays. Delayed completion. Loss of productivity. Acceleration. Attempts at fast-tracking.
- *Instructions*
  Delay in providing, or failure to provide instruction; the owner may not have programmed its managerial tasks or have systematic contracts administration practices in place. Late approvals.

- *Inspection*
  Failure to make tests. Opening up and testing work.
- *General*
  Unreasonable administration of the contract. Late or inconsistent decisions. Interface and interference problems with others. Adversarial relationships. Problems in contract formation. Role of owner's representative.

External influences (possibly non-contractual)

- *Changes in statutory requirements*
- *Late approvals by outside bodies*
- *Injunction proceedings*
  Proceedings brought by others.
- *Bankruptcy or financial difficulties*
  Including nominated subcontractors.
- *Inclement weather*
  Delays.
- *Strikes*
  Delays.
- *Claims of no real substance to make up for losses*

The cause of the dispute may be associated with a claim within the contract or outside the contract. Those outside the contract may be complicated and take considerable time and effort to settle; settlement may partly rely on the owner's goodwill.

1.4 EXAMPLE DISPUTE SCENARIOS

1. Claims and records

A contractor makes a claim. The owner then discovers that its records are inadequate, and cannot refute the claim. There may be a reluctance by the owner's staff to admit that their documents are inadequate, or that their administration of the contract was not as tight as it should have been.

The contractor is then perceived as trying to 'rip off' the owner, and animosity develops between the owner and the contractor. The owner believes that the contractor is trying to build up enormous profits.

The animosity builds up at the workface, such that each party becomes committed to its own position and less trusting of the other. The dispute becomes harder to settle. An independent third party may be considered to assist the resolution. However if either party wants to fight and does not want to settle a claim, it is headed for a long and costly exercise.

Non-adversarial contracting and dispute avoidance would appear an attractive alternative.

2. Scope management

Scope management focuses on agreeing on the extent of work in the project, and controlling changes to it, to stay within cost and schedule limits. Disputes arise over changing scope.

One particular software project experienced significant problems in scope changes. The initial scope was for a solution based on commercial off-the-shelf applications, with minimal owner input. As time went by, it became clear that it was not possible to integrate over twenty off-the-shelf applications without a larger than expected degree of owner input. Off-the-shelf applications went from comprising 80% of the work to 40% of the work, while owner input went from 20% to 60%. This was as a result of the realisation that off-the-shelf applications were a solution to only the simplest of problems. The vast majority of off-the-shelf applications needed at least some (and in some cases, substantial) modification to incorporate the owner's business processes. Off-the-shelf applications were not the cost and technical risk panacea that the owner had thought they might be.

This change in the technical direction gave fertile grounds for disputes with the project's main contractor. The contractor found it difficult to adapt the design of the system to keep pace with the change in owner content.

These changes in the project's scope required the drafting of a new contract. The new contract included several new provisions for the resolution of disputes. It accepted that disputes were likely to occur in the normal course of a high risk development project. Heavy emphasis was placed on informal dispute resolution mechanisms.

1.5 DISPUTE MANAGEMENT SKILLS

Dispute management skills have application across all industries. A requirement for such skills exists in small contractor companies through to large multinational companies, as well as in the public sector at all levels. Regardless of the size of the project, disputes will arise between the contracting parties and those with which they have dealings. As a general rule, the larger the contract value, the more complex the relevant industry regulations and the greater the degree of technical risk, the greater will be the need for effective dispute management skills.

To manage disputes effectively, it is suggested that, amongst other things, each of the following actions should be addressed:
- Understanding the typical patterns and appearances of disputes.
- Awareness of opinions and alternatives for dealing with disputes.
- Awareness of one's self in terms of typical behaviour in dispute situations.
- Adoption of appropriate attitudes and beliefs to handle disputes.
- Skills development in dispute avoidance, minimisation and resolution in a variety of situations.

Dispute management may be thought of in terms of:
- *Avoidance*
- *Mitigation*
- *Resolution*

This includes elimination, suppression and reduction.

Generally, the preferred practice is to *avoid* or *minimise* disputes wherever possible.

Suppressing disputes. Suppressing a dispute may not settle the dispute; it may remain hidden, and the parties may remain unsatisfied, and possibly hostile under the surface. This might manifest itself in various forms of *undesirable behaviour* either against individuals or an organisation, ultimately being reflected in poorer project performance.

Autocratic people have a tendency to stifle discussion on matters and not to consider the possibly legitimate views of others. This may be indirectly reflected in one party only providing minimally required effort, and not performing in a team spirit.

Where one party ignores a dispute, doesn't take a position on a dispute, pretends to be unaware of a dispute, refuses to deal with a dispute, or doesn't accept a third party's view on an issue, indirect expressions of an unresolved dispute may still occur as in the case where the dispute was suppressed outright.

Dispute elimination. It may be possible to eliminate disputes altogether. This might be able to be done in a number of ways, for example, by carefully selecting people, improving the co-operation and co-ordination between the parties, and getting the documentation right.

Responses

Suppression, elimination etc. may be thought of as examples of responses to the cause or existence of disputes. Responses to the presence of a dispute might be classified as being either *passive* or *active*.

Example *passive responses* include denial, sidestepping and capitulation.

The *denial* of the existence of a dispute could be expected to lead to increased tension. The issue needs to be investigated, discussed and adequately addressed in order to avoid frustration to the party that perceives the issue as important, and a gradual withdrawal of that party's cooperation.

Sidestepping, with a desire to maintain the peace, can have long-term consequences. Failure to confront and deal with problems could result in a 'no go' situation and reduced commitment to project goals.

Capitulation to the demands and threats of the other party may bring an incorrect perception that a dispute has been resolved when in fact it has only been unwillingly suppressed. It is suggested that effort be made to bring the matter to a head, and have it resolved.

Example *active responses* include domination, bargaining and compromise.

Domination is an aggressive response. With a perceived weaker party, unreasonable demands are made and one-sided solutions are imposed on that party.

When people are backed into a corner with threats of excessive additional costs and delays, they tend to retaliate with excessive demands of their own, making resolution a lot more difficult.

One party may be obsessed with winning in an attempt to 'drive a hard bargain'. All this ego driven behaviour usually achieves is the withdrawal of cooperation and support

from the less powerful party. The solution to a dispute then comes, not through cooperation but, through litigation or similar.

Bargaining
Competitive bargaining can result in the opposing party withdrawing cooperation, and
setting about defending its adopted position. Such an environment is not conducive to
obtaining workable solutions. *Cooperative bargaining* might be classed as a creative
response. The emphasis is on identifying creative and workable solutions which can
satisfy the needs, and dispel the fears of the parties involved. The defensive attitude is
removed and a wide range of possible solutions are developed.

Active v passive approaches
Being *passive* might encourage 'industry bullies' to drive even harder, leaving the
weaker party unsatisfied. It distorts the situation in favour of the party which 'over
inflates' its initial demand to end up with what that party considers reasonable for it at
the end, with little regard to the other side. Passive approaches may be effective in the
short term, but not necessarily in the long term.

Of the various *active* approaches, domination might end up forcing a one-sided solution on the weaker party. One party might 'win the battle but not the war'. The other
party may look for opportunities to 'even the score', and that will remove all build-up
of goodwill, lead to a loss of trust, lead to a withdrawal of cooperation, and produce an
environment not conducive to finding workable solutions.

The active approach of 'integrative' bargaining is suggested as being effective. Parties are encouraged to participate, and have a positive attitude which helps in identifying
creative and workable solutions that can satisfy the needs and dispel the fears of the
parties. At the least, applied early enough, it can lead to the parties accepting their losses
and the parties losing motivation to escalate the dispute.

1.6 DISPUTE MITIGATION

Once a dispute arises:

• *Speed*
As soon as is practical, the dispute is brought to the attention of the other party. There is
an early intervention of the project managers of both parties.

Disputes are resolved as quickly as possible and as cost effectively as possible,
causing minimal disruption to the project as a whole (Antill, 1978). The most opportune
time to reach a resolution is when the people involved in the project are still available.
It is generally in the interests of both parties to reduce the time of resolution to a
minimum.

• *Commitment*
Resources, time, and organisation support are committed towards this aim.

• *Cool off*

Initially it might be wise to take one step back from the dispute to cool off, or to remove personalities from the dispute. This avoids the people involved becoming emotional. Emotion can also be used against a person, in later situations.

The issues of the dispute are focused upon and mitigation pursued, rather than reinforcing claims, which prolongs the time to reach agreement.

• *Negotiation*

Informal negotiation by the personnel involved in the project is used to resolve disputes. Genuine discussion and negotiation is encouraged. An agreement need not be reached there and then. Sometimes an extra couple of hours or a weekend may give the chance to review the dispute and come up with a better resolution.

Nevertheless it may be necessary to argue the case when the time comes. If a dispute arises from genuine misunderstanding, it may be beneficial to have a disagreement. This gives both sides of the argument a chance to air their views, and as long as there is a commitment to resolving the dispute, then a reasonable outcome may still be achieved.

If necessary, the dispute is raised to a higher authority level in each party's organisation; concerns are removed over 'lack of authority'. Senior management then makes the necessary commercial judgments.

If necessary, a third party neutral is used to assist the negotiation. The dispute is prevented from evolving into an arbitration or litigation matter, where the environment is more formal and adversarial, the resolution costly, and both parties may be unsatisfied with the outcome.

If negotiation fails, common ground and matters that can be agreed are identified.

• *Dispute course*

The methodology for resolving disputes, in place in the contract documents, is followed, or agreement is made to follow a course of action. An example progressive stepwise approach is:

(i)   Nominated (project) people to address (negotiate) each issue as it arises.
(ii) Dispute referred to senior management from both parties in the hope that a 'big picture' stance of the issues, and the removal of personalities, will see a solution quickly.
(iii) Mediation or similar to be attempted.
(iv) Arbitration or litigation as a last resort.

1.7 OUTCOMES

> *Another victory like that and we are done for.*
>
>   Pyrrhus

> *In a philosophical dispute, he gains most who is defeated, since he learns most.*
>
>   Epicurus

What is best for one party in a dispute outcome is not necessarily best for the other.

The aftermath of a dispute may lead to both *constructive* and *destructive results*. However the continuity effects of a dispute should be noted, in that the outcome of a dispute with its subsequent rearrangements may contain seeds which may mature into another dispute.

Three broad results can follow from the resolution of a dispute between two parties:
- *Win–lose* — the dispute ends with a winner and a loser. The loser may or may not be determined to win the next time they clash.
- *Lose–lose* — nobody gets what they want. Both sides may finish the dispute more determined to win the next time they clash.
- *Win–win* — both parties believe they have obtained what they want. This is the best outcome as it usually ends the dispute.

A dispute resolution which results in a win–lose or a lose–lose result may lead to a continuation of the dispute. That is, the dispute is unfinished. To avoid this, dispute management strives as its goal to achieve a win–win result.

Example

The classic textbook example of a win–win outcome is where two people both want the same orange. A compromise solution cuts the orange in half. Deeper questioning of the reason why each person wants the orange reveals that one person wants the juice to drink, while the other wants the rind to make a cake. Both people can be accommodated in this win–win outcome.

Fisher and Ury, 1981

Of course, finding win–win outcomes in real disputes is not that simple. In fact some commentators, with good reason, are very critical of the simplistic win–win, lose–lose and win–lose terminology, which is pushed quite heavily by conflict resolution consultants.

Example

When one contractor was introduced for the first time to the notion of win–win as a possible outcome of a dispute, his reply was that he was totally in favour of a win–win outcome as long as the other party's outcome was lose–lose.

A *win–win* result might be achieved by using one or more of a number of techniques. These include determining the real (may not be the stated) grounds for the dispute and establishing a situation with which both parties are happy, setting a common goal for both parties, and by introducing an outside party to resolve the dispute from an impartial viewpoint etc.

Compromise

> *Every human benefit and enjoyment, every virtue, and every prudent act, is founded on compromise and barter. We balance inconveniences, we give and take; we remit some rights, that we may enjoy others; and we choose rather to be happy citizens than subtle disputants.*
>
> Edmund Burke, 1775

Some form of *compromise* may be involved unless a solution can be found extra to the matters in the dispute. Compromise involves trying to find some middle ground that is acceptable to both parties. Both parties accept 'part of a loaf' in the interests of the common good. Both parties get something of what they want but not everything; the needs of the parties are partially met; in that sense it is a lose–lose outcome, though perhaps only mildly. The parties can *live with the result*.

> Juhani's Law:
> *The compromise will always be more expensive than either of the suggestions it is compromising.*

The disadvantage of thinking in terms of a compromise is that initial demands may be inflated. A lot of time and resources may be spent arriving at a solution which at best is only partially acceptable to the parties.

It can be distinguished from *cooperation* where the intent is win–win. An industry push for dealing with disputes or issues on projects is to adopt a cooperative approach, however it commonly manifests itself as a compromise approach, with people failing to realise the distinction.

> The 'Stay Cool' Rule:
> *You cannot sink someone else's end of the boat and still keep your own afloat.*

Provided the compromise is reached by agreement of both parties, then the issue in dispute may disappear. There is the potential for a long lasting agreement. If the compromise is forced on the parties, then similar manifestations as occur in suppressing a dispute may occur.

The parties might be encouraged to work together to find a solution, and modify their adopted positions in the interests of the project.

1.8 CHANGES

Following the resolution of a dispute, the parties to the dispute will undergo certain changes. These changes can be summarised as follows:

Changes which may occur within each party

Each party may:
- Be united by the dispute, and close ranks. The members of each party may become more closely knit.
- Become more formal in its behaviour and structure.
- See the other party as an enemy rather than another legitimate entity.
- See only the best parts of itself and only the worst parts of the other party. (The other group does not 'play fair'.)
- Only listen to its side of the argument and tend to ignore the doctrine of the other party.

Changes which may occur in the winning party

The winning party may:
- Become more cohesive.
- Become complacent.
- Become selfish.

Changes which may occur in the losing party

The losing party may:
- Deny the reality of losing by inventing excuses such as the referee was unfair, the rules of the game are ridiculous, or it suffered from bad luck.
- Find a scapegoat to blame for its failure from within its own ranks.
- Become bitter and isolated.
- Eventually regroup and renew the dispute with more bitter feelings.

It is generally accepted that continuing disputes are undesirable and are detrimental to project effectiveness and efficiency.

1.9 DISPUTE ANALYSIS

To resolve a dispute in a proactive rather than reactive manner, it assists if an analysis can be made of the dispute, including exploring each party's needs, wants, interests, power, relationship to the other party, values and emotions related to the issue.

A first step in any analysis is to collect *information* about the dispute and this may involve lateral thinking on top of the immediately obvious material. For each dispute identify the problems and needs of each party, and relevant history.

An attempt may then be made to *assess* or *evaluate* this information.

With this in hand, *options* for resolution may then be considered.

The approach is a measured one rather than being spontaneous, and like in most matters, time spent on preparation is rarely wasted.

> *In all matters, success depends on preparation.*
>
>   Confucius

With any dispute there may be matters on the table (*visible* or *manifest issues*) and hidden agendas (*invisible* or *non-manifest issues*). The non-manifest issues include those that are unconscious and those known to the parties but unstated.

A hidden dimension may be suspected if a relatively minor dispute leads to an extreme emotional reaction, if an objectively minor problem seems incapable of solution, or if a relationship is characterised by repeated minor disputes. ... Attempting to deal with the manifest [dispute] when it is but a symptom of something larger is unlikely to be successful other than in the very short term. [Parties] to a [dispute] may, however, be unwilling to deal with, or even to recognise, the deeper dimension, particularly if consciously or unconsciously they feel that it is overwhelming and threatening.

                                                                    (Tillett, 1991)

Disputes also have a *time dimension* and a present argument may be the result of something in the past or an accumulation of things, perhaps minor, over time.

The *needs and values*, both real and perceived, of the parties, and the *constraints* on the parties have to be recognised. Constraints include financial, time and resource matters, managerial matters and regulations.

Disputes may not exist in isolation. Their *context* (personal, organisational, geographical, legal, ...), is important in understanding them. Disputes, themselves, may be part of larger disputes, and may also have subdisputes.

1.10 DISPUTE RESOLUTION METHODS

The framework for resolving disputes can be as varied as the types of disputes. Nevertheless there are accepted good practices, that hopefully may work but are not guaranteed.

Increasingly the world seems to be making more use of *litigation*. This is worrying many people in the engineering and like professions who are aware of the costs and time involved in legal proceedings. *Arbitration* has always been a favoured alternative to a court hearing in many technical projects. Now there is a further push for *alternative dispute resolution (ADR)* methods such as *mediation* and *disputes boards of review*.

Disputes commonly involve emotive issues, and they become intertwined with the facts. If the two could be separated, many disputes might be readily settled. Employing an independent third party can assist in isolating the emotive issues. Using such third parties, either as judges, mediators or other title, appears to be as old as the occurrence of disputes themselves, or as old as recorded history at least.

Negotiation, of course, has always been used in settling disputes and will continue to be so used. A successful outcome involves reaching an agreement or understanding.

There might be a presupposition that the parties will work together to avoid or minimise the amount of disputation and adopt procedures that will avoid disputes or minimise their adverse effects. But this may not always be the case. Often, the compromise positions of the parties are too far apart, and instead of the negotiation process being fruitful, it often turns out to be an information gathering exercise. The parties are more interested in preserving their legal/contractual rights and entitlements.

Until there is an industry wide realisation that the resolution of a dispute is inevitable, and people act with goodwill and honest intent in the interests of both parties as quickly and efficiently as possible to achieve resolution, then alternative dispute resolution procedures are unlikely to flourish, no matter how desirable. Most court cases settle before they get to trial. And so the question is asked, why spend large amounts of money and time as a prelude to an inevitable settlement that could possibly have been achieved at the outset?

If prior to 'seeking the high ground', adopting a 'legalistic' stance and pursuing formal arbitration or litigation, project personnel were to consciously identify the likely overall cost inclusive of all legal fees, loss and/or distraction of employee time, and the likelihood of a more expensive resolution with the uncertainty of winning, then most disputes would reach that inevitable settlement a lot quicker.

With the benefit of hindsight, very few project personnel who have embarked on or been forced into a major litigation or formal arbitration would be enthusiastic about the prospects of such further involvement.

Cost of a dispute

In seeking a method for resolving any dispute, a full analysis of cost, time and risk should be undertaken; there may even be commercial wisdom in walking away from a dispute and 'cutting your losses'. Properly given advice by legal experts or technical consultants is advisable. The true cost of any action includes any time spent by organisation representatives in carrying out the preparation, and attendance at any meetings or hearings.

The most cost effective approach in many cases is initial dispute avoidance. All forms of dispute resolution cost the project dollars and frequently both parties suffer. There is skill in avoiding disputes and resolving disputes.

The cost of a dispute includes:

Direct costs
- Cost of staff time in preparation and attendance.
- Legal, third party, expert witness and venue costs.
- The time value of money (where there is a long time to settlement).

Indirect costs
- Lost production.
- Delay and time costs.
- Loss of esteem; lower staff morale.
- Loss of trust in future negotiations.
- Loss of goodwill, business relationship and future work; drop in sales.
- Drop in company share value, credit rating.
- Loss of public confidence.
- Lost opportunities.
- ...

Some of these may take many years to recover to their pre-dispute levels. As well, some settlements may take many years.

A contractor known to be dispute-oriented by owners may lose business to other contractors, in a competitive environment.

A project's scarcest resource is commonly time, and a significant premium is attached to it. Time is money. In most industries, staff time is measured in dollars, and when the cost of running the dispute exceeds the value of the dispute, it becomes attractive to simply give in and move on. In some public sector organisations, attaching a dollar cost to the time of project personnel can be at best vague, and the dispute resolution process can often be drawn out and bureaucratic. Few organisations realistically cost a dispute, commonly ignoring many of the indirect costs listed above, and some of the direct costs such as wasted managerial time.

> *Time is waste of money.*
>   Oscar Wilde, Phrases and Philosophies for the Use of the Young, 1894

The cost of resolving a dispute can bear no relationship to the amount being disputed, and there have been many farcical cases where the disputants have ended up arguing over which disputant pays for the resolution, with the original dispute being long since forgotten. All these costs are non-value adding to the project; they increase the cost of the project while adding nothing extra to the finished product. As well, there is all the anxiety and ill-feeling accompanying a dispute. And there is ultimately a cost to the community at large, whether this is through a non-productive activity like dispute resolution, using the public court system, or other.

The amount spent on dispute resolution should take cognisance of the return on the investment.

1.11 NEGOTIATION

Negotiation is the art of reaching an agreement or understanding through bargaining. There are no formal rules, though there are culturally accepted styles. To ensure smooth conduct, and an avoidance of excessively adversarial behaviour, some recommended practices are available.

Negotiation involves a contest between two parties in that each will take a position (or side) and then bargain from its strengths to achieve a mutually acceptable final position, using a mixture of assertiveness and cooperation. The process can be clouded by emotions which may severely impair logic. Negotiation may not be a pleasant experience. Gaining concessions may have to be hard fought with the other party.

Negotiation is driven by discontent. It involves one party wanting something the other has, or two parties wanting something from each other. For a negotiation to proceed, it is necessary for both parties to be able to pressure each other in some way. If one party is powerless, there can be no negotiation.

To be successful, a party to a negotiation should be aware of which areas of the dispute it must remain firm, and which areas it is willing to sacrifice or make concessions.

Negotiation involves:
- Each party presenting its position on a particular matter to the other party.
- Each party then analysing and evaluating the position of the other party.
- Adjustment of each party's position in an attempt to move closer to agreement.
- The establishment of a final position which is acceptable and agreed to by both parties.

During the negotiation, both parties strive to achieve the 'best deal' for themselves. In order to do this, each party needs to be aware of the position of the other party and be knowledgeable of its own position. Because of the competitive aspect of negotiation, the parties to a negotiation may not openly disclose their full position to the other party.

Third party assisted negotiation

Where negotiation between the parties stalls, an independent/neutral third party might be employed to progress the negotiation.

The *stages to the resolution process*, assisted by *a third party neutral* (for example in the *alternative dispute resolution method* known as *mediation*), might commonly involve with respect to the third party:
- Defining the *issues* and *needs* of the parties.
- *Processing* and *exploring* the issues.
- *Resolving* the issues.
- Finalising an *agreement*.

Defining the issues and needs
Defining the issues and needs of the parties involves learning from the parties their views, and the history of the dispute, as well as:
- Defusing any defensiveness or hostility.
- Gathering facts.
- Clarifying discrepancies.
- Identifying the parties' positions.
- Avoiding making judgments.

Processing and exploring the issues
Processing and exploring the issues involves, generally in private discussion (*caucus*) with each party:
- Clarifying facts, priorities and positions.
- Looking for trade-offs.
- Trying to understand the real needs and interests of each party.
- Generating possible solutions.
- Creating doubts to soften each party's position.
- Testing the validity of positions.

Resolving the issues
Resolving the issues involves:
- Searching for areas of common ground.

- Narrowing the differences between the parties.
- Using leverage and external pressures to assist.
- Acknowledging the fall back position if no agreement occurs.

Of course, this is coupled with various tactics that the third party neutral can use to progress the resolution.

1.12 CULTURAL INFLUENCE IN SETTLING CONTRACTUAL DISPUTES

Markets are extending beyond national bounds to become global markets, projects are extending across borders and project players are operating globally. This inevitably brings different cultures into contact.

A *definition* of culture might be: a frame of reference consisting of the assumptions, values, beliefs and traditions shared by a majority of people, usually within a defined geographic area.

Culture is the way of life for a group of people and it encompasses every aspect of living, and thus affects how people think and behave. It forms values, creates attitudes and influences behaviour.

Con Petropoulous (Eric Bana) and Tracey Kerrigan (Sophie Lee) had recently returned from their honeymoon in Bangkok, and were involved with Tracey's father, Daryl Kerrigan (Michael Caton), in a Kerrigan family dinner table discussion at the Kerrigan holiday house at Bonnie Doon (Victoria, Australia), with all its 'serenity'.

Daryl: "I'm curious. Now, I know it's unfair to compare any place to Bonnie Doon, but why would I want to go there [Bangkok] instead of here?"
...
Con: "It's the culture, Daryl. The place is full of culture."
Tracey: "Chockers!"
Daryl: "Yeah?"
Con: "Something for everyone."

The newly married couple then go on to talk about a twenty-four hour kickboxing television channel in the honeymoon hotel, five dollar satay meals, and the cheap purchase price of electrical goods.

From the movie The Castle, Frontline Television Productions, 1997

Different cultures have different approaches to resolving disputes. Cultural differences contribute to the way people interact with each. Disputes involving different cultures can appear very complex. A lot of this is due to misunderstandings and inadequate communication.

Each culture has a different perspective that it brings to a dispute. An understanding of culture, and how it contributes to the dispute resolution process, will assist a person approaching multicultural disputes from an informed and advantageous position.

With an awareness of how different cultures vary, dispute avoidance and resolution might be more achievable.

The following is not intended as a comprehensive treatment of the topic, but rather to alert the reader to some of the issues. The notions of cultural values, cultural dimensions, cultural groups and social and cultural determinants are discussed. The discussion is in general terms, though obviously individuals within a given group may behave differently.

Cultural values

Each culture has values, prioritised according to the importance the society places on them, and developed over time. The values are reflected in work practices and personal life, and are brought to any dispute resolution process.

As an example, in a recent survey, Elashmawi and Harris (1994), gave participants a list of twenty general cultural values that could apply to any country. They were then asked to reorganise the list of values from most to least important. Table 1.1 presents the top ten values listed from Japanese, American and Malaysian respondents.

The results show that each culture could be expected to bring a different set of background values to a dispute resolution. Assuming these results were generally applicable values of each of these cultures, the following conclusions might be drawn.

Japanese find a higher value in belonging to a group, and achieving group goals. In resolving a dispute, they might possibly attempt to achieve group success rather than personnel gain. With their emphasis on relationships and privacy, they are unlikely to be as direct or open during discussions of the circumstances of the dispute. They may try to avoid embarrassment and be as tactful as possible when confronting the issues under discussion. The Japanese may also bring a more patient and less demanding attitude due to their long-term view of relationships.

Americans value freedom and openness of expression. Sometimes their forthright discussion of the exact issues, without reference to group, or social issues, may appear very confrontational to other cultures, but merely reflects their individual values. Personal achievement may be an important issue, and Americans might expect to have direct personal involvement in any dispute resolution process. They may bring a competitive

	Japenese	American	Malaysian
1	Relationship	Equality	Family security
2	Group harmony	Freedom	Group harmony
3	Family security	Openness	Cooperation
4	Freedom	Self-reliance	Relationship
5	Cooperation	Cooperation	Spirituality
6	Group consensus	Family security	Freedom
7	Group achievement	Relationship	Openness
8	Privacy	Privacy	Self-reliance
9	Equality	Group harmony	Time
10	Formality	Reputation	Reputation

Table 1.1. Priorities of values (Elashmawi and Harris, 1994).

attitude and expect input from all levels of responsibility in the dispute due to their value of social equality. A shorter-term viewpoint may be likely.

Malaysians, with their emphasis on family and religion, would bring a different perspective to any negotiation. Cooperation and hospitality between parties would be important.

Asians are said to consider family relationships and friendships more importantly than business, a trait coming from their ancestors. They view keeping these relationships, a good reputation, and loyalty as the essential elements of a successful business. On the other hand, Westerners separate relationships from business.

Many Asian people dislike trouble and wasting time fighting over an affordable loss. Although they may be morally in the right, issues of reputation and relationships arise.

The differing cultural values of the countries provide an explanation of why each culture approaches a dispute in a different manner, and how there will be varying motivations to achieve an outcome.

Cultural dimensions

Underlying all cultures are basic assumptions about the world. Hofstede (1980) calls these *dimensions of cultural variability*.

The dimensions can be used to explain why cultures approach dispute resolution differently, and to define the particular components of a group's culture that contribute to this difference. The following is after Pendell (1995).

• *Context*
Hall (1976) differentiates cultures on the basis of the type of communication that predominates within the culture — high context v. low context. *In high context cultures, the information or content of the message is actually in the physical context surrounding the communication or internalized in the people participating in the communication. Little of the meaning of the message is in the coded, explicit, transmitted part of the message. Most Asian cultures, including Japanese, Chinese, and Korean, are high context cultures.*

In low context cultures, on the other hand, the information or content of the message is in the code. The meaning is found more in the symbols than in the context or the people. Dominant Australian culture, U.S. American culture, and Scandinavian cultures are low context cultures.

High context cultures tend to be indirect and group oriented, whereas low context cultures tend to be direct and individualistic oriented.

• *Collectivism and Individualism*
High and low context communication are the predominant forms of communication in collectivistic and individualistic *cultures, respectively.*

Collectivistic cultures, such as Japanese culture, are dominated by groups. People prefer a tightly knit group framework in which relatives, friends, and coworkers look after them in exchange for loyalty. Individuals are motivated by the need to belong, and there is a definite 'we' v. 'they' orientation. Members of the in-group get better deals than members of the out-group; you employ, do business with and socialize with members of your in-group. Friendships are intertwined with family and business relationships. Group goals and accomplishments are important, not individual ones. And the maintenance of proper forms of behavior and of harmony is strongly desirable; loss of face is more painful than physical pain.

Opposite to these characteristics of collectivistic culture are the characteristics of individualistic cultures, typified by dominant Australian culture and U.S. American culture. Individuals are supposed to take care of themselves and their immediate family. They are motivated by the need for self actualization, and the focus is on 'I'. Business is separate from family and friendships, to a large extent, and people should be treated on the basis of individual worth. Individual goals and accomplishments are important, and openness, directness, and confrontation are virtues.

• *Power Distance*
... the extent to which members of a culture accept that power in relationships is distributed unevenly. High power distance cultures accept significant differences in power and status, whereas low power distance cultures consider such differences undesirable. Following the pattern, Japanese and other Asian cultures tend to be high power distance, whereas dominant Australian and U.S. American cultures are low power distance.

High power distance cultures have visible differences in status and are paternalistic. The norm is loyalty and respect, and the power of a superior is absolute. Age is held in high regard, and wealth is inherited, usually involving ancestral property.

Low power distance cultures tend to downplay status differences and are individualistic. The norm is independence and consultation, and power is negotiated. Age tends to be negatively evaluated, and wealth is earned, involving happiness, knowledge, and love.

A culture's conception of power distance is reflected in all relationships: superior/subordinate, parent/child, and, of course, teacher/student.

• *Uncertainty Avoidance*
... the degree to which members of the culture dislike ambiguity. Asian cultures tend to be high uncertainty avoidance cultures; dominant Australian and U.S. American cultures are low uncertainty avoidance cultures.

In high uncertainty avoidance cultures, behavior is rigidly prescribed by written rules and unwritten social codes. Procedures are standardized; structure is formal; precision and punctuality are required. Deviant ideas and behaviors are not tolerated, and conflict is avoided.

In low uncertainty avoidance cultures, written and unwritten rules are considered more a matter of convenience. What works counts more than what is prescribed; structure tends to be more informal; things are more relaxed. Deviant ideas and behaviors are more frequently tolerated, sometimes even encouraged, and conflict is confronted.

• *Sexrole Differentiation*
... what Hofstede (1980) refers to as masculine/feminine or the way a culture assigns social roles to the sexes. The parallel structure among dimensions breaks down with this one. High sex-role differentiation (masculine) cultures include dominant Australian, U.S. American, and Japanese cultures. Low sex-role differentiation (feminine) cultures include Norway, Sweden, and The Netherlands.

High sex-role differentiation cultures emphasize competitiveness, and rewards are based upon performance. Things are important, as are titles and other symbolic expressions of the importance of one's career. Finally, sex roles are strongly differentiated and concrete.

Low sex-role differentiation cultures emphasize social justice, and rewards are based upon need. Quality of life is important, as are one's work group, friends, and family. Sex roles are not clearly differentiated and are fluid.

General
While I have discussed the five dimensions of cultural variability in terms of opposites, in reality cultures fall somewhere on a continuum between the extremes. Gudykunst and Ting-Toomy (1988) compiled a chart from Hall and Hofstede's work which lists a number of countries and their scores on the various dimensions — countries have relative strength in each dimension. But discussing the dimensions in terms of opposites enables us to see clearly the characteristics of cultures strong in those dimensions.

All relationships and social organizations of a culture are grounded in these five dimensions ...

To Hofstede's and Hall's dimensions of power distance, individualism versus collectivism, masculinity versus femininity, uncertainty avoidance and context, Trompenaars (1997) has added:
- *Neutral versus affective relationships*
  This concerns the way different cultures express emotions. The potential for misunderstanding between neutral and affective cultures is significant.
- *Specific versus diffuse relationships*
  This refers to the balance between family and working life, to people's sense of private and public space and to the degree to which people feel comfortable sharing their feelings with others who are outside their immediate family. Specific cultures tend to keep their private life separate from work and guard it closely.
- *Achievement versus ascription*
  This relates to the degree to which cultures see social status as a birthright or something to be earned through merit.

Cultural groups of the world

Some writers simplify the cultural issue by grouping culture into seven broad groups that dominate the world:
- Northern Europeans.
- East–Central Europeans.
- Arabs.
- Developing Countries.
- Asians.
- Latins.
- Anglos.

It is reasoned that, although there is a great deal of cultural variation in the world, some nationalities are more compatible than others. Of course, within every group there are exceptions to the rule and it is dangerous to stereotype everyone from a particular group into the same mould.

Social and cultural determinants

The forms of conflict resolution chosen by a society will reflect basic beliefs about the nature of that society. The rights and duties of individuals towards one another, towards the community within which they live and towards whatever authority is placed over them; the influence of religion; the relationships between social classes; attitudes towards property: all these will play their part.

More practical considerations will also shape systems of justice. If a nation's communications are slow and poor, a centralised 'King's Justice' will have little effect in isolated communities which will have to devise their own ways of keeping the peace. Isolation is also likely to make communities introspective and determined to maintain on-going stability whatever sporadic conflict may occur.

Contemporary Western cultures can trace their dispute resolution processes back to Graeco-Roman law, strongly influenced by Christianity and more recently by the rise of capitalism. During most of the second millennium AD, most European nation states were physically quite small. All these influences combined to create mostly highly centralised, formalised and coercive societies with forms of justice in both the civil and criminal fields that reflected these characteristics. In the present context perhaps the most important differences within Western law developed between the adversary approach of the Anglo-Saxon common law countries and the inquisitorial approach of the civil law countries. It is significant that the latter group have always had stronger traditions of non-coercive, non-adversary dispute resolution outside their formal legal systems.

A much greater difference can be detected between European and non-European cultures. Asian and Oceanic cultures — non-Christian and (until quite recently) non-capitalist — have almost all been based on harmony, a strong sense of personal dignity and the restriction of overt conflict to precisely defined 'appropriate' situations and forms. In these societies both the formal courts and the less formal community dispute resolution processes have almost always put great emphasis on settlement by agreement or consensus rather than by the imposition of a decision from a third party. (This is the case with civil disputes between individuals or groups within a society: treatment of crime or defiance of authority may not be so harmonious.) In fact, what we are calling alternative dispute resolution has in much of the world always been the means of first choice for most disputes. To a great extent it remains so even though the pressures for more centralised, coercive justice are likely to grow.

(Pears, 1989)

1.13 CASE STUDIES

1.13.1 CASE STUDY — RELATIONSHIPS AND DEADLINES

This case study relates to the management of a road construction project with a foreign-trained contractor. The owner was a public sector organisation.

The construction contract was awarded to a relatively small company with minimal experience in projects within the country. The owner and management staff of the company were all recent immigrants. The company's major roadwork and contracts management experience had been gained in the immigrants' home country.

Relationships

There are a number of relationships within any project that influence contracts management. Culture influences relationships in not only how they are formed, but the expectations that each party has of the relationships. The following is a discussion of some of the relationships within the project and how they were influenced by culture, and of cultural differences and how they affected the resolution of disputes.

Contractor's staff relationships
Each member of the upper management within the contractor's company was in some way blood-related to the managing director. Appointing only family members to responsible positions rather than being based on professional competency resulted in the problem of personnel being appointed beyond their skill level. Non-family members were not given any responsibility, or control of the project, even though they appeared to be better suited for the positions held by family members. This resulted in friction and dissatisfaction within the contractor's company, as some staff had the view that there was no reward for good service if you were not a family member.

Relationship with subcontractors
Some individuals may impose stereotypes and discriminate against people because of their ethnicity. One example involved a subcontractor that would not undertake additional repair work, because a previous experience with a foreign contractor resulted in the subcontractor paying directly for the work. The subcontractor stereotyped the head contractor, assuming that 'all ... are the same'. Another example was a crane operator (a subcontractor) who stopped all cranework on one day, because the head contractor's senior manager could not inspect the work. The head contractor's senior manager was unable to wear a safety helmet due to a religious practice. The owner was required to liaise between the contractor and subcontractor, causing great embarrassment to the contractor. The solution to the problem was quite simple, however the subcontractor used the safety excuse to embarrass the head contractor for other non-related issues.

Relationship with the owner and other organisations
The contractor had difficulty in dealing with the owner due to the contractor's preconceived ideas. These ideas had been based on experiences with government work in its

home country on how a contract should be administered. There were also various problems with the contractor attempting to make deals outside the contractual requirements. It was inappropriate for a public sector body to partake in such deals, because it was accountable to the public, and it also had to act with probity. Additionally the contractor had difficulty understanding the power held by the workers' union within the local construction industry, and the politics of a unionised labour force. This resulted in delays on site as union members stopped work until some issues relating to union fees were resolved.

Meeting deadlines

Another important issue that arose was the perception of the importance of deadlines. The project ran for approximately three times the length of its contractual period. Any step taken by the owner to attempt to communicate the importance of completion of the project on time was not addressed by the contractor. The contractor came from a culture that was relaxed about meeting deadlines.

Exercise
1. Which contractual problems in this case study would you attribute to cultural differences?

2 How important is it for project managers from both parties to understand the impact of different cultures on relationships within the project? How do different cultures lead to different management styles? Once a project manager identifies any cultural differences, is it then possible to implement strategies to minimise or address problems that could result from these differences, or is the best a project manager can do is be aware of the differences, and perhaps understand the reason for the differences?

3. As the contractor was new to the local culture and business practices, could the appointment of family members be seen as one way of increasing potential trust, or insulating it from the local culture?

4. Issues of corrupt behaviour arose, though the contractor didn't perceive it as such, but rather acceptable business practice. Is this a case of the contractor not being aware that it was operating in a different culture, or is there something else of issue here?

5. Should the owner's organisation have been expected to adapt to the culture of the contractor, or the contractor to the owner's culture, or both come half way?

6. Based on anti-discriminatory legislation, it is possibly not allowed to have 'culture' as one of the evaluation criteria in choosing between tenderers. Yet differences in culture between the owner and the contractor contributed to poor performance. This poor performance would presumably exclude the contractor from winning future work, but how do you address cultural issues in tender evaluation?

1.13.2 CASE STUDY — BUILDING LOCATION

A firm decided to build an office/apartment complex in a foreign country. The initial site chosen was occupied by a state-run store. To tear down and relocate the store, a large amount of money was demanded. The foreign firm resisted. The deadlock was only broken when a high ranking politician brought up this matter on a visit to the country. The firm was then given permission to negotiate for a new site. The new site was occupied by a computer-parts factory. Negotiation dragged on for two years, despite the intervention of prominent people from the firm's home country. Finally, the firm engaged a consultant, who had taken part in other inter-country relations negotiations, to represent it. The consultant met with the local mayor, who said he would give instructions that the contract should be signed within one month. True to his words, it was. The computer-parts company became a joint venture partner for the project.

Exercise
1. Is it possible to collectively group a culture as 'Asian', 'African', ... or are such descriptions too broad? Is classifying culture by country, and then have groupings within a country, more informative? Is the culture homogeneous across Asia, Africa, ... or do you have to look at region by region?

2. Why is it that potential investors don't take time to understand the different thinking and culture of their foreign partners, when it would result in a better understanding of each other, reduce surprises, and avoid unnecessary failures that would ultimately lead to commercial disputes? Doing business, not only involves financial matters, but cultural matters as well.

3. People learn how to conduct themselves, how to treat others, how to distinguish between right and wrong etc. from their immediate group. Rules of behaviour make sense within their own cultural boundaries, but can be regarded as strange, illogical, frustrating and sometimes wrong by others. Do these views apply to resolving commercial disputes as well? If so, is this the only explanation as to why different cultures have different approaches to resolving commercial disputes? Is there such a thing as culturally diversified methods of dispute resolution?

1.14 EXERCISES

Exercise 1
What would happen, in the case of a dispute, if no dispute resolution clause was included in the contract?

Exercise 2
How does industry perceive disputes?

Exercise 3
Are parties satisfied with compromises? What is the difference between compromise, collaboration and consensus?

Exercise 4
Suggest ways of obtaining win–win results in disputes.

Exercise 5
How predictable are the changes that occur in groups in going through a dispute?
    Analyse the changes that occurred in the parties to a dispute that you are familiar with. How predictable were these changes?
    Would you expect similar changes if the circumstances were repeated?

Exercise 6
Examine a dispute. Analyse it in terms of parties, problems, needs, issues, history, values, context, ...

Exercise 7
A disagreement on some simple matter may escalate to a ridiculous size. To start, the simple matter may broaden to more general issues, and new issues unrelated to the first matter are raised. Even when the original matter is settled, other issues raised may still exist, personal animosities may still exist and so on.
    Examine a dispute. Describe the stages that the dispute went through.

Exercise 8
Is the analogy with a competitive sporting contest the right one for negotiating? Does one side win and the other side lose? Do people use defensive and attacking tactics?

Exercise 9
It has been said that over 90% of all project disputes are settled through direct negotiations between the parties themselves. For the projects in which you have been involved, how true is this statement?

Exercise 10
In negotiation parlance there are things called BATNA (best alternative to a negotiated agreement) and WATNA (worst alternative to a negotiated agreement). For both parties to a negotiation, where would BATNA and WATNA fit into any negotiation strategy?

Exercise 11
Why is it that as the level of a dispute rises, the parties engage in more and more destructive acts, and constructive interaction becomes less and less possible?

Exercise 12
Suggest some tactics a mediator could use to assist the resolution process.

Exercise 13
Given the same dispute, different cultures will approach it in different ways. For example, some cultures may use strategies from history as a mirror to resolve the dispute; they believe history can help them in their understanding of the dispute (as well as events in their lives). However other cultures tend to resolve disputes with their knowledge and

professional experience, making judgments according to current issues and the current environment.

Does this mean that the outcome of the same dispute will be different if between people from the first culture type or if between people from the second culture type?

Exercise 14
Power distance refers to the way that societies deal with human inequality. Low power distance societies minimise inequalities, while high power distance societies accept inequalities, and institutionalise it by using hierarchical structures (Hofstede, 1980).

How would you expect high power distance to be reflected in the method of negotiation in any dispute? Would you expect a senior member of an organisation to make all the decisions?

Does a low power distance society imply numerous people may be called upon to offer specialist advice, and may in fact feel dissatisfied if all aspects of the dispute are not addressed?

Exercise 15
Individualism 'describes the degree to which individuals are integrated into groups' (Hofstede and Bond, 1988). High individualism societies expect people to act primarily out of their own self-interest, and low individualism societies are those where people are born into, or associate with in-groups, and are expected to act in the interest of that group (Hofstede, 1983). A person from a high individualism society may consider personal achievement as an important part of commercial negotiations.

Would you expect that a member of a low individualism society be more likely to be concerned with group goals, and may not be able to make a decision without consulting other members of the group?

Exercise 16
Confucian dynamism deals with the time perspective of particular societies. A high Confucian dynamism society looks long-term, while a low Confucian dynamism society looks short-term (Hofstede, 1993). Would you expect people from a relatively low-Confucian dynamism society to approach dispute resolution from a short-term perspective, hoping to achieve solutions that will satisfy immediate goals? Would you expect people from a high Confucian dynamism society to take a longer-term perspective in attempting to resolve a dispute, and to be less satisfied with shorter-term solutions?

Exercise 17
Australia is a multicultural country. However, behaviour is still mainly influenced by Anglo-American thinking. Therefore when Australians approach a matter, they are very direct and short-term oriented. Why haven't other immigrant groups influenced Australian culture as much as Anglo-American culture?

Exercise 18
In terms of authority, and relationships between people, some cultures prefer to have as little difference in status between people as possible, with few hierarchical layers; disputes are to be worked out by direct confrontation, and participative approaches to

leadership are best. Conversely, other cultures have a highly stratified set of relationships with formal rules about who can talk to whom and about what, and the accepted norm is for the leader to be like a father or a benevolent autocrat. Does it then imply that, because of this conformity to formality, in times of disputes, such (second mentioned) cultures would like to arrive at a consensus before a decision is taken? And because of this, there appears to be indecision and endless meetings? Therefore must people from the first mentioned cultures have the patience to wait until all those involved agree on how to proceed? Or is this an incorrect implication?

Exercise 19
Some cultures are such that people expect to accept responsibility for their work and to rectify any problems using their own initiatives; they expect to be rewarded personally for their achievements and to receive feedback on their performance. This component of working culture is well suited to the task and output focus of project management, but how might it run into difficulties in more collectively oriented societies? Would individuality in resolving disputes run counter-productive to a communal perspective of some other cultures? Does this explain why some people (from the first mentioned cultures) wonder why such second mentioned cultures appear indirect and why their behaviour seems to avoid saying no in resolving disputes? Why do people (from the first mentioned cultures) interpret this as being unassertive or even misleading, while for the second mentioned cultures disagreeing directly with somebody may be seen as rude and disrespectful? Should therefore people (from the first mentioned cultures) find discreet ways of giving feedback, and collectively rather than individually addressing people from the second mentioned cultures?

Exercise 20
One needs to have patience when dealing with some cultures; cultures that have been around for thousands of years have taught that things will always change, and that life holds endless possibilities that are better awaited than forced. This is to be compared with other cultures which believe that events are controlled through planning and respecting deadlines. Things have to be ordered. Time is something tangible, it is limited and can't be wasted. How has such thinking contributed to different approaches to dispute resolution? Does it mean that some cultures will resolve their disputes through an emphasis on establishing personal relationships and informal contact? How important is it to create an atmosphere of trust, built by patiently spending time on social contact? How might informal social events be more productive in resolving disputes than the officially scheduled meetings that are favoured by other cultures?

Exercise 21
Some cultures tend to like everything spelt out in detail, assuming that unless things are stated they become woolly and a source of disputes. Because of this, they require the use of contracts, with each aspect of the agreement stipulated in writing, and if any disputes arise, they have the authority of the courts to deal with them. Does this necessarily mean that such cultures are adversarial because they concentrate on areas that may cause disputes? Does this necessarily imply a litigious culture, or are there other contributing factors? Can such cultures adapt to taking a different perspective, and

focus on common project interests? How can a balance be achieved between spending sufficient time to build contractual infrastructure to provide security and a way of working, with time building relationships, that can be tapped to assist in the resolution of disputes and realise a project's goals?

CHAPTER 2

Dispute Avoidance Practices and Non-Adversarial Projects

2.1 INTRODUCTION

It is often said that 'prevention is better than a cure'. So it is with disputes. Disputes are non-productive and can be stressful, and the parties may not get, in the end, what they want or think they deserve anyway.

> *In this world, there are only two tragedies. One is not getting what one wants, and the other ... is getting it.*
>
> Oscar Wilde

People are aware of the benefits of non-adversarial projects, however projects frequently reduce to having 'them and us' attitudes between contractor and owner, and a win–lose scenario. Generally there are advantages in avoiding disputes compared to assuming disputes are inevitable and administering the work accordingly.

A major dispute on a project has the potential to be extremely detrimental to the project's end result, with respect to completion times, cost, quality and rapport between the parties. The notion of 'nipping it in the bud' is very relevant. Dispute avoidance has the potential to save money and time, and prevent a lot of bad feeling and loss of future business relationships. Dispute avoidance is the ideal goal. Should a dispute not be able to be resolved, then third parties (often with little technical background) have to be paid large amounts of money to give opinions, awards and/or judgments which may turn out to be disagreeable to both parties. Dispute avoidance is something to be practised from start to finish of a project.

Where possible, the best method of resolving a dispute is to remove the basic cause of the dispute. This course of action, however, may not always be easy, because the parties tend to interpret their rights and obligations very differently, and time is required to appreciate these respective positions.

Of course, it may suit one party to encourage disputation, if for example that party is stalling for time or wishing to extend a project. Dispute avoidance in such circumstances is obviously not applicable. And there is the view that disputes are inevitable, and it is only a case of minimising disputes or maximising avoidance.

When owners are evaluating tenders, an important evaluation criterion is the level of disputation that the tenderers had on previous projects (even though the disputation may not have been caused by the tenderers). Accordingly, short-term gain from adversarial contracts management may in the long term lead to pain, with the owner overlooking that tenderer's proposal and favouring a contractor less prone to taking issues to dispute.

All disputes cost the project money (and often time) without adding value to the project. The time and money spent on resolving disputes is counterproductive and could be better spent on developing dispute avoidance practices. Energy can be redirected to concentrating on the project work, rather than being wasted on disputes. Whilst some disagreements are unavoidable, it is essential that both parties realise the stage at which the project begins to suffer.

Experience with past disputes and their causes can assist in avoiding similar situations in the future. Over time, over a large number of projects, and considered from the viewpoints of the various project stakeholders, it is possible to reflect on how different approaches impact on the overall success (or otherwise) of projects and how effectively disputes can be minimised.

There are a number of procedures and practices that can be adopted to minimise adversarial relationships developing. One particular dispute minimisation formula will not produce the perfect result in each and every contract because there are too many variables to be taken into account, but a line of thinking can be developed.

Learning

Information on good and bad practices, problems and solutions should be taken forward to future projects. Both the owner's and contractor's organisations become learning organisations.

Lael's Law:
Hindsight is always 20/20.

Fagin's Rule on Past Prediction:
Hindsight is an exact science.

General

The relationship that develops during the course of a project is invariably influenced by personalities, egos and money, and as a result the ambience on some projects will be adversarial in nature. Differences of opinion, different goals and motives, unclear guidelines, misconceptions and errors and a breakdown of personal relationships add further fuel.

In almost all cases, the ability to avoid or diminish disputes is within the hands of the contracting parties. Adopting some common sense practices will help. The minimisation of the occurrence of adversarial relationships can be achieved, amongst other things, by:

- The education of the parties.
- Careful selection of consultants and contractors; qualifications and motives.
- Adequate resourcing of projects with experienced professionals.
- Careful attention to documentation, and its on-time delivery.
- Ensuring that contracts administration procedures are in place that minimise or avoid the generation of disputes.
- Good communication, understanding and rapport between the parties.

Up to 80% of all construction litigation is a result of the personality types that are all too common on a construction team — the low bid owner, the change order contractor, and the ivory tower designer.

The low bid owner *wants more than it is prepared to pay for, and will accept the low bid without any vetting, not taking into account that the bidder may have discounted something to come in below a fair market price, or may have made an estimating error.*

The change order contractor *will chase variations and perhaps even price the job, based upon the assumption that there is some exploitable ambiguity, inconsistency, mistake, ...*

The ivory tower designer *is not prepared to entertain suggestions to vary from an impractical design, even though this approach may cost the contractor much and not deliver any improved facility to the owner.*

B. Symons, Civil Engineering, February 1993, pp. 68, 69

The range of *experience* of individual owners and contractors can vary greatly, from those which have successfully completed many projects, to those which are embarking on their first foray, and accordingly knowledge of their rights and obligations also varies markedly.

Knowledge about contracts also varies greatly, with many people knowing very little. Staff are quite often 'thrown in at the deep end' without education in administering contracts, learn 'on the job' and often pick up bad practices and wrong information. Poor practices are followed because the people 'don't know any better'.

Teamwork

Experience shows that it is a much more pleasurable work environment where a team approach is encouraged to bring about the project's goals. The owner's requirements with respect to time, cost and quality are met, the contractor makes a profit, and both parties are happy to do it again should a future opportunity arise. Projects with a 'them and us' attitude put strain on all involved. As time goes on in projects, there can be a tendency for each party to concentrate more and more on its own goals, and give less and less concern for the other party's goals; each party focuses on short-term issues and benefits. For example, an owner may be interested in minimising costs, while the contractor in, say, a cost plus arrangement, may be interested in maximising costs. Competing interests between the parties cause conflicts.

... there is the tendency to speak of the owner's [representative] and the contractor as being 'on opposite sides of the fence'. A far better attitude is to consider that there is no need for a fence at all, then they will both be in the same paddock, with a common [goal]: to see the project grow successfully under their joint guidance.

(Antill, 1979)

On larger projects, a project logo may be developed and used to generate a feeling of shared responsibilities.

Attitudes

Good relationships between contracting parties develop around the parties having mutual respect for each other, and being aware of and respecting the rights and obligations of the other party. That is, attitudes are important.

Mistrust can give rise to adversarial attitudes. Each party suspects that the other is trying to maximise its benefit at the expense of the other. One effect of the suspicion may be a retreat by the parties to a strict enforcement of the contractual words, and the threat of legal enforceability of these words. Each party must feel that its interests are protected, it is not exposed and more importantly that should there be some exposure, this will not be exploited by the other. Partnering, and like thinking, attempts to establish a trusting, cooperative environment.

Project stakeholders

Contractors compared with owners
Contractors are often regarded as *the cause of conflict*. They are seen as being shrewd business people, greedy, using underhanded tactics, misleading, the ones that bite the hand that feeds it if there is a profit in it, and attempting to cut as many corners as possible, and generally a breed that should be watched and controlled with a big stick. There is mistrust and a belief that the work ethic of 'near enough is good enough' is driven by the contracting industry culture and commercial pressures.

Contractors are seen to consciously underbid competitors, and this gives the owner a false impression that the tendered price bears little resemblance to the cost. Contractors are assumed to take the art of *'claimsmanship'* one step further by setting out not only to recover legitimate unforeseen costs through claims, but also allocate their effort and resources to build strong claims so that they may tender at a loss, but make a profit on the claims.

> *Example*
> An owner, against recommendations, accepted a bid worth three quarters of the esti-mated project cost. Immediately the owner's representative took an approach of always looking out for 'tactics' that the contractor might use in order to make up the cost by claiming for extras.

Contractors are assumed to engage in *'programmanship'*. This is where a contractor deliberately submits an overly optimistic program, thereby placing the contractor in a situation where it can more readily claim for disruption due to owner-caused delays.

On the other hand, *owners* are seen by contractors as wanting a 'Rolls Royce' product for the price of a 'Mini Minor', and making life as difficult as possible for the contrac-tor.

Both these perceptions of contractors and *owners* have an element of truth in them. Such impressions and lack of trust influence the way contracts are written and adminis-tered, and hint at why solid working relationships may not exist on many projects.

Some owners think that they are 'employing' the contractor, since the owner is the one paying for the work. It is when the owner takes the 'employer' attitude, then there is an 'employee' response from the contractor. The emergence of an 'employer–employee' attitude leads to conflict, in common with much in industrial relations.

Many *owners* are unable to totally describe what they want, and in protecting their interests, quite often become overly cautious of contractors. At the same time, the con-tractor's interests of maximising profits and minimising exposure, conflict with the own-er's interests.

Consultants

Other stakeholders such as consultants, and their prides, often cause adversarial relation-ships to occur. Often consultants portray an arrogant attitude and appear to consider themselves above contractors. They may also see contractors as adversaries. Credit is not given where credit is due. They design in 'ivory towers' and do not become involved closely with the construction. They believe that what is shown on the drawings or in the specification is the only and best way, when a suggested approach from the contractor deserves consideration. On the other hand, consultants might be able to offer solutions to contractors' problems because of their close involvement with the design. By keeping an open mind, consultants can help to keep projects running smoothly.

Professionals have a strong pride in their work, and will defend it against all attacks. When a mistake is encountered in the contract documents, the professional will com-monly attempt to down play the mistake to the contractor.

> *Example*
> A mistake was discovered in the specification, but no immediate attempt was made to reveal this mistake to the contractor. The consequences of this were costly, as the owner's engineer and contractor argued their cases.

> The specification called for concrete aggregate to have a minimal quantity of deleterious secondary minerals, because of a splash-zone salt water environment, where swelling and shrinking could cause cracking and allow salt water penetration and steel corrosion.
>
> The specification read:
>
> 'The secondary mineral content must not exceed five percent of the total aggregate mass. Of this, no more than two percent must be clay.'
>
> That is, no more than two percent of five percent (or 0.1%) should be clay. Obtaining locally an aggregate such as this was an impossibility. What was intended was that no more than two percent of the total aggregate was to be clay.
>
> The project had already begun, but the contractor had yet to locate a source of aggregate. Under the contract, it was the contractor's responsibility to locate an aggregate source and send samples to a laboratory for testing.
>
> The contractor was unable to find an aggregate that complied with the specification. The owner's engineer attempted to locate a source of aggregate, but eventually concluded that an aggregate such as specified, could not be found, and nominated an aggregate source, close to the project work, that the contractor could use.
>
> Much time and money was spent by both parties trying to deal with a mistake in the specification.

Many owners are reliant on consultants to get a working end-product. It may not be the cheapest option, because the consultant is not the one paying the bill and there are liability issues. Rather, the consultant protects its reputation by using a bit of padding, a 'belt and braces' approach. The consultant's goals do not match the owner's goals, and the contractor's goals are different again.

Owner's representative
In conditions of contract, a person (variously named as owner's representative, superintendent, engineer, architect, ...) is usually given a certifying role. Initial claims from the contractor are channelled through this person. This person's role as a certifier requires him/her to exercise judgment, rather than, for example, recording an indisputable quantity. Even measurement of progress requires judgment. Typically the conditions of contract require this person to act impartially in the treatment of claims from the contractor. Therein lies a conflict of interest — impartiality in the treatment of both owner and contractor versus being an employee of the owner, or consultant employed by the owner. Rulings (or failure or refusal to give rulings) on delays, variations etc. find the owner's and contractor's interests opposed. Even if unintended, and despite the integrity of this person, a perception of bias still exists. ('Justice must not only be done, but be seen to be done.') Contractors naturally believe that this person makes determinations in the interests of the owner, and hence show no trust in decisions made. This conflict of interest, or perceived conflict of interest, promotes adversarial thinking and can be the start of many disputes. The petty rejection of 'or equivalent' substitutions for materials and products fuels the contractor's negative attitude towards the owner's representative.

The initial claim by the contractor and its treatment is the start of a dispute. Having a person with a conflict of interest escalates a claim assessment into now being a dispute to be resolved.

Example

On an international project, a major problem resulted from the inflexible attitude of the owner's representative towards the conditions of contract. The contract was written by the owner and was heavily biased towards it. One clause was particularly onerous for the contractor. This clause required that, following the issue of a variation, the contractor must notify the owner's representative of the value of that variation within ten days. If the contractor failed to give such notice within this period, the contract stated that no additional payment could be claimed for the variation work.

Many of the variations that were issued during the course of the work had major additional cost implications. Also these variations involved the contractor in obtaining materials and components from various sources throughout Europe; for example marble from Italy, and wall cladding from Germany. Consequently the period for obtaining base prices from suppliers could be as long as four or five weeks. The contractor was unable to submit its price for the variations within the ten days as stated in the contract. This was explained to the owner's representative.

It was not evident until later in the project that the owner's representative would strictly impose the conditions of contract. This meant that the contractor did not receive any additional payment, because it had not submitted the estimate for the work within the stated time period.

This decision created a series of major disputes between the owner and the contractor. The owner's representative unfortunately was a man with only one main goal — to complete the project within budget, even if it meant a total disregard for what was fair. The owner's representative was under pressure from his head office overseas to keep to budget.

The dispute went to arbitration.

Suggestions for eliminating this cause of disputes include appointing a truly independent certifier, or have an independent third party (for example, a disputes review board) to which appeal can be made following a perceived biased determination. In each case payment to these persons would be jointly by the parties. With an independent certifier, the contractor would perceive less risk, but the owner may perceive greater risk. As well, the owner would have to appoint another person to look after its contractual interests, and the contract would have to be written to reflect this division of responsibilities.

With appeal to an independent third party, this person's decision could be made binding for the duration of the project, pending possible arbitration or litigation at the end of the project should one party feel aggrieved.

Another option, that has been suggested, is to select a certifier from a list of possible candidates that are acceptable to both parties, though this doesn't eliminate the conflict of interest, only unsuitable people.

A further compounding issue is the remuneration that the owner's representative receives. Commercial realities drive this person to balance the resources committed or consumed in performing his/her function against the remuneration received. If a consultant, there is also a keenness to derive ongoing or future work from involvement in the current project; this leads to a balance between devoting sufficient resources to the tasks for them to be adequately performed and profit from the work — minimum resources versus maximum profit. When quoting for the work and with a view to winning future work, the consultant will necessarily have to offer a low fee to win the work.

The standard of the performance of owner's representatives is dependent upon the expertise of the individuals performing the function, and the support resources they have. The likelihood of disputes could be expected to increase as the standard of performance of this person decreases. Commonly mention is made of owner's representatives not administering contracts with understanding and flexibility, and not fully appreciating the possible ramifications of their decisions.

External and unavoidable influences

There are some external influences, such as politics, labour union activities (for example, site agreements), and environmental issues (for example, indigenous people's land rights), that contribute to disputes. Regulations and requirements may change after a project starts.

It is important to recognise that some problems, that could lead to disputes, cannot be anticipated. As such, the project management needs to be carried out in such a way that both parties are kept aware of what is happening, so that problems can be resolved before either side faces substantial losses.

Example

In the construction of early nuclear power plants, where each typically took a number of years to build, governments became more aware of the dangers from nuclear power plants, and continually introduced many regulations. These changes occurred while many plants were under construction, causing disruption, redesigns, altered construction schedules and contractual disputes, as costs blew out excessively. Costs of supplies and equipment also increased as the companies producing them changed their activities to meet with the new guidelines. Clauses for contingency payments and unforeseen payments were tested, as contractors sought to avoid bankruptcy caused by the price rises. Some projects were cancelled, while others were significantly delayed. Projects were only completed through a joint commitment of the contractor and the owner. Both parties had to compromise. The contractor lost profit, while the owner paid for the extra construction costs. Parties that pursued assertive dispute resolution through formal means (such as litigation) were believed to lose more than they gained, and neither party achieved what it wanted.

Example
The regulatory environment, that projects must operate within and imposed by governments, can be a major source of disputes. A newspaper recently reported of a contractor, that was prosecuted by a government department because a small defined percent of its four employees were not women. Subsequent compliance costs with retraining and partially employing an employee as an Equal Employment Opportunity (EEO) officer, were a cause of dispute with the owner. This was due to an increase in the labour costs through employing a fifth employee to cover the workload.

Example
An owner contracted with a builder to construct a house. The owner was unaware that the standards adopted by the builder were less than that required by the local government approving body. The owner had to accept the large cost of variations to bring the house up to a suitable standard.

Chronological approach

Some people take the view that adopting the following suggestions on dispute avoidance might be interpreted as a weakness by the other party. Accordingly a strategy, midway between dispute avoidance procedures and allowing disputes to occur and then resolving them, might be adopted. The writer doesn't share this view.

Activities that contribute towards minimising adversarial relationships start pre-project and arise at all stages of a project. The following adopts a *chronological approach* in conjunction, for definiteness, with a *traditional delivery method* (Carmichael, 2000). A lot of the practices may be criticised for focusing on the negative side and a 'protect your back' approach (because they may not inspire confidence with the other party), and rectifying this.

Dispute avoidance actions start well before any contract exists. It is a matter of recognising or anticipating the sources of potential disputes and responding to them. In effect, a *risk management* approach could be followed.

'In all matters, success depends on preparation' (Confucius). Generally, effort spent in the early stages of a project could be expected to be rewarded in terms of positive project outcomes.

2.2 LONG TERM

The establishment of longer-term working relationships between owner and contractor, even though the nature of projects may work against this, and the gaining of trust between owner and contractor have the potential to smooth matters in the workplace. Strategic alliances attempt such relationships. Owners and contractors with like attitudes and philosophies are linked. Parties look beyond short-term financial gains that can cause a deterioration in relationships.

In the 'lean production' approach, the relationship is the most important aspect regardless of the issue at hand. At the outset of a new project, the owner and the contractor discuss each other's financial and technical goals with a view to accommodating as much as possible. There is a commitment to work together in achieving the maximum cost savings throughout their activities. In conventional relationships, that are often conducted at arms length, a dispute may often be resolved by simply terminating the relationship and acquiring the same services from someone else. In the 'lean production' approach, once a contractor has been selected, there is an ongoing commitment to see the relationship work. Both parties examine each other's accounts and methods of operation, to identify more efficient ways of meeting targets and building their relationship. Commonly firms might assess the performance of their contractors and partners over a period of years rather than months. Although there might be questions over competitiveness, longer-term relationships could be expected to result in fewer disputes, meaning less down time and more efficient progress.

Example
A contractor bought into a holding company (the project owner), and accepted a share of the profits over 25 years instead of a direct profit on that project.

The desire for a long-term working relationship promotes the willingness to see each other's point of view and can lead to a mutual understanding of each other's needs and requirements.

Relationship development should also occur in the short term.

Too often a contract represents a fixed-term arrangement, whereby the owner 'gets' what it wants and the contractor in return 'gets' its money. Because there are no longer-term prospects, players adopt an adversarial, short-term 'winner-takes-all' attitude. If two strangers were brought together for only a short period, knowing that future contact won't exist, it is doubtful whether a friendship would develop. Often a contractor, knowing that it only has to work with an owner for a fixed period, does not bother to develop a relationship, as it sees no gain in such effort. Additionally, and often more the case, it is easier to be adversarial if such a relationship does not exist.

Gaining repeat business for a contractor also has commercial sense to it, in that the risks associated with working with unknown parties are avoided. The owner also gains through having a skilled and experienced workforce, that is the contractor, for future projects.

Unfortunately, however, many owners are 'one-off' purchasers.

In the same way that individuals have personality conflicts, and do not develop close friendships or relationships, it is inevitable in contracting that incompatibilities and conflicts can exist between organisations, and these will render an amicable working relationship impossible. Organisations may be forced into entering into contracts through such things as market pressure or process monopoly, but usually both the owner and contractor can be selective in their choice of the other contracting party.

Factors which influence relationships between people are also relevant to the relationships between organisations. Factors such as trust, respect, and like interests and attitudes create compatibility and strengthen relationships, while others such as dishonesty, abuse, over familiarity, and cultural extremes can make relationships unworkable. The time of exposure to these factors is generally much greater over the period of contract work than is the case with personal life, and hence the sensitivity to negative factors is arguably greater in a contracting relationship.

It follows that organisations should be as careful in choosing contracting 'partners' as people are in choosing friends in personal life. People have to feel comfortable with whom they are about to entrust an investment and expectations, and with whom they are going to communicate, negotiate and possibly spend a considerable amount of time.

Trust must be present. It can be either built up over a period of time through past collaboration, accepted by reputation or be established in an intensive effort of gaining the confidence of the other party. Trust leads to the respect of the other party's abilities, management honesty and professionalism.

Compatibility of the parties in such things as similar management philosophies, work ethics, work practices, quality standards, and cultural and industry background is desirable in order to minimise conflict and create a more efficient work environment, where management effort is directed to production rather than resolving differences.

The greater the number of uncertainties that exist in a contract, the greater the chance of conflict; the greater the number of uncertainties the greater the importance of entering into a contract with parties that have an established working relationship.

Project review

As part of a longer-term strategy, recommended practice is for the parties to a contract to conduct a project review, definitely at the end of each project and also possibly during projects, with a view to managing disputes better on the next project. The review would look at, for example, work practices and irregular events that occurred, and also look to how improvements or changes could be made.

2.3 PRE-TENDER STAGE

2.3.1 CONTRACTOR'S POSITION

A contractor might be advised to target owners with which the contractor is able to work successfully. The contractor's marketing and relationship development would be aimed at such owners.

Successful projects are used as a means of developing credibility and trust with the owner, and establishing the owner's faith in the abilities of the contractor.

The contractor does its homework on the owner, the owner's representative, and the owner's project team. Learning something about the personalities involved can help to establish strategies to avoid conflict and to resolve problems as soon as they occur. The contractor selects its project manager and team to match the owner's staff.

A contractor will tender and conduct a project according to how it feels that it is at risk over not receiving payment or payment on time from the owner. A small contractor might be wary of a large and strong owner. It is not unheard of for a large owner to refuse to pay a smaller contractor for work it has undertaken. The smaller and weaker contractor can potentially be put in a difficult position and quite possibly bankrupted by a belligerent owner, especially one which sees some financial advantage.

2.3.2 OWNER'S POSITION

Feasible project

The feasibility of a project needs to be well established. A project of marginal feasibility at the outset is likely to heighten pressure to achieve a least cost outcome from the consultants and contractor. This is likely to have a consequent effect on relationships under the contract as stricter interpretations of contractual meaning are made by the owner.

The tender process

Appropriate tendering practices are required. The whole tender process of inviting and evaluating tenders must be seen to be fair, equitable and unbiased and, for public sector bodies, must be able to withstand an independent audit.

A goal of the tendering process is the selection by the owner of the tenderer most likely to complete the work, in accordance with the contract, at the lowest overall cost to the owner.

For any significant project, industry opinion is that *short list* or *selective tendering* is preferable to public or open tendering. Public tendering might be reserved for low monetary and standard items.

Consultants and contractors are carefully vetted to ensure that they have the capability and track record to undertake the work. For short list or selective tendering, this vetting may be carried out before requesting tenders (prequalification); public or open tendering requires this vetting after the submission of tenders.

In selective tendering, the number of tenderers invited to tender should be as few as possible. There is a balance between getting competition and wasting tenderers' resources. Some sources suggest a maximum of five or six tenderers, but the debate is open as to what is an optimum number.

Prequalification, expressions of interest

There are some contractors with an apparent main aim in life, once awarded a contract, to exploit to their maximum financial advantage any weakness, in the documents or project administration, that they find or allege by colourful interpretation, or any alleged changes in the project. These contractors with an overzealous claims mentality become known over time. To minimise the risk to the owner, tender prequalification processes, which among other things include a review of past performance, may be used. Information sought in prequalification includes:

- Organisation details.
- Ability and experience.
- Personnel resumé.
- Financial details.
- References and examples of past work.
- A history of claims and disputes, and exclusions from contracting.
- Quality assurance systems.
- Occupational health and safety information.
- Industrial relations information.
- Equipment ownership.
- Current work.

Quality accreditation by itself may be meaningless. Procedures and checklists are of no use if they fail to be translated into good attitudes and work practices. In principle, though, *quality assurance* ensures that the contractor 'gets the job right first time', reduces the cost due to defects (nonconformances), reduces the number of supervisors and ultimately removes 'shonky' contractors from the industry. Quality assurance gives the contractor the opportunity to provide objective evidence to the owner that the work is being completed to the specified standard and quality. If the owner selects a contractor with an established and proven quality assurance system, then a high level of trust and faith can be established knowing that the contractor will produce a finished product to the desired quality. The number of disputes over quality will also be reduced to a minimum, especially if the expected quality is defined.

Robertson's Law:
Quality assurance doesn't.

Whilst prequalification schemes are used mainly by public sector bodies, which award many contracts each year, a similar purpose is served by an owner inviting prospective tenderers to submit expressions of interest based on an owner-produced document, which briefly describes the scope of work to be performed. After considering all submissions, the owner selects a short list of tenderers to be invited to tender based on a full set of project-specific tender documents. In this way, owners can feel more secure in awarding contracts to the lowest bidders, because all the invited tenderers have the prerequisite experience, resources, financial capacity and more desirably, a cooperative team attitude, necessary to successfully complete the project.

Whilst the above process is used in the selection of the 'right' contractor, the same process can be used earlier in the project to select consultants to join the project team. Contractors are not the sole source of claims and disputes on projects. Contractual relationships, which at the start of a project are based on goodwill and cooperation, can be soured by the attitudes of supervisory, administration and technical personnel engaged by the owner. In the evaluation of expressions of interest or proposals for professional consultancy services, the owner should consider the consultant's past performance on similar projects to ascertain the likelihood of the prospective consultant generating or inflaming disputes. The level of document-related claims that arose on previous projects is another indicator of a consultant's performance.

Prequalification and expressions of interest approaches minimise the risk to the owner associated with the 'change order contractor' and the 'ivory tower designer' personality types.

Advice

Consideration may be given to involving a contractor early on in a project as a consultant. Such a contractor could provide input on constructability of the design, methods of construction, cost and time estimates, and selection of the main contractor and subcontractors. For some projects, the consultant contractor may end up becoming the main contractor, but there are both positives and negatives for such an approach. The positives include improved efficiency, contractor involvement, less reasons for claims and disputes, and quicker delivery. The negatives include the contractor's extra fee, the owner's vulnerability to the contractor having inside knowledge, and the contract having to be negotiated rather than tendered.

Professional indemnity insurance

Although professional indemnity insurance is not compulsory, conditions of contract often require this of consultants, and it is attractive to owners. The practice implies a lack of trust in the skill and integrity of the consultants. As well, the inclusion of the insurance requirement brings in the insurance company in the case of a dispute, and can interfere in the dispute resolution process.

Contract documentation

Defective documentation
Many disputes and claims can be attributed to poor quality, inadequate, ambiguous, defective or vague contract documentation (including drawings, specifications, conditions of contract, design revisions, project schedules, contract procedures, and quality assurance documents). Some writers go so far to say that errors, contradictions and ambiguity in documentation are the greatest cause of claims and disputes. Correct documentation is fundamental to a successful project, and is as important as any other project item. Over- and under-documentation can lead to troubles just as much as poor documentation. The contract documents set the scene for nearly everything that follows. The responsibility for the quality of the contract documents lies with the owner, and the owner should be fully aware of the possible consequences of proceeding with poor quality documents. The courts favour the party which was not the author of the contract documents, if there are discrepancies and ambiguities.

The commercial reality is that contract documentation, despite the best of intentions, is always open to differing interpretations. And both parties are seen to deliberately manipulate the interpretation of the documentation for their own purposes.

Example
The standard of required and expected workmanship needs to be clearly stated. Merely quoting, for example, that workmanship shall conform to Standard XXX results in a minimum standard of workmanship, sufficient to satisfy the standard. This may lead to a dispute because of the owner's dissatisfaction. As with the intent of the contract, the standard of workmanship should also be included in the discussions at the tender stage.

To gain an advantage and to submit the lowest bid, some contractors exploit inconsistencies, errors, omissions, ambiguities, contradictions and discrepancies in the contract documents. Any conflicting information or interpretation can provide fertile ground for claims; hence the need to clarify documents where the potential for misinterpretation may arise. Contractors, at tender time, are encouraged to point out errors or anomalies in the contract documents, although unscrupulous contractors choose not to do so. In fact, the extreme contractor engages legal officers and others to look over the tender documents with a view to potential claims down the track and balancing this against a cheap, price-competitive tender; this is sometimes called 'loophole engineering'.

Polis' Attorney Law:
Any law enacted with more than fifty words contains at least one loophole.

As well, there are predatory owners which use contractor misunderstanding of the documents and the consequent tendering errors to their advantage.

The owner, ideally, notifies the tenderers (later the contractor), as soon as possible, of any errors or anomalies in the documents; the tenderers (later the contractor) are asked to do likewise in informing the owner.

Document preparation
Disputes and claims resulting from poor quality documents, whilst being almost impossible to totally eliminate, can be minimised by the owner adopting some common sense practices during document preparation:
• Have support of senior management.
• Allocate appropriate time to the task (drafts, reviews, approvals, checking, printing, ...).
• Use competent personnel; establish continuity by using the same people throughout the documentation process, with some of these people available for the whole duration of the project to deal with documentation issues as they arise.
• Have a system of independent checking in place, with checking by experienced personnel.
• Use grammar and spell checking capabilities of word processing software.
• Coordinate the assembling of the various documents (conditions of contract, specification, drawings, ...); cross-check and establish consistency between documents.
• Have individual specifications which are either performance-based or prescriptive-based, rather than mixing the two (state the method or the result but not both).

- Develop the documents for any project based on that project's individual requirements; resist the temptation to make cosmetic changes to, or 'cut and paste' from, a previous set of documents; documentation should be specific to the project.
- Use standard general conditions of contract where possible; this has an added advantage of parties being familiar with their rights and obligations; don't mix different standard conditions of contract.
- Use an instrument of agreement.
- Provide an order of precedence to the contract documents.
- Nominate one person as a single point of responsibility.

Many of these activities would be picked up in comprehensive quality assurance practices.

Example

While preparing the contract for the laying of concrete pipes, the owner's engineer overlooked the issue of who was to supply the bedding sand in the trenches. The value of supplying the sand was approximately a quarter of a percent of the entire budget. Rather than create a confrontation between the owner and the contractor, the owner decided to pay for the sand. The project was on a strict time scale and a dispute was the last thing that the project needed.

The involvement of experienced contract legal officers to provide advice on the suitability of special conditions of contract and modifications to standard conditions of contract, and to generally vet the contract documents, can also assist in the avoidance of troubles down the line resulting from poor contract documentation.

Nevertheless, errors, ambiguities etc. in contract documents will still occur.

Bukota's Typographical Truths:
- *Typographical errors will be found only after the final copy is bound and mailed.*
- *Typographical errors appear in inverse proportion to the number of syllables in the misspelled word.*
- *Engineers can catch misspellings only in words written by someone that's not an engineer.*
- *The incidence of missed typographical errors increases in direct proportion to the number of people who will see the copy.*
- *The incidence of missed typographical errors increases in direct proportion to the size of the letters in the copy (about 1.3 errors per point size, but that proportion isn't proven beyond doubt).*
- *Success in finding typographical errors is in inverse proportion to the finder's income and number of years of education. Ask the janitor.*

The Secret-of-Success Law:
Discover all unpredictable errors before they occur.

Assumptions
The nature of some project work leads to many assumptions being made in the contract documents. For example, the level of the water table, the location of rock beneath the ground surface, the hardness of the rock and similar factors can only be determined after tender and during construction. The result of this can be a build up of claims, and this in turn leads to conflict on the project.

Standard conditions of contract
Conditions of contract are chosen as appropriate for the project. Where possible, industry standard conditions of contract, and 'tried and true' specifications, rather than inventing unfamiliar 'home brew' documents, are preferred. This will minimise misunderstanding and misinterpretation of the owner's expectations. Standard general conditions of contract contribute to the parties understanding their obligations. Greater familiarity and comfort with the routine and form of the contract conditions will leave more time available for dealing with the content of issues, rather than having time consumed developing and learning procedures within a new and unfamiliar framework.

Standard contract conditions which have been prepared by industry interest or pressure groups, and which are promoted for use by members of those groups are to be avoided, because of their usual bias in favour of those members, and against the other party. Such promotional practices are to be expected because the groups exist to serve the interests of their members. Unscrupulous group members may use the contract terms to their advantage.

Example
Builders may push for contract conditions developed by a home builders' organisation when contracting with a consumer wanting a house built. Given the difference in power between the builder and the consumer, and given that for the owner this may be the only time that it contracts for a house to be built, projects using such conditions start off with the possibly ill-informed consumer feeling intimidated, and believing that the contract exists solely for the benefit of the builder.

Standard contract conditions that have been developed by a broad spectrum of interests have more chance of being fairer.

Some recommended inclusions in the conditions of contract include:
- Realistic notice provisions such that communication and relations between the contractor and owner are not adversely affected.
- Any float or contingencies (time reserves) in the contractor's program belongs to the contractor.
- For delays caused by the owner, but not by the contractor, the contractor is entitled to an extension of time. How delays caused by neither the owner or contractor (neutral delays) are to handled.

Complexity
Conditions of contract can be made unnecessarily complex. What the owner wants may be a straightforward product of a given quality, at a given price, in a given time. In

some instances, there may be more clauses in the special conditions of contract than in the standard general conditions of contract. Or the standard general conditions may be changed extensively. Trying to cover every eventuality and concentrating on what might go wrong, may not necessarily work in practice. Complex contract conditions may not be understood properly and may result in disputes. Complex contract conditions may contain hidden issues. Complex contract conditions may not necessarily contain less loopholes, compared to standard contract conditions, for the contractor to exploit. In fact, greater complexity may create more loopholes.

Forde's Third Law:
The longer the letter, the less chance of its being read.

Bad owner experiences have led to contract conditions being modified over time, such that they have become convoluted, user unfriendly and adversarial in nature. This has increased the difficulty for both parties administering their sides of the contracts.

Extent of documentation
Documentation needs to be accurate, clear and complete, and include a well-defined extent/scope of work (in one document rather than being spread over many documents), standards of workmanship and materials, technical requirements, contractual/commercial conditions, time for completion, and criteria for the evaluation of contractors. There should be no room for the parties having different perceptions as to which party is responsible for what. Such perceptions may not surface until a problem occurs. Opinion is divided over the need to develop a systematic breakdown or bill of quantities (BOQ, BQ) of the work that the contractor is required to do because, although useful in clarifying the work, errors in the breakdown could be interpreted against the owner. The alternative is for the owner to provide a bill of quantities, but word the conditions of tendering and conditions of contract such that the risk associated with pricing the work is placed on the tenderer. Where quantities are provided, these are checked. Changes in bid item quantities after contract award may disadvantage the contractor with consequent expected contractor behaviour. Changes may also advantage the contractor with consequent expected owner behaviour.

A *standard method of measurement* should be used in the bill of quantities.

Minimal use of *provisional sums* is recommended in order to reduce the incidence of variations later on. Where provisional sums are used, reasonable care is required in their estimation. (And when the provisional sum item is being carried out, owner verification can be undertaken.) Undervalued *prime cost items* need to be guarded against.

Some humour on prime cost (PC) items:

PC items are when the owner puts on a blindfold, takes out his/her wallet and asks everyone to help themselves.

External influences, such as politics, labour union activities (for example, site agree-ments), and environmental issues (for example, indigenous people's land rights), that might lead to disputes, have to be anticipated and allowed for in the contract.

Simplicity in stating obligations assists with clarity, and with this comes certainty.

Example
In information technology (IT) projects, the end-users are often unsure of their requirements. As well, their requirements change mid-stream. This usually leads to time delays and added costs, resulting in disputes.

There can be a mind set that links the value of the project and the quantity of paperwork. There is no basis for such a correlation.

Language
Plain language (rather than what has been described as a 'farrago of obscurities') is preferred to the use of legal jargon (legalese). Legalese in contracts should be avoided; it is at the expense of clarity. Legal terminology may generate fear among the users of the contract documents. Legalese is not understood by people implementing the contract. It may give the owner the perceived necessary protection, but may not result in getting the work done. The presence of legalistic terms leads the contractor to seek advice from a lawyer which tends to increase the chance of the parties becoming adversarial because of the fundamental training of lawyers.

Murray's Law:
If written correctly, legalese is perfectly incomprehensible.

'When I use a word,' Humpty Dumpty said in a rather scornful tone, 'it means just what I choose it to mean, neither more nor less'.
'The question is', said Alice, 'whether you can make words mean so many different things'.

Lewis Carroll, Through the Looking Glass, Ch. 6

The contract documentation acts as protection for each party, and a basis for negotia-tions if things go wrong. However, this focuses on the negative side of the contract. The 'protect your back' approach to contracts is not effective in inspiring confidence in the other party. There is a view that a contract should be viewed more as a road map and emergency guide rather than as an enforcement document.

No matter how good the documentation, there is nearly always some way of interpret-ing part of it in a way that was never intended. There are grey areas. It is frustrating to specification writers just what bizarre and creative interpretations can be given to what appears to be simple, clear and obvious statements. Problems with specifications can come from errors in the source (owner's input) or from errors in interpretation (contrac-tor's input).

Klipstein's Law of Specification:
In specifications, Murphy's Law supersedes Ohm's.

Example
Some humorous examples from poorly written actual contract documents (definitely not to be emulated):

1. The contractor shall submit his claim to the owner's representative, and his decision shall be final ...
2. The contractor shall be paid for not more than a trench the width of which shall be less than the diameter of pipe plus 18 inches (450 mm) and more than the diameter of the pipe plus 12 inches (300 mm).
3. If the owner's representative deems any employee of the contractor to be inefficient or dishonest he shall be dismissed forthwith.
4. The contractor shall leave the site in the same condition when the work is finished as it was before operations were begun, and shall be cleaned of all rubbish to the satisfaction of the owner's representative.
5. The material for the embankment shall be obtained as far as possible from the road.
6. When completed, the contractor shall leave the job in a neat and orderly fashion.
7. After the pipe is laid the trench shall be backfilled with bulldozers.
8. The contractor shall make monthly claims for extra work with invoices attached to the owner's representative.
9. The ready-mix subcontractor shall be in position at the forms 20 minutes before the concrete placement begins and shall remain in position until 20 minutes after the concreting is finished.
10. All the water for the concrete in the marine structure shall be passed by the owner's representative.

Example
Specification writers commonly use the terms 'or equal' and 'as approved by the owner's representative'. These terms lead to uncertainty. Ideally what is required, or the standard of performance expected, should be specifically stated. Research time spent on establishing what is available in the market, while compiling the specification, can save much debate later on.

Example

When a contract has three volumes of technical specifications, it is no wonder that there are disputes about interpretation. It tends to be the case that consultants feel that the more words there are in the specification, the better. Unfortunately they do not write each specification from scratch, but recycle old specifications. The result can be hard to digest, without necessarily clearly spelling out the requirements.

Example

A project specification may make reference to a standard, certain legislation or a previously published specification relating to particular workmanship or a particular material or piece of equipment. Such standards, legislation or specific specifications may not be readily available to the contractor, especially on a project site. Although economical in words, such referencing may be of little value if its content is inaccessible to the contractor.

Dispute resolution mechanism

Necessary mechanisms are included in the contract documents to promote genuine discussion, cooperation and negotiation aimed at resolving problems as quickly as possible, should problems arise. The contract is structured to encourage or force the early identification and notification of issues likely to be disputed.

Incorporation of an appropriate *dispute resolution clause* (outlining procedures to be followed) is an important step toward dealing with disputes. The contract clause needs to make it clear that it is there to facilitate the resolution of disputes in the simplest and most impartial ways possible. The important emphasis of the clause is to bring the parties together, rather than provide instruction on how to have a dispute. It is aimed at dealing with the dispute at the earliest opportunity where resolution is more probable; each party has a greater opportunity to make adjustments in ongoing expectations when matters are dealt with early.

The clause itself, and the approach it embodies, is able to give tenderers a feel for the likely general conduct of the owner on the project. For example, a dispute resolution clause that has onerous obligations mostly upon the contractor may well lead the contractor to expect the owner to be harsh in its conduct. This would tend to precipitate disputation rather than contribute to its avoidance or speedy resolution.

All project contracts

There should be no contradictions in the contract documents between the various project players — owner, contractor, consultants, subcontractors, suppliers. All contract documents should be consistent and compatible.

Unfair contracts

Extremely one-sided or unfair contracts are not recommended, although superficially they may appeal to owners, and owners may be persuaded to adopt them by their adversarial-minded lawyers on the pretence that they are looking after the owners' interests. Contractors, desperate for work, may accept such contracts, but then use any loopholes as a basis for claims.

The Golden Rule:
Whoever has the gold makes the rules.

Dan Green's Rule:
What the large print giveth, the small print taketh away.

Unjust contract conditions and unachievable demands have no place.

There exist a number of fair standard conditions of contract, yet owners still produce contract conditions that are unfairly biased in favour of the owner. Such owners have the attitude that, if the contractor wants the work, it will accept the contract at any cost.

Contract conditions may also be forced upon contractors in a buyers' market. Some project industries are fiercely competitive, with many contractors trying to win work of any nature.

Some owners have contract conditions structured in such a way that contractors cannot find a way around them. Whatever happens, the contractor is in the wrong. Such lack of fairness does not promote non-adversarial relationships, but rather becomes a challenge to a contractor to try to 'beat' the owner somehow.

The contract documents should be presented in a way that does not present an adversarial image, for example by not 'hiding' important information on one matter within something of a different and less important nature. A layout which doesn't mix the fundamentally different document types is preferred. A suggested distinction between the contract documents is:
- All the tender documents:
  - All notice(s) to tenderers.
  - All conditions of tendering.
  - The form of tender.
  - The general conditions of contract (to be).
  - The special conditions of contract (to be).
  - The specification.
  - The drawings.
  - The schedule of rates (prices)/bill of quantities (if any).

Together with:
- The tender and its acceptance.
- Instrument/form of agreement (optional).
- (Possibly) correspondence between the parties, post-tender meetings minutes, post-tender negotiations.

Natural justice ideals should be guiding the contract writers.

Risk

Many project industries carry high risks, for example due to industrial relations problems, adverse weather or latent conditions.

The contract documents need to adequately communicate the risk sources to the parties. Risk allocation should be to the party able to best manage it, rather than, say, being onerous toward the contractor. For example, a contractor might be penalised for schedule delays when owner-supplied equipment is late. A contract that assigns all risk to the contractor will create a favourable environment for claims and disputes later in the project. Passing an undue amount of risk to the contractor, through harsh conditions and penalties, causes the contractor to find areas of fault belonging to the owner as an offset.

Each party generally tries to minimise its exposure, often by trying to shift responsibilities to the other party, or it will accept a specific risk if the reward for accepting the risk is sufficient. The key to this is the early identification of risk sources, and an analysis of their consequences.

For example, the absence of a latent conditions clause can be disastrous for the contractor if the ground conditions are not what was anticipated at tender time. Losses due to wet weather is another risk which the owner could make the contractor responsible for. Such practices lead to problems between the owner and the contractor. Owners need to realise that off-loading risk to the contractor results in unnecessary friction between the contractor and the owner, and also may lead to higher tender prices.

Open and frank discussions between the owner and the contractor can assist in addressing unfair risk allocation, and the avoidance of disputes down the line. It may take both parties working together to, for example, mitigate a risk. Risks associated with disputes may be best addressed by up-front open discussions between the parties. For example, design-and-construct delivery might require the contractor to bear the risk associated with development application (D.A.) approval, but this would be inappropriate because of the likelihood of disputes over problems arising during the application assessment.

Minimisation of disputation is best served when the obligations of parties generally conform to industry norms. Unless there are overwhelming benefits of structuring things in a radical fashion, risks associated with latent conditions, rise and fall and so on should be borne by the owner. While risk shedding has its merits, it is likely to contribute towards greater disputation during project execution.

Delivery methods, payment types

Selection of the most appropriate delivery method (traditional, design-and-construct, ...), as well as payment type (lump sum, schedule of rates, cost plus, ...) and tendering procedures can help reduce some adversarial elements (Carmichael, 2000). For example, if it is difficult to define the quantum of work, then a schedule of rates contract would be more appropriate than a lump sum contract.

Example
Work started with a lump sum contract. The design was not fine tuned or advanced enough at the time of contract award. As the project progressed, scope changes were made, and resources were wasted on costing and negotiating the ever changing scope. With payment based on a schedule of rates, all resources could have been concentrated on doing the work.

Example
Due to government regulations, all funding for the construction of some targeted roads had to be spent by the end of the financial year. This meant that the project had to be completed in a six-month period. To satisfy this time constraint, the project had to be fast-tracked, and design-and-construct delivery was chosen so that construction could start before the design was complete. If traditional delivery was used, the time constraints would have been hard to satisfy, and the chance of conflict would have risen.

A project can be dissected into packages and an appropriate delivery method and payment type selected for each package. The owner then, however, has to be capable of performing a greater coordinating function, and this may not be possible. But even then, care has to be exercised in the usage of payment types. For example, disputes in prime cost contracts arise when it is unclear what constitutes reimbursable costs. These need to be defined explicitly in the contract along with the means of measuring these costs. As another example, an owner accepting an unrealistically low tender price in a guaranteed maximum price (GMP) contract could expect trouble.

Consideration might also be given to incentives/bonuses for delivery under budget or ahead of time, as these tend to be well received and also generate creativity. Properly thought through incentives encourage cooperation rather than competition, and teamwork rather than fragmentation.

Partnering and project alliances

Consideration might be given to using partnering or alliance thinking on a project as a way of embodying trust, cooperation and goodwill, and developing a unified project team. Such thinking has the potential to prevent many disputes from occurring, or if disputes do occur then their resolution is streamlined.

Nominated subcontractors

Owners wish to see their projects completed to the appropriate standard, and believe that nominated subcontractors can achieve this. However, the use of nominated subcontractors may cause conflict. Contractors take a view that owners are trying to impose themselves. There may be conflict between the contractor and nominated subcontractors over various different practices. With nominated subcontractors, there is a blurred responsibility which can cause friction, claims and disputes.

Rather than nominating a particular subcontractor, the owner may be better advised to provide a list of acceptable subcontractors, from which the contractor can choose.

Reasonable objections by the contractor to any nomination should be considered by the owner, and the subcontract amended to reflect these objections. If nominated subcontractors are used, the head contract conditions of contract and the subcontract conditions of contract need to be compatible.

Investigations

Appropriate resources need to be allocated to ensure thorough preliminary investigations — surveys, environmental impact studies, traditional land owner negotiations, public opinion trends, geotechnical investigations, service locations, potential supply difficulties etc. Saving money through inadequate investigations before starting design, leads to uncertainty and unexpected eventualities, and contributes to confusion and disagreements.

Example
The notion of saving money is illustrated best in the cartoon showing two people looking at the drawings for the Tower of Pisa prior to it being built, with one person describing to the other how much money could be saved if geotechnical investigations weren't carried out. The tower is now known as the Leaning Tower of Pisa.

Design

Many design problems can be eliminated with a carefully thought through owner's brief. Poor documentation and design may flow from an inadequate design brief.

Any design needs checking, perhaps using an independent person:
- For adequacy. A designer, for example, may argue that problems are construction related rather than design related. For example, a suspended slab sags following construction; here, the designer argues that it is poor construction, when in fact it may be inadequate design.
- To ensure it fulfils the owner's brief, which too needs checking so as not to permit the designer to claim for additional work. The design brief needs to be clear and concise.
- For accuracy, precision and proper drafting of the associated design drawings; for errors or omissions in the design.

Deficient design may be due to lack of adequate basic investigations, which in turn may originate from financial constraints or urgency requirements. The general feeling is that money invested in investigation and design is recovered later in a project (and during the lifetime of the end-product). Design effort should not be started too late or unrealistically limited by expenditure or time constraints. The consequences of calling tenders with an incomplete set of drawings should also be apparent. Once a contract is awarded, the cost and time effects of changes to requirements become largely outside the owner's control, and are governed more by the contract.

One person should be nominated as a single point of responsibility for design.

Adequate budget has to be allocated for the design process. Paying designers at unrealistically low levels is counterproductive; substantial savings can occur during the construction process and life cycle costs as a result of a properly considered design.

Late provision of design documentation will cause claims. Also, it may be unclear whether documents provided for construction involve variations, compared to the documents provided for tendering purposes.

Constructability

Whenever the design and construction functions are separated, there is the constructability (buildability, value analysis) issue.

Drawings and specifications may be developed by design staff who have only limited exposure to actual construction activities and methods, and may not understand the way contractors like to work. As such, they may not be totally suitable. Some aspects may be impractical, impossible or uneconomical to construct. This leads to on-site changes with the associated administrative work and potential for disputes. To counter this, design requires input from experienced construction personnel. Specialist subcontractor personnel, who in many cases do the bulk of the work on projects, could also assist in this area. This contributes to ensuring constructability of the design, and that the design can be constructed safely, and within budget. Early involvement in the project of persons with construction knowledge is recommended to avoid as many documentation and coordination problems as possible.

Example

On a building, the holding down bolts for the first floor structural steel frame were designed to be positioned over concrete columns at the intersections of reinforced concrete beams integral with the first floor slab. The result was that the holding down bolts could not be positioned exactly as per the design because of the amount of beam, slab and column reinforcement. Each set of bolts was positioned as best it could be and an 'as-built' survey was performed after the slab was poured. A drawing of each set of bolts was done in relation to where they should have been, and each column base plate had to be individually made.

Consequently there were increases in costs because of:
- Increased time in setting up bolts and moving some reinforcement.
- The need to do a highly accurate 'as-built' survey.
- Increased time in fabricating base plates and completing structural steel erection.

The whole problem could have been solved at the design stage by making the concrete beams slightly wider. The cost of the extra concrete would have been small compared with the above mentioned extra costs.

Alternatively, tenderers may be able to provide valuable input on construction methods, and the period set aside for construction. This can only really occur in a preselect or short list tendering situation. If discussions are held with several tenderers, then confidentiality of the tenderers' ideas, particularly those involving innovation, needs to be preserved.

Consideration can be given to allowing the contractor to be free to vary the method to achieve the required end-product. This may produce cost savings to the contractor, cost savings that could be balanced against losses elsewhere in the contract work.

A conflict may arise because people don't like change — 'We have always done it this way, and we don't want to change.' There is a need to demonstrate the advantages of change, and reach acceptance with its introduction.

New technology

Where new technology is being used, the owner needs to be aware of the potential issues associated with using the technology.

Budget

The owner should ensure that all items in the project are correctly allowed for in the budget, else the final cost may exceed the owner's budget.

Often, owners have only a fixed amount of money within which to carry out a project. Often, the effect of a cost overrun can be politically sensitive. One result of this is that owners establish burdensome and time-consuming procedures for the approval of variations. This may have the effect of contractors shelving legitimate claims until the end of the project because they do not want to ruin the contractor–owner relationship; and the storing up of claims contributes its own problems. A realistic contingency allowance should be included in the project budget for variations.

Fixed price contracts are not absolutely fixed, only 'nearly' fixed (Carmichael, 2000), and if an owner wishes to absolutely set the upper cost of a project, then a guaranteed maximum price (GMP) contract is recommended as the closest possible alternative. Owners may be reluctant to pay more than the initial contract price if they believe that a fixed price contract means an absolutely fixed price.

It may be worthwhile including in a contract the ability of an owner to terminate the contract where unforeseeable variations (those not allowed for in a contingency sum) exceed, say, 10 to 20% of the original contract price, and the associated project duration extends beyond, say, 50% of the original contract duration; this would apply to owners on strict budgets and not having the capacity to pay beyond this. The contractor, of course, would be paid for any work done up the termination point.

Should disputes arise, then there is a cost for resolving these. Owners (and contractors) generally do not budget for anything more than the finished product. A contingency to cover the cost of dispute resolution is not included. Few owners and contractors realise how costly the resolution of disputes, particularly by arbitration and litigation, can be. Even by other dispute resolution methods, there are significant costs.

Alternatively, for disputes over small amounts, it may be decided that the cost of pursuing a resolution is not worth it. In which case, this amount in dispute becomes an additional project cost, which also perhaps has not been budgeted for. The precarious balancing act often involved in budgeting becomes disturbed.

Competition

Tendering, by its very nature, implies competition. Contractors may be competing in a tough business environment, and this attitude can be carried over into the project. Contractors are already adversarial in their thinking of their competitors; they seem to never have anything nice to say about their competitors, and many won't even talk to a competitor (with consequent detrimental flow-on effects to the development of an industry at large).

There is often no long-term association between contractors and owners, and hence each might try to make the most of each opportunity. Alternatives to tendering, such as direct negotiation with a potential contractor based on the contractor's qualifications, could bring the contractor and owner to work as one.

Competition can seed conflict between the owner and the contractor, and this may be manifested in various types of disputes, such as the interpretation of the contract documents and/or the performance of the contractor.

2.4 TENDER STAGE

First impressions

In many cases, the tendering process is the first point of contact between the contractor and the owner. Many relationships are built or doomed on first impressions.

Much of the form that the initial relationship takes is influenced by the nature of tendering itself. There will only be one winner, and the other tenderers will have spent often a lot of money only to be disappointed. The knowledge of such looming rejections causes many owners to maintain a detached involvement, seemingly indifferent to the problems faced by the tenderers.

2.4.1 CONTRACTOR'S POSITION

Capabilities

The contractor checks that the work is within its capabilities, and has the resources available for the anticipated duration of the project.

Tender documents

Appropriate time and resources have to be allocated to compile a complete tender. An eager contractor may not analyse the completeness of a design and may limit its tender to a narrow, obvious scope.

Special attention is given to studying, clarifying (technical, legal and administrative issues) and understanding the specification in particular, and the scope of work with all its peculiarities and odd or unusual parts. All the contract documents have to be read and understood. All negotiated and agreed items have to be verified that they have been included.

Any clarification, for example in the owner's requirements, is done in writing. Unclear items in the tender documents are removed from the tender; unclear items are 'tagged out' of the contract. Where alternative materials and products are permissible, proposed alternatives are checked for acceptability with the owner, to eliminate any possible misinterpretations. Materials and products from suppliers are checked to ensure that they match the specification or are acceptable alternatives.

Example

The situation involved the purchase of a used motor vehicle. After some bargaining, the seller and buyer agreed on a price. A small deposit was paid, at that time, by the buyer to show her intent on buying the vehicle, with the remainder of the purchase price to be paid on pick up of the vehicle. On arrival to pay the outstanding amount, it was noticed that the roof racks had been removed from the vehicle. When the seller was questioned about this, he implied that it had been agreed that the selling price did not include the roof racks. However that was not the way the buyer saw it; she believed that everything on the vehicle, at the time of agreeing on a price, was included in the price. A dispute arose from this and it became a matter of one person's word against another's word, since nothing had been written down. In the end, it came down to not what was actually said (as there was no commonality on this), but how much the buyer wanted the vehicle. Eventually the buyer decided to accept the vehicle at the agreed price, but without the roof racks.

Possibly at the initial meeting with the owner, the owner is alerted to any queries or anomalies in the tender documents. Such queries and anomalies are clarified as soon as possible. Consideration might be given to guiding the thinking of the owner into accepting the contractor's interpretation of the requirements of the tender documents.

Regardless of how well or how poorly the contract documents are prepared, the appropriateness or inappropriateness of the specification, tolerances, and so forth, if the contractor agrees to do the work as presented, it may be risking its commercial future on the assumption that 'common sense will prevail'. If something should go wrong, the contractor will be finally judged on the words in the contract. Ambiguities aside, this means that the contractor will be assumed to have read and understood the contract documents, and priced accordingly. Contracts involving unrealistic conditions of contract and specifications are best avoided.

Owner expectations

The contractor attempts to understand the expectations of the owner, and before accepting work, ensures expectations can be met, especially with respect to quality and scope of work, and staffing of projects. ('The personnel who we thought we would be contracting with are not who we ended up with.') There is a potential source of conflict if the owner's expectations and contractor's expectations do not match, or if one party's expectations are not met.

In preparing a competitive tender, it is not unreasonable that the contractor identifies the cheapest solution to meet what it sees would be its obligations under the contract. The expectation of the owner, however, may have been based on a quite different interpretation of the obligations.

There may also be regulatory requirements that a public sector owner may have to conform with, and which the contractor should be aware of.

Owners compare quotes from contractors on the assumption of identical quality. Owners often expect the highest quality, even though they may only be paying a cheap price. Owners and contractors fall out over what constitutes an acceptable standard of work.

At the tender stage, there is an opportunity to get to know the owner's personnel, and what part of the project they are likely to be fussy and particular about, and what part of the project they see as not being too important. For example, an owner may insist that monthly invoices are presented in a particular way, but may not be concerned whether or not they are submitted on time. In this case, extra effort should be put into the presentation even if it means submitting the invoice one day late.

Contractual obligations

There is a need for the contractor to understand the contractual obligations, possibly through discussion with a legal officer, to scrutinise the contract conditions, and negotiate if necessary. Some owners request the contractor to initial each document page and drawing sheet, in the belief that this makes the contractor aware of everything in the contract.

Some specific issues that need to be examined and understood include latent conditions, time limitations/bars for notices and claims, terms of payment and variations procedures.

Risk analysis

The contractor ideally performs a risk analysis on the tender documents and the project. A checklist is suitable for this purpose. This will include consideration of wet weather, delay in equipment and material supply, access to site, site awards, other contractors present, latent conditions, statutory requirements and so on.

There can also be an upside of any risk analysis, that is opportunities, good project performance, additional profit, ... However most people focus on the possible downsides, and if upsides occur then this is regarded as a bonus.

Many contractors are guilty of not sufficiently identifying risk sources, and quantifying risk, or deciding not to include something in their bids covering risk, for fear their bids may be non-competitive.

A contractor's planning, work sequencing, resource allocation and cost estimates may provide little room to accommodate additional requirements introduced once the work commences. Such unforeseen requirements resulting from ongoing design and/or procurement efforts can interfere with the orderly and efficient performance of the work. Disputes are commonly over unanticipated extra costs and responsibility for those costs.

Some owners hold a risk management meeting with tenderers, where potential uncertainties are discussed. This gives the opportunity to resolve issues before they occur, or commence action to resolve issues before they develop into later disputes. It also gives the owner the opportunity to review risks that have been imposed on the contractor, and if the current approach is appropriate.

Each party may try to minimise its exposure, or may accept a risk if the associated rewards are sufficient. Each party can establish what its own risks are, but it may take both parties acting together to mitigate a particular risk. Discussions pre-tender help this mitigation process.

The contractor's bid basis

To avoid disputes over quantities and scope of work, the contractor can include in the tender a schedule detailing the basis of the tender. This also provides a basis for establishing what is extra work (for example, resulting from revisions, as may occur in fast-track projects), and if a schedule of rates is also included in the tender, extra work can be costed simply. This can avoid a problem, later on, with non-acceptance of the cost of variations. The rates need to be adequate for all circumstances. Rates would be necessary for different categorisations of equipment and labour. Values for overheads and profit would also be necessary.

Example
Where a prime cost sum allowed in the tender turns out to be lower than the actual cost of the installed item, the contract sum is increased accordingly. As well, a contract may also allow the contractor to charge an additional amount to cover margins. If a basis for calculating such margins has not been agreed at the time of entering into a contract, a disagreement will almost certainly ensue.

For delays and disruption, pre-agreed delay costs may be difficult to establish accurately, and may have to be submitted as approximate only; actual delay costs may not be known till a particular event occurs. Tenders can include an indicative resource schedule in order to assist delay cost arguments. Disruption affects the contractor's productivity, and actual disruption costs can only be worked out at the time.

An extension of this type of thinking is contained in the idea of *escrow bid documents* (EBD) which contain the contractor's calculations and information used in preparing its bid. The EBD are consulted if they would facilitate the resolution of a claim or an issue. The presence of EBD facilitates a more honest and cooperative relationship on the project.

Project planning

Unrealistic planning/scheduling may lead to later modifications having to be made. Sometimes planners can be overly optimistic. For example, they may not consider the impact of delays in materials supply, climatic conditions, or machinery breakdowns. The initial plan provides a basis for tracing the progress of the project, but requires continuous adjustment according to actual project progress.

Cheop's Law:
Nothing ever gets built on schedule or within budget.

Materials and equipment orders are scheduled to allow timely delivery and so as to not hold up the project.

The program

The submitted program may take on contractual status or may be merely an outline of intentions. Ideally, the program should not be used as a statement of contractual obligation. When a dispute arises, it is often the contractor's program which is produced, but only taken seriously if it suits. Those who do not wish to see the contractor's program in the contract argue that the owner's consultants and the owner cannot be expected to endorse a particular work sequence at the outset. However, there is no reason why the contractor should not be asked to express its intentions clearly so that the owner's consultants can view them. The program can be cross-checked against the drawings, bill of quantities and the specification. This ensures that the consultant's intentions and the contractor's understanding are compatible, removing some uncertain ground as a basis for disputes.

   The program is a tool that can be used by the contractor to verify failures to meet targets. But if the program is unrealistic, it can be used by the owner against the contractor. The program can be a 'double-edged sword', to both owner and contractor.

Price

The price developed by the contractor needs a realistic contingency (cost reserve) included. This contingency allowance is available to resolve unanticipated problems.

Some tendering humour:
A tender is a poker game in which the losing hand wins.
The lowest bid is a wild guess carried out to two decimal places.
The successful tenderer is a contractor who wonders what has been left out.

The tender

Before submitting, the contractor reads the tender, and checks for errors. A checklist is suitable for this purpose. On submission, the contractor brings to the owner's attention any qualifications, exclusions and nonconformances.

Example
For the construction of a treatment works clarifier, three tenders were received at approximately the same price, while a fourth tender was priced at half this. The fourth tenderer had made a major estimating error due to its lack of knowledge of market prices. Acceptance of the fourth tender would most certainly have resulted in problems down the track.

Chisholm's Second Law:
Proposals, as understood by the proposer, will be judged otherwise by others.
Corollary:
If you explain so clearly that nobody can misunderstand, somebody will.

2.4.2 OWNER'S POSITION

Tenderer information

Any unusual, different or non-standard contractual requirements are brought to the attention of the tenderers by the owner. As well, the owner ensures that the scope of work is understood.

The owner provides or tables whatever technical information (site data, ground conditions, condition of existing equipment, ...) is available, even if qualified, to prevent liability issues and to assist the tenderers. The owner needs to be open, and not withhold information. To give the tenderers the best opportunity to prepare proper tenders, the owner should ensure that as much relevant information about the project is made available to the tenderers.

An extension of this thinking is contained in a *geotechnical design summary report* (GDSR) which clearly and specifically outlines the subsurface conditions anticipated by the geotechnical engineer (engaged by the owner), and their likely impact on the construction. It is a baseline of anticipated conditions by which differing site conditions can be compared. Should the conditions be worse than the baseline, the contractor could expect some compensation. It is incorporated into the contract without any associated disclaimers for accuracy or completeness. It is not a guarantee of ground conditions, rather a professional engineer's informed judgment. It is argued that the engineer is in the best position to assess the reliability of site data, being able to spend much more time reviewing the data, compared to the short tendering time available to the contractor. Both owner and contractor are able to establish their risk levels relative to this baseline. With an ill-defined statement of ground conditions, unrealistically high and low bids may be obtained, with the latter possibly leading to extensive claims in order to recover losses. With a well-defined baseline, more informed and competitive bids would be expected, with lower contractor contingencies. The potential for disputes over what constitutes differing conditions is reduced, and possibly eliminated.

A *meeting of tenderers* is recommended, where questions from the tenderers are fielded. These are general questions, and would not relate to something, for example an alternative tender, which would give one tenderer some commercial advantage. However, the meeting may lead to alternative submissions that may be of mutual benefit to the owner and the contractor. The meeting is minuted and signed by all parties present.

It should be mandatory for the tenderers to inspect the project site, though this may not be practical for some projects, for example a project involving the construction of a long pipeline. The tenderers should be accompanied by the owner's staff who are sufficiently conversant with the project so that all site features, peculiarities and risk sources are identified for the benefit of the tenderers.

There may be owner requirements not mentioned explicitly in the contract documents, for example alternative phasing of the project to facilitate early access or sales or to keep part of an existing facility functioning while the new work is being carried out. If these kinds of requests can be accommodated by the contractor, it can be of great benefit to the working relationship between the parties.

The owner may also wish to communicate aspects of its budget and other project constraints. Contractors are generally not privy to information on the owner's budget. They are unaware of the funding arrangements, and how much money has been allocated to the project stages. They also may not be aware of the absolute time constraints on the project or how the project fits into some master plan. The contractor does not know how much was budgeted for the project, and how much contingency is available. It might be desirable for a relationship between the contractor and owner to be developed where some of the information about the owner's budget could be communicated to the contractor. The contractor may be able to provide suggestions to improve the value or decrease the cost of the project, or forewarn the owner of any possible problems that may incur more cost. The notion of an incentive might be introduced to bring this about. This informing and involvement of the contractor promotes the recognition of a common cause in the project. If the contractor can understand the pressures and constraints experienced by the owner, then it is less likely to look for avenues to confront the owner. Detractors of such an approach, however, would argue that the contractor would charge up to the limits of the owner's funds, and work accordingly, and operate on the owner's vulnerability, and hence it is better to keep the owner's budget and constraints secret.

Parkinson's Second Law:
Expenditures rise to meet income.

Example

A project involving the refurbishment and fitout of a city office building. During the tender preparation period, a site inspection required the attendance of all tenderers. Tenderers had received the tender documents well in advance of the site inspection and therefore had ample time to review them.

The site inspection was effective in familiarising the tenderers with the overall extent of the work. The tender documents were reviewed and discussed. There was genuine intent shown by the owner to enable the tenderers to fully understand the project goals, and there was openness in discussing the tender documents. This injected tenderer confidence and the tenderers responded with goodwill. The tenderers brought inconsistencies, vague issues and discrepancies to the attention of the owner. These were rectified immediately and all tenderers informed.

Tender documents

Making amendments to the tender documents should be avoided during the tender period. A freeze on changes to the documentation allows tenderers to develop a tender in a structured way.

Any amendments or addenda released to tenderers require the tenderers' acknowledgment of receipt.

Example

The following example demonstrates bad practice calling for tenders for a project involving civil and building work. Prequalification preceded the tendering. Repeatedly asking for new tenders based on updated information, and extending the tendering period alienated the tenderers. Poor preparation and an indifferent owner attitude left the tenderers reluctant to participate in the project. Fast-tracking the project put a strain on the owner's resources and eliminated any cooperative attitude towards and consideration for the tenderers.

Submission [1]

    Original submission date 25 November

    Tender documents, 100 drawings, $10k cost

    Tender bond (extended twice), $1M

    Addenda, 4 in 4 weeks

Submission [2]

    Submission date for revised tenders 29 December

    Meeting, revised drawings (12 December)

Experience submission

    4 February

Submission [3]

    Submission date for re-revised tenders 8 March

    New specification, 300 drawings

    Negotiations (19-26 January)

    Submission extension date (granted 2 March) 29 March

    Addenda, 6 in 4 weeks

    New 300 pages of specification (23 March)

The tenderers

The owner should understand the expectations of the tenderers, and before offering work, the owner ensures expectations can be met, particularly with respect to profit and cash flow. There is a potential source of conflict if the owner's and contractor's expectations do not match, or if one party's expectations are not met.

The owner should understand the needs and requirements, particularly timetabling of activities and resource allocation, of tenderers, and where these clash with the needs and requirements of other project stakeholders. Conflicting needs and requirements have to be reconciled. The owner should consider the flexibility, adaptability and willingness of stakeholders to accommodate others' requirements. Promised undertakings are minuted.

The motives of tenderers are important issues. For example, a contractor may have geared up for some expected industry work boom, and become over-resourced. When the boom doesn't eventuate, the contractor engages in extremely competitive tendering, or bidding at a loss in order to win work, and then engages in aggressive chasing of claims. The fault of the owner in this case is to accept such a tender.

The main motives of tenderers are:
- Profit.
- Prestige.
- Owner relationship.
- Expansion in present field.
- Expansion in new field.
- Utilisation of plant and personnel.

Tenders have to be viewed in the light of which motives are present.

The tender

The owner reads the tender, and particularly reads 'between the lines' in case the contractor is setting the scene for later claims. Some contractors make a practice of winning work at low prices with the knowledge that they will make their profits on the variations and disputes over poor tender documentation or through cleverly worded tenders.

Owners should carry out own their own tender calculations, and ask 'why?' if a tender is different (particularly in price) to that anticipated. Tenders are closely reviewed to ensure that the tenderer is capable of doing the work for the price. This may avoid the contractor later submitting ambit claims because it had underestimated the cost of the work.

The contractor's proposed work method and its feasibility are reviewed. This may indicate, amongst other things, whether the contractor understands the extent of the work.

Lowest tender price

Accepted practice in many places, both public sector and private sector alike, is to accept the lowest tender price. Low cost is often associated with low quality. The policy of accepting the lowest bid can lead contractors to submitting unrealistic bids. As well, the cheapest price tenderer may not have fully understood the project requirements. The end result is financial trouble for the contractor, the contractor 'cuts corners' to contain costs, or the contractor recoups additional money through claims and extras ('claimsmanship'); the contractor later seeks excuses for time extensions and extra costs in an attempt to recover financially. This practice is supported by a 'claims industry', sometimes working on contingency fees. Over time, this has led many contractors to devise strategies to assist them recoup additional money from projects, and within each project the inevitable outcome is conflict. Accepting the lowest bid may also result in poor workmanship, lesser standard materials, greater supervision, longer completion time, cessation of work due to the failure of the contractor's business, or increased cost of the work. The contractor may focus on recovery of costs rather than the timely completion of the work.

The acceptance of the lowest bid in open competition for major public works has resulted on many occasions in the selection of a contractor [which] suffered from incompetence in estimating, inefficiency in production, and inadequacy in financial backing.

(Antill, 1970)

It's unwise to pay too much, but it's worse to pay too little. When you pay too much you lose a little money, that is all. When you pay too little, you sometimes lose everything because the thing you bought was incapable of doing the thing it was bought to do. The common law of business balance prohibits paying a little and getting a lot — it can't be done. If you deal with the lowest bidder, it is well to add something for the risk you run. And, if you do that, you will have enough to pay for something better.

John Ruskin (1819–1900), English writer and critic

One of the early American astronauts was asked on his return to earth what went through his mind as he hurtled through space. He is reported to have said that a recurring thought was that every part of the spacecraft had been made by the lowest bidder. He was able to rely on NASA's tendering procedures, documentation and quality control systems being equal to the task. The success of the mission depended on NASA being prepared to pay the market price for the equipment and in making sure that it was getting what was specified. Some later astronauts were less fortunate when quality control was allowed to slip.

Cooke, 1991

> *Example*
> A subcontractor submitted a tender, for the supply and installation of external fascias
> and soffits to a building, which was 35% below its nearest rival and 25% below
> the contractor's original budget estimate. The warning bells were sounding from
> the outset, but still a contract was signed. What followed on site was a string of
> delays caused by the subcontractor's complete inability to perform. It was unable to
> staff the work sufficiently, it was unable to keep a supply of material ahead of the
> work, and it was not able to honour its contractual obligations. Its original program
> duration was five weeks. Its actual completion time was twelve weeks. As a result,
> there was an extra seven weeks scaffold hire, the cost of which was passed on to
> the subcontractor. There was also the problem of not being able to carry out any
> external work near the building because of the scaffolding.

The notions of buying work and low bidding are not uncommon in some industries.
A contractor, which buys work, has committed itself to a course of, and has anticipated, claims and disputes in order to make a profit and survive commercially, right
from the time of tendering. Such claims thinking also comes from a contractor that is
losing money on a project, even though the work may have been realistically tendered
in the first place.

There are unethical owners (and contractors in contractor-subcontractor relationships)
which shop around for prices (*bid shopping, bid peddling*). No price is bottom price. One
price is used as a lever to get a lower price from another contractor. *Ethics* is an issue
that needs to be addressed if projects are to be non-adversarial, though some people see
bid shopping as nothing more than good business practice.

Large sums of money are involved in many contracts, and many owners, because they
are making large profits, assume contractors are making similarly large profits. As well,
a spread of tender prices might imply that large profits are involved. In reality, many
project industries operate with contractors only receiving a few percent profit, and curiously the spread in tender prices might actually exceed the profit in the project to any
one contractor. This imbalance and misunderstanding promotes an uneasy relationship
between owner and contractor.

Awarding a contract on the basis of cost alone, often curiously goes hand-in-hand
with an owner's expectation of high quality and service. Conflict might result if a contractor submits a tender with little room for error or change, the contract documents do
not spell out the quality requirements precisely, the quality provided is at a low level,
and/or the contractor's work quality is below expectations.

> Drew's law of Professional Practice:
> *The client who pays the least complains the most.*

Awarding a contract on the basis of cost alone, may also prevent the contractor from
offering something more or different, that could be to the owner's advantage, for fear
of being non-competitive.

Some owners accordingly speak of selection on the basis of 'value for money' rather than 'lowest price'. Selection systems, that use evaluation criteria (track record, claims history, project team, method, management support, quality, environmental record, ...) other than price alone, are preferred. These selection systems send a message to the marketplace that best practice is as important, or more important, than price. It also requires owners to consider these facets when developing project budgets.

Example

A medium-sized civil engineering contractor works with margins of 3% to 5% when winning work by tender, and if successful, expects a growth on the contract value of approximately 10% in variations. Based on these figures, the potential to absorb cost overruns is small, and therefore encourages the contractor to pursue every avenue possible to increase its return and maintain an already small project margin. This leads to disputes and an unsatisfactory work environment.

The contractor believes that a better way to avoid disputes occurring would be for owners to hand pick a small number of tenderers from a preregistration process, have them price a given scope of work, and then negotiate with the preferred tenderer on not only price but other factors which are important to the project, such as quality assurance (QA), occupational health and safety (OH&S), and industrial relations.

A detailed scrutiny of the bids by an estimator should be undertaken. If a low bid is unrealistic, including a fair profit margin, it would be prudent to discuss the tender with the tenderer. Otherwise, acceptance of a bid close to the estimated cost (including a fair profit), rather than the lowest bid, would be recommended.

Nevertheless, there are legitimate contractor motives (to enter a new field of work, prestigious project etc.) which would allow the lowest bid to be accepted.

Unbalancing bids

In a balanced bid, a contractor allocates project indirect costs to activities (bid items) in direct proportion to the direct cost of each activity. In an unbalanced bid, this proportioning is varied so that some bid items carry a greater or lesser amount of the indirect costs. Contractors unbalance their bids to stay competitive, to improve cash flow, to increase profit, or to protect themselves against possible fluctuations in quantities and the time needed to complete work activities. Contractors argue that they have no choice but to adopt such a practice.

In contrast, owners view this practice with distrust. To increase trust, owners need to recognise and understand the indirect costs incurred by contractors. Where possible, indirect costs and particularly major indirect costs, should be separate items in the owner's schedule and the contractor's bid. Fluctuations in quantities will then not have as much effect on the indirect cost spread. Any perceived underhandedness of the contractor should disappear.

Limited time

A common and valid criticism put forward by contractors is the limited time which they are allocated to prepare tenders. Projects, involving design time of months or years, may require contractors to digest the tender documents and submit a tender in a matter of weeks. Insufficient time is allowed to do the planning, estimating, developing a work method, studies etc. necessary for a properly thought through tender. An appropriate tender period is required in order for contractors to develop thorough tenders.

By rushing the process, more advantageous solutions may be missed. The compressed time frame only allows the mechanics of the solution to be tackled, and value adding is missed. Short time frames don't provide time to think and reflect.

Ideally, the owner indicates an intended date for the awarding of the contract. Repeatedly asking for new tenders based on updated information, and extending the tendering period alienates the tenderers.

Van Roy's Law:
Buy in haste — repair at leisure.

Tendering costs

The costs associated with the preparation of a tender can be substantial. The cost of unsuccessful tenders is incorporated into any successful tender as an overhead. Tendering 'on the cheap', because of the reluctance of the tenderer to invest in a tender which it is not certain of winning, can lead to incomplete or inadequate tenders. For example, special contract conditions or certain activities may not be properly costed and an unrealistically low bid results. Paying preselected tenderers to prepare tenders would assist in maximising the up-front understanding of the project and this would have a beneficial impact on the project implementation.

Consideration should be given to reimbursing unsuccessful (genuine, not unscrupulous) tenderers (at least part of) the expenses incurred in preparing a tender, and making this known up front. However such an idea usually meets with resistance from owners. (The downside to the owner is the possibility of paying, but the contractor not putting in commensurate tendering effort. This includes the practice of submitting cover bids.) It has more applicability for large projects where tendering costs are significant, or specialist projects, and with a short list of reputable, reliable and prequalified tenderers.

Tender evaluation, contract negotiation

The range in prices received in tenders can be attributed to a number of factors. For example, a tenderer may have recently won other work just prior to the closing of tenders, and may submit a high bid with no intention of winning, but with the intention of being considered for future work. In other cases, low bids may be the result of tenderers unwittingly omitting sections of the work. Uncertainty in the tender documents can lead tenderers to interpret documents differently, leading to a range of prices.

In order to obtain clarification and resolution of matters in the periods during tender evaluation and prior to lodgement, procedures for contact with tenderers should be developed and strictly controlled. Clarifications sought by tenderers prior to lodgement should be coordinated by a single nominated person, and any necessary addenda or information to tenderers should be sent to all which have received tender documents. Many public sector bodies, due to the need to be seen to be impartial in the tender evaluation process, do not allow any contact with tenderers between lodgement and the letter of acceptance. Often communication channels are cut to reduce the possibility of unethical practices. Without contact for clarification purposes, contracts can be let with unresolved matters that could have been economically or conveniently dealt with in the pre-award negotiations.

Some typical issues discussed would be:

- Access to site.
- Existing ground conditions, services, structures.
- Discrepancies, ambiguities and differences in interpretation.
- Financial matters such as progress payment assessment, payment terms, variation pricing methods (including how overhead and indirect costs are to be handled), cash flow, and a bill of quantities.
- Claims and disputes procedures (not in the sense of encouraging claims and disputes).
- Dates for the submission of outstanding information.
- Program and method.

Provided interviews/*meetings with tenderers*, for the purpose of seeking clarification, are carried out in a structured business-like manner, with a well-prepared agenda, and provided tenderers are not given any indication of the status of their tenders, contact can reduce the uncertainty and assist in the eventual recommendation of a preferred tenderer. Complete records of the meetings are kept (Figure 2.1). Confidentiality of tenderer's information is maintained. A memorandum of understanding is useful to record mutual intent with respect to how a contentious issue is to be handled.

Meetings also serve to help each party understand the expectations of the other party, and provide a forum where inconsistencies and errors in the contract documents can be brought to light. The contractor's involvement in meetings may also start the process of ownership of the project by the contractor; this implies a non-adversarial climate to the meetings.

Some public sector owners may have difficulty holding such meetings and being involved in negotiations because of public accountability and probity reasons.

Tenderer interviews should not merely review that the tender conforms with the tender documents. Tenderers must be able to explain their understanding of the owner's technical and commercial requirements, their implementation procedures, the risk sources that they have identified, and how they have assessed and allowed for the associated risks.

Where prequalification has not been carried out, then the same issues that prequalification covers need to be focused on during tender evaluation. As well, there will be project-specific items such as the construction program and method, and price.

Project	
Meeting type:	
Meeting number:	
Purpose:	
Date:	
Time:	
Place:	

Attendance		
Name	Position	Representing

Item No.	Item/Question	Action/Answer	By	Date

Agreed as a true and accurate record:

Name:..   Signature:..

Date: ....

Figure 2.1. Meeting record.

Tenderers, if allowed, may submit non-conforming tenders as well as conforming tenders. These non-conforming tenders may be based, for example, on some change to the contract documents, design change, method change or timetable change. A non-conforming tender may be advantageous to the owner, but its implications and repercussions need to be studied.

Contractor losses

The main goal of most contractors is to perform the work in accordance with the contract documents, at the price tendered. However, a sequence of unexpected events (for example, bad site conditions or extended wet weather) can change the contractor's attitude, especially if the contractor is making a loss on the project. A contractor may take shortcuts to minimise its losses or at least recover part of its costs in such a situation. Checks need to be made to ensure that the contractor has adequately allowed for risk in its tender price. Often, in order to win work, contractors may ignore risks or believe that by incorporating something to cater for risks, their tenders will be non-competitive; tender selection on other than lowest price can help to eliminate such thinking.

The owner should recognise the need of a contractor to remain in business and make a profit.

A contractor's living is made from the completion of a continuing series of projects. Each project is unique. The reputation and continued existence of a contractor is dependant upon the successful completion of these projects. The profits of the contractor's company are dependant on each project making a profit. The effect is that, if a project does not make money for the contractor, then it is hurting financially. A severe problem during one project could be enough to drive the contractor into liquidation. Each project represents a potential threat to the existence of the contractor. The issue can be that serious.

Completion time

Unrealistic time periods for the completion of the project, and the application of liquidated damages, to safeguard the owner's interests, can introduce adversarial thinking. Owners need to be aware that unrealistic time periods may also lead to substandard workmanship, which may go unnoticed until the project is completed and the asset is handed over to the owner. Pressure on the contractor to perform within an extremely tight time scale, may lead the contractor to claim delays for anything done by the owner.

Ideally, the project completion time should be tendered by the contractor, though it is realised in some circumstances that the owner may have its own commercial or accountability reasons for designating a completion date.

2.4.3 COMMON CONTRACTOR–OWNER POSITION

Pre-contract planning

The failure to do pre-contract planning by both parties, for example, not allowing adequate time to carry out adequate site investigations, to properly investigate construction methods, or to prepare a tender, contributes to disputes, and reinforces the Confucian view that 'in all matters, success depends on preparation.'

Project conditions will change, and this will introduce its own problems, but without prior planning, such problems will take on an added dimension.

Codes of practice

There are a number of codes of practice to guide the behaviour of owners and tenderers. These are largely based on what is considered fair and ethical behaviour for both parties. Conforming to these practices provides a platform from which a good working relationship might be developed.

Communication

Communication lines are facilitated to be open. Breakdowns in communication are to be avoided. A single person in each of the owner's and contractor's organisation should be nominated as a contact person, and all communications channelled through these people.

A restriction to the free flow of information can come from each party trying to protect its bargaining power (both pre- and post-contract award). An owner may feel, for example, that letting a tenderer, or subsequently the contractor, know too much about susceptibilities or weaknesses, may lead the contractor to exploit these.

All communications of importance are confirmed in writing, while good records of owner–tenderer discussions, meetings, telephone conversations and correspondence are maintained, along with reports and programs. Records can be used to support actions taken.

The other party

It is important to recognise the other party's commercial and financial requirements, as well as understand and respect the other party. Insolvency of either party has severe consequences for the project.

A contractor will conduct a project according to how it feels that it is at risk of not being paid, or being paid on time.

The agreement

Both parties have a need to know and understand the contract (to be). A working knowledge of the contract removes the many origins for disputes deriving from ill-informed people.

A dispute is certain where there is a disparity of views by the parties as to their respective obligations under the contract. The scale and complexity of obligations that form part of a contract on a large project make it difficult for there to be an unanimity of view by all individuals involved in the project as to their respective rights and obligations. In reality, contract documents for example, can never be complete, devoid of inconsistency and in need of no interpretation. It is important to attempt to reduce the incidence of opposing views. The greater the mutual understanding of the obligations under the contract, the greater the likelihood of a smooth contractual relationship.

The parties should try to agree on each party's responsibilities and the scope of work, rather than make assumptions as to what the other party will do. '... but I thought you were doing that.' A good rule of thumb is to anticipate the other party to do the bare minimum, until confirmed otherwise. Attention to detail makes it possible for any deviation from the scope to be readily identified and then either rectified or treated as a variation to the contract, whichever is appropriate.

Rates tendered by the contractor need to be agreed as being reasonable to use for changes in quantities.

An instrument of agreement can be used to tie together the contract documents including evidence of amendments or elaboration made during the tender review period, whether by correspondence, memoranda of understanding, minutes of meetings etc.

Disputes Review Board

For larger projects, a disputes review board (DRB), where the cost is shared equally by both parties, and which keeps informed of project progress, is effective. The DRB is a panel, of three experienced people acceptable to both parties, that gives resolutions to disputes as soon as they occur.

This can be taken a step further where the board also looks for potential disputes. The referral of potential problems to the DRB means that plausible arguments are identified at an early stage, the quality of the presentation of claims and their validity are more carefully established, and justification for their rejection is more carefully considered.

The presence of a DRB may in itself lead to an attitude of compromise from both parties in their day-to-day dealings, and act as a deterrent to disputes. Regular appearances by the DRB encourages openness and fair play by both parties and may lead to claims being settled before the need for further action.

On smaller projects, a dispute resolution person (DRP) can act in the same way as a DRB does on larger projects.

Neutral certifier

A neutral person is recommended to administer fairly the contract provisions (as opposed to looking after the interests of either party) — issuing instructions, dealing with variations, resolving errors or discrepancies in the contract documents, dealing with matters involving site conditions, other contractors and suppliers, complaints etc. Preferably the neutral person is chosen by both parties.

Conflicts can be a direct result of this person's inability.

The neutral person has to be seen to be fair and impartial, and reasonable in determining claims, and recognise legitimate differences in the interpretation of work. Harsh decisions may lead to disputes. The neutral person has to display an understanding of both sides of an issue.

Where the person is not neutral, but rather appears to act with partiality or unfairly, the person may not understand the role of the 'neutral administrator', or may have a conflict of interest (for example, the person is also the designer, or the person is an employee of the owner). The person is expected to be even-handed and display judicious qualities, as well as being honest, reasonable and fair.

The certifier should not have a *dual project role*, because this promotes a conflict of interest. For example where the certifier is also the designer, whenever there is ambiguity of interpretation or error in the design, this person is unable to act impartially, and given human nature will interpret the issue in his/her own favour, and against other project participants.

Approvals

It is important that all necessary government and private approvals have been obtained. Times taken for such approvals cannot be guaranteed, and planning allowance has to be made accordingly.

2.5 POST-CONTRACT AWARD STAGE

2.5.1 CONTRACTOR'S POSITION

Scheduling

A schedule is regarded as essential, and would contain, as a minimum, a breakdown of activities, correct logical sequencing, timing (including milestones and end date) and resources. A schedule is not only an important management tool but is also an important tool in avoiding disputes with respect to time performance.

Any delay to critical activities leads to a delay in the end date. Other activities which are delayed and which may become critical can be easily identified. If the schedule is updated regularly, then delays can be identified and attended to early. This may either lead to an extension of time which can be discussed and resolved at that stage, or the work accelerated to allow the completion date to be met.

Not only does a good schedule allow potential problems to be earmarked, it also attaches responsibilities, and can lead to the matters being sorted out at the time before they become disputes.

Example
In an effort to speed the building work up, certain work was performed out if its usual order. For example, the ceiling grid was installed prior to the in-ceiling electrical, data and security cabling being completed. The result was that the ceiling grid was pushed out of shape and damaged.

Monitoring, recording, reporting

Cheop's Law:
Nothing ever gets built on schedule or within budget.

To engender trust from the owner, the contractor is encouraged to monitor and regularly and accurately report progress and events, without bias. Exception reporting backed up by information is useful. Clear and comprehensive *records* of productivity, procurement, variations and incidents affecting technical and financial performance are maintained. The information needs to be easily retrievable in case any question is raised. Many potential disputes can be cleared quickly by providing proof of documentation, correspondence etc.

Reporting needs to be accurate and to reflect the status of the project. Under-reporting can lead the owner to take actions to increase productivity; over-reporting can result in a false sense of well-being if, say, the project is behind schedule.

All project participants are advised of critical dates and activities as the project progresses.

Formal documentation of meetings, actions, issues, milestones, exceptions, delays etc., is kept and copied to the owner.

The owner is kept informed of progress and any problems that can be foreseen. Time/ cost overruns are forecast early, such that the owner does not get surprised with claims later on, and/or apply penalties at the end of the project.

Delays should be reported as soon as they occur or can be identified. Extensions of time should be claimed as soon as the extent of the delay can be quantified. Responses by the owner's representative to the claims should be within a reasonable time. For prolonged delays, interim claims are encouraged. Any change to the timing of an activity may affect the cost. Delays are a common source of disputes.

Site management

Poor contractor site management results in the lowering of the standard of work, and affects the progress of the project. A site manager has to be aware at all times of the materials and resources required to maintain progress, safety and quality of work. Poor site managers might use the owner as an excuse to cover for personal limitations as a manager. A project which is failing to maintain quality or progress raises the ire of the owner's representative, and continual failing of site management ultimately leads to 'show cause' notices being issued and subsequent termination of the contract.

Site management cannot operate to efficient levels without the support of head office. Whilst poor head office management is not a direct cause of disputation, the failure to provide adequate resources and equipment to maximise the performance of the project team, leads to a chain effect of non-performance at the workface, and hence a drop in safety standards, quality and progress. The poor performance of the head office has a consequential effect on the development of disputes.

Subcontractors

Subcontractor interfaces can be a crucial part of any project. Faults on a project commonly show up at interfaces. Checks on interfaces need to be done. Coordination of subcontractors, including access issues, need to be addressed.

Disputes between the owner and contractor may originate in the relationship between the contractor and a subcontractor. Even a dispute on another project can spill over. Programming and coordination of subcontractors by the contractor requires special effort; this may lead to building in extra subcontractor idle time as a result of building in some float to the program, and taking the pressure off all subcontractors to perform to a strict timetable.

Owner awareness

Sometimes the owner and the contractor are mentally building different things.

Example
Ceramic bathroom wall tiles were being fixed using a single dab of adhesive. The owner assumed that the tiles would have their backings totally sealed.

The contractor is sensitive to and aware of the needs and expectations of the owner. Involving the owner in project discussions, gives the owner a sense of involvement and 'ownership' of the project, rather than alienation. The owner is kept informed. Good rapport is established.

The owner's 'hot buttons' are recognised, that is those activities and matters that the owner perceives as important. Examples include reporting, consultation and technical standards.

The owner's expectations are managed; these include project outcomes (budget, completion date, ...), facility/asset evaluation criteria (aesthetics, functionality, long life, ...), and inter-company relationship. The contractor raises the profile of successful tasks and completed milestones within the project.

The contractor ensures that the owner is happy with what it has paid for.

Hill's First Law of Salesmanship:
Treat the customer like a mushroom; keep him in the dark and spread manure on him at frequent intervals.

Quality assurance; Safety; Industrial relations; Environment

The work is executed in keeping with what was offered to be done in the tender, for example, to maintain the promised quality program with auditing. Any *quality* control procedures and quality plan are adhered to. Quality concerns that arise at the end of a project can be costly to rectify.

Example
It was a requirement of the bricklaying subcontract that the bricklayer do all its own setout. After a particular brick wall had been cement-rendered and tiled, it was found that one end of the wall was 50 mm out of position resulting in a tapered cut in the floor tiles. This was unsatisfactory and the wall was subsequently demolished and rebuilt. The cost of the demolition, re-erection, re-rendering and re-tiling was placed on the bricklayer because of the contractual clause regarding setting out of the work. Money was recovered via reduction of the retention sum. The contractor could feel partly to blame here for having inadequate supervision on the project and insufficient quality checks.

Example

After some suspended concrete slabs were poured, the 'as-built' levels were found to be way out of tolerance, resulting in large sections of slab requiring scabbling and topping to either lower or raise finished levels respectively. It was part of the subcontract that the concreter set out levels from a benchmark provided by the contractor. All costs were borne by the concreter.

Mesikimen's Law:
There's never time to do it right, but there's always time to do it over.

Drazen's Law of Restitution:
The time it takes to rectify a situation is inversely proportional to the time it took to do the damage.

Wright's First Law of Quality:
Quality is inversely proportional to the time left to complete the project.

The work is executed using appropriate high standards of *safety*-procedures and equipment; concern for *industrial relations* issues; and with consideration of *environmental* requirements.

A general feeling is that the head contractor should have overall responsibility for industrial relations matters, but seek input from subcontractors, which in many cases are the primary employers of labour on projects, before making decisions. This is so because the head contractor is responsible for the overall performance of the project.

Many conflicts relate to safety standards and work site conditions. Most of these conflicts can be prevented from escalating by maintaining regular communications between project participants, and addressing items before they become strong issues.

For each project, employment awards/agreements are usually jointly developed. Changing or not adhering to an accepted award/agreement without unilateral consent only causes problems for one of the parties, namely the party that ends up worse off, and this leads to discontent and a potential conflict.

Projects progressing without troubles and being effectively managed in these areas tend to have fewer disputes. Clear goals for their management, and plans for their implementation are required.

Variations and changes

If possible, variations should be kept to a minimum. Urgent variations should be avoided, because work generally has to start before they can be properly planned and meshed with other work, and before agreement can be reached on their valuation. All variations disrupt the orderly carrying out of the work, and involve a disproportionate amount of effort and resources.

Changes may come to the contractor by way of owner instructions and revised contract drawings. Instructions need to be explicit as to the extent and nature of the work. Having values agreed relies on communication, honesty and integrity along with a few necessary negotiating techniques. On receipt of an instruction, the contractor should advise, as soon as possible, whether it is a variation and also advise of the cost differential, and possibly give the owner options to make further scope alterations to stay within budget. Full cooperation with the owner, by providing complete and accurate breakdowns of the costs of variations and their time impact, backed up with supporting evidence, such as quotations, assists the negotiation to reach agreement and promote confidence between the parties of a mutual trust.

Variations and changes may also be instigated by the contractor. They are checked for approval, before proceeding. If there are no rates of pay covering the work, agreement should be reached on costs prior to performing the work. The contractor should not perform work outside the agreed scope without prior approval of the owner.

Undocumented, unsigned and undated variations can lead to exploitation by either party.

Example

In landscaping/irrigation work, the owner verbally increased the scope of work and the contractor passed this on to the irrigation subcontractor. But after considering the budget and a two-day delay, the owner decided not to vary the original scope. However in the meantime the extra pipework and sprinklers had been installed by the subcontractor. Negotiation then ensued between the owner, the contractor and the subcontractor. The subcontractor threatened to pull out the extra pipework and sprinklers. The contractor tried to establish the subcontractor's minimum price that would not justify its pulling the fittings out, but would still satisfy the owner. Everyone had acted too quickly on a verbal instruction.

Where there are significant changes, it may be necessary to manage the changes as a separate project. Where the owner requires acceleration (project compression), this is a similar case.

2.5.2 OWNER'S POSITION

Contractor awareness

The owner recognises the possible vulnerability of the contractor to delays, risks beyond its control, timing of payments and so on.

Access

Many disputes start because of the contractor's lack of access or only partial access to the site, by the contracted date. Prior planning on the part of the owner will overcome this.

Monitoring

Monitoring and the accurate reporting of progress and events are encouraged. Project documentation, including changes, minutes of meetings etc., are recorded faithfully, without bias, to engender trust from the contractor.

Prompt *inspection* of any work requested by the contractor allows the contractor to proceed with minimal delay. Any delays brought about by the owner in the approval of work prior to it being covered up, is an area where problems, claims and frustration occur. A detailed knowledge of the contract's stated quality levels enables the owner to act promptly to enforce the stated levels. Such an approach reduces conflict, particularly later in the project.

Example

On a wharf construction project, the specification for the concrete reflected the salt and wet environment requirements, and the presence of additives and plasticisers. An on-site slump test, just prior to concreting, produced poor results which led to the rejection of a batch of pre-mixed concrete. The same result followed in two further batches of concrete. Although the rejection decisions brought heated debate, they did have the good effect of setting the standard of material quality which was to be maintained throughout the project. In setting this standard at the start of the project, the owner's representative was not put in the vulnerable position that would have occurred should s/he have stepped up the material quality requirements midway through the project, after having accepted lower quality at the start of the project.

Timely inspections also prevent the contractor claiming delays due to tardy performance on the part of the owner. Delays in inspecting work may also result in the contractor covering up the work, inevitably forcing the owner to accept the work, poor or otherwise.

Variations and changes

The number of changes ideally should be minimised. Ordering extensive changes, after contract award, is undesirable. This includes the introduction of untimely design revisions (with or without allowing commensurate time extensions for the completion of the project or recognising the contractor's right to impact costs), and introducing changes under the disguise of correcting deficiencies. Urgent changes should be avoided.

Changes disrupt the orderly performance of the work, and demand a disproportionate amount of management time and resources.

Example

In the construction of a building, half of the internal walls were cement-rendered brickwork. After painting was complete on these walls, skirting boards on, carpet down and ceiling completed, the owner decided that the finish on these walls was not good enough for the function of the building, and requested that all the walls be plaster set. This meant a two to three week delay on job completion, with skirtings having to be removed and replaced, walls repainted (three coats) and carpet covered with plastic for protection. The painter in particular had to put extra staff on to complete this extra work.

In this instance, as it was a legitimate variation, the problem was not of payment for work done, but more of a logistical problem in getting the work done within expected time constraints. The variation impacted on the project schedule and on other trades.

Alterations, brought about by the owner's actions, to the program and their effect on the cost of the project, are communicated to the owner. Claims might be made for employing additional people and for working outside normal working hours, attracting penalty payments, where the program is compressed or acceleration is required. *Acceleration* and *suspension of the work* may generate disagreement over claimed extra costs.

Owners should consider paying for variations as soon as possible after their completion. Alternatively owners should consider making without-prejudice payments for variations that have to be carried out before a price has been agreed.

Example

On a large infrastructure development project, the owner made design changes in the name of public welfare and safety with little justification. This imposed additional costs that the contractor was forced to absorb, because the contract was fixed price and did not include provision for increases due to public safety. Poor owner management of the project was corrected through the public safety screen, and the extra cost allocated to the contractor. Numerous delays by the owner in the approval of changes and design specifications, even well into construction, considerably escalated costs. A dispute obviously followed.

Example

Sometimes an owner orders work (without stating that it is extra work), if it thinks that the contractor won't object. This order may be slipped in among genuine contractual orders, hoping perhaps that it will not be noticed, and language may be used to disguise the real nature of the work.

Payment

Disputes commonly occur over payments.

Both the owner and the contractor face cash flow problems from time to time. It may be possible to arrange a flexible payment procedure and credit system, such that both parties can work together to balance their finances. Contracts administration procedures could allow for this. Cooperation will increase and adversarial attitudes will decrease.

Prompt handling and payment of progress and approved variation claims is an excellent catalyst for a good working relationship. Late payment is not. Recognition that the contractor is entitled to payment, and quick payment enables the contractor to plan its work, and order and pay for further materials. The goodwill established with such an approach is a further step towards maintaining a non-adversarial relationship.

Example

Commonly, lump sum contracts provide for periodic payments to the contractor. These payments are 'on account' only, and do not imply that the owner believes that the work has been done satisfactorily; the owner's legal rights are still open. The last payment occurs on 'practical completion'.

Commonly, contracts do not make any provision for the withholding of payment by the owner, should the owner consider the work unsatisfactory. Also, progress payments may require payment for materials and equipment delivered but not yet installed, and this may cause concern and hesitation from the owner.

Late payment may be a breach of the contract. This may set in train a sequence of events. The contractor might stop work until the matter is resolved. Subcontractors might not be paid. The matter might go to arbitration and the contractor's costs start to accumulate. If the owner is unsuccessful in arguing against its breach, then the owner will pick up the contractor's costs.

Time extensions

Time issues are examined continuously. If there are justifiable extensions of time, these should be approved progressively; to hold off approving them until the end of the contract work invariably leads to conflict.

Interference by the owner with the sequence and timing of the contractor's work, whether directly or indirectly, for example through delays in the delivery of owner-supplied equipment/materials, contributes to claims and disputes.

Unnecessarily long time delays may affect the borrowing and loan repayments of either party.

Liquidated damages

The provision of liquidated damages in a contract is for the owner to recover money from the contractor should the project be delayed by the contractor beyond the date for practical completion. The amount of liquidated damages (for example, dollars/week) nominated in the contract is not a penalty to the contractor for not completing the project on time, but is a true estimate of the owner's likely costs as a result of not having the project completed on time. However, too often owners specify an amount for liquidated damages in the hope that it will be enough of an incentive for the contractor to finish on time. As well, most contractors will make every effort to complete the project on time as a matter of pride. Inevitably, as soon as liquidated damages are hinted at, the contractor will look for ways of making up the lost money, for example by seeking errors and omissions in the contract documents, or by taking short cuts (in time and workmanship). Liquidated damages should be applied as a last resort, and only when the owner has genuinely incurred a loss due to late completion of the project. Liquidated damages amounts stated in the conditions of contract should be commercially realistic.

Rather than impose a remedy, such as liquidated damages, an amicable agreement is far more preferable.

Some humour on liquidated damages:
Liquidated damages are the penalty for failing to achieve the impossible.
The completion date is the point at which liquidated damages commence.

2.5.3 Common contractor–owner position

Contract documents

Experience suggests that (for each party) a (signed) complete copy of the contract and a copy of the current program (design, construction, maintenance, ...) needs to be maintained in a single location, within easy reach, and accessible on a needs basis, and senior project team members be well versed with both the content and intent of the contract clauses.

It is not uncommon that one or both of the parties' contract administrators have not read or do not understand the contract documents, and only read them when a problem occurs. Part of the problem here is a result of a lack of suitable education and training. Advice is not sought on rights and obligations until something goes wrong.

A drawing register should be kept, and different issues of drawings clearly identified by date and purpose. Electronic versions of documents place a further challenge on document management.

Records

Contracts are great generators of paperwork. Thorough records during a project protect both parties and ensure that disputes do not arise from misinformation. Complete and auditable documentation systems are set up. At least, if disputes occur, they can be kept to dealing with agreed facts and not distorted memories. Complete records provide for more efficient and meaningful dispute resolution, should a dispute not be able to be prevented. Items that need to be recorded, documented or issued include:

- Transmittal records of drawings, memoranda, letters.
- Site instructions.
- Contract modifications, changes of scope.
- Requests for approval.
- Test results.
- Records of delays from rain and other unforeseen circumstances.
- Records of receipt of owner-supplied materials.
- Receipts for insurance, levies, securities.

Each significant communication is committed to paper, and signed and dated. Efficient filing procedures are necessary. Adoption of the use of diaries may be either a contract requirement or a general good practice. A document control system should be implemented to ensure that correspondence is processed, distributed, answered or acted upon and filed. The document control system also ensures that the latest revisions of all documents are being incorporated in the work.

Claims

'Claims' here generally implies claims under or within the contract (that is, those covered by the provisions of the contract), but may also contain elements in tort and under statute (extra or outside the contract).

Commonly, for contractors engaging in claimsmanship, claims will be based on the owner not administering the contract in a timely fashion (leading to delay cost claims), and on the quality of documentation (leading to a variation claim); the claim documentation may look like an ambit claim, but is directed at extracting something from the owner (perhaps through compromise in order to avoid the costs of litigation) and contributing to the contractor's profit.

Some humour on claims:
A claim is the contractor's guess at the amount of money needed to transform a net loss into a gross profit.

Example
Contractors may submit claims for extra work, real or alleged, and not worry unduly about the lack of any formal orders for extra work. Owners commonly don't reject claims on the grounds of there being no order, but rather treat all claims seriously. Where the owner does not take a stance on the claim, this represents profit to the contractor.

Outside of claimsmanship, genuine claims in many project industries appear inevitable. Careful management can prevent them turning into disputes. The contract documents should address time limits on claims and the early response to claims; both factors force the parties to confront problems early while the issues and events are still clear in the minds of all. By resolving claims as they occur, allegations can be compared with prevailing facts, when the impact of delay, disruption or loss can be seen first hand. When contractors delay making claims, the late claims can cause surprise and hostility to the owner; claims need to be 'on the table'. Unresolved claims should not be allowed to build up.

Notice provisions can work against good feeling on a project. They force the contractor to notify the owner early of troubles, but by so doing may give the owner the impression that the contractor is claims-oriented and not interested in a harmonious relationship. Some owners, quite naturally, react defensively to any notices issued by the contractor.

The owner's representative is kept informed at all times of the measures being undertaken to overcome problems. This avoids the owner's representative rejecting a claim because it had 'not previously heard of it'.

Contractual claims are discussed whilst in 'draft' form. An opportune time to discuss claims is during the regular project meetings, which might be fortnightly or monthly. Additional meetings might be scheduled if progress is not made during these regular meetings. Contractual claims are itemised along with their value at the time, claims to date, and revised contract work value, in order spell out the rolling financial status of the work.

Claim notification should identify the broad reasons (including contractual) for the claim, so that the other party has the opportunity to mitigate any effects. A goal is immediate agreement of as many components of a claim as possible. Early payment of claims should occur for the agreed parts. This assists the contractor's cash flow, and enables the owner to adjust its budget before problems arise. Claims resolution should occur when the people directly involved are available, and haven't moved on to another project. The legal process should not be used to delay or avoid meeting justified claims.

Claims should be well supported by facts, on which the claim is based, so that the credibility of the contractor is not undermined. (This implies keeping adequate records.) The practice of making ambit claims is discouraged. If a contractor claims only what it honestly feels is justified and can support these claims, then an element of trust is developed. Unrealistic and unsubstantiated claims promote ill feeling, and shouldn't be used as bargaining 'chips'. The temptation to pursue claims aggressively should be avoided.

When presented with a claim, people should not take the claim personally, but accept it as something which is following contract procedures, whether that claim is sustainable or not. Frequently owners blame contractors for claims. While the claim is under review, work must proceed, and issues currently on the table should not influence the outcome of other issues that may arise during the currency of the review.

Modern society emphasises the pursuit of profit and the 'holy' dollar. Costs are scrutinised, dissected and then re-examined over and over to maximise profits. There is the possibility that both the contractor and the owner will apply such thinking when it comes to claims, instead of viewing claims in a more just and fair way.

The valuation of claims may lead to disagreements between the parties should the claims be seen as excessive. There may be a tendency for contractors to 'load' their claims as a means of recovering money not allowed for in a competitive tender price. The administration of such claims should be treated in a proper and sensitive manner by the owner. An outright rejection attitude by the owner can only exacerbate the 'them and us' attitude between the parties.

Example

A pipeline contractor working for a local government body encountered more rock than was anticipated. He completed the work and put in a claim for the extra work, which was quite legitimate. The contractor was reasonably new to working for this owner and stated that he was not aware that an engineer must measure the quantity on site and both parties were to sign for this agreed quantity. The owner refused his claim for a very small sum in extras because the conditions were clearly stated in the contract. The owner believed that was the end of the story, and the contractor would not make the same mistake again. However the matter was put into legal hands by the contractor at great cost to both sides. Both parties had received bad legal advice. The owner's decision stood.

Owners should view claims fairly, and recognise the need for a genuine return to the contractor operating a business.

A checklist might be used when reviewing claims:

* Completeness of work.
* Standard of workmanship.
* Valuation acceptability.
* Compliance with contract documentation.
* Completeness of supporting documentation.

Example

Many events cause delays on projects. Notices of likely delays are submitted by the contractor to the owner, as and when relevant events arise. Every possible delay event is covered by a notice. The owner responds to such notices by challenging the event notified as not likely to cause delay. The contractor reacts to such correspondence. Animosity develops between the contractor and the owner.

The owner assesses the contractor's time extension claims on a monthly basis and issues a formal notice of determination. Almost invariably there are differences between the time claimed and the time allowed. Claims are disallowed mainly on the ground that the event in question does not delay the critical activities.

The gap between the contractor's claimed time quantum and the allowed time quantum grows wider and wider as the project progresses.

Example

As a project progresses, the paper war between the contractor and the owner escalates. Day-to-day administration degenerates into a trial by correspondence on many significant issues. The relationship between the contractor's project team and the owner's project team becomes polarised.

The contractor's claim strategy is simple — claim for everything. Not surprisingly, many claims are rejected. No serious attempt is made by either party to resolve the substantive question of the contractor's entitlement to an extension of time.

The contractor suffers loss on the project. The strategy is to recover as much loss as possible through claims.

The owner believes the contractor isn't doing what it contracted to do and is putting in unjustified claims for delays which do not affect the critical path. It appears to the owner that a number of the contractor's problems are caused by either its lack of management or coordination of the work, or that of its subcontractors.

Dispute source identification

It is rare for a project to be completed without unforeseen problems occurring. The associated question that arises is which party should pay.

It is suggested that the parties be proactive rather than reactive where anyone can see a potential issue arising. Concerns are notified to the other party.

A forum is provided for project personnel to offer early warning signals, and suggestions/improvements that could enable management to head off potential disputes, perhaps by resolving the problem at 'grass roots' level.

Problems are responded to early. As soon as is practical, any problems encountered are brought to the attention of the other party. An attempt is made to understand the problem, and to identify the problem's stakeholders and ramifications for the rest of the project. Problem denial, and delay in dealing with the problem are avoided. The air is cleared. Many disputes arise from relatively insignificant issues which are not identified or correctly responded to when they first occur. Unaddressed apprehension festers, leading to problems being larger than they otherwise should be.

When a problem occurs and a project starts to run behind schedule, the parties should consider the likely other effects on the project, on other projects in the future and on the people involved in the project. The process of solving the problem at its core is kept as efficient as possible. There may be a tendency to focus too much on the particular problem and lose sight of the project.

Where a dispute may arise out of the action of some other party or from a situation that has arisen, the possibility is highlighted and the two parties work toward the elimination of the dispute occurring. The matter is not to be taken personally. It does not help the situation to confuse a problem with personal relationships.

Authority to resolve problems is delegated to the lowest level possible, or to the level best able to deal with the problems.

Each party works in a manner such that neither party is considered to be the cause of a dispute.

The parties have to be realistic, and understand the limitations of what can be achieved.

Mueller's Law:
You only have a problem if you think it is a problem.

Nef's Law
There is a solution to every problem; the only difficulty is finding it.

Application of project management procedure and bureaucracy can obscure creativity, leading to inflexible and ineffective approaches towards dispute resolution.

Cooperative joint problem solving, where a wide range of possible solutions are developed, and a 'best' solution selected, is encouraged. For this to occur, an understanding of the other party's goals is needed, such that options from both points of view can be considered and explored in the resolution of a problem. If a problem is only looked at from one perspective, only some of the possible solutions will be found, and these may not meet the needs of the other party.

The solution that is finally accepted by both parties must be accepted without any lingering resentment by either of the parties. If any resentment of the solution exists, this has the potential to affect future problems, as the resentful party may feel that it again will be hard done by, and this will erode the trust and respect for the other party. If any concern is held about a proposed solution to a problem, this must be communicated and not left unresolved.

Negotiation

Problems are resolved through negotiation by the personnel involved in the project. Genuine discussion and negotiation is encouraged. This, at least, means the parties are still talking to each other. There could, however, be the perception that compromise, implicit in much negotiation, could override legitimate claims. This is countered with the advantages of pragmatism offered by negotiation - negotiation offers speed, low cost and maintenance of relationships between the parties. Lateral thinking to find a solution is encouraged.

Resources and time, and organisation support are put towards this aim. The problem is prevented from evolving into a dispute.

Each side will not want to acquiesce to the other by giving an early concession to the other's view. The personalities of the people involved have a large bearing on negotiation proceedings. If necessary, a third party neutral is used to assist the negotiation. In some instances, the parties are amicable and keen to work together in finding a mutually agreed solution. In other instances, the parties are unwilling to cooperate, and formal dispute resolution procedures may be the only available option.

When people are backed into a corner with threats of excessive additional costs and delays, they tend to retaliate with excessive demands of their own, making resolution a lot more difficult.

Each party should consider whether a dispute is worth pursuing. There is little point in having a Pyrrhic victory that leaves the project worse off in terms of time and cost.

Both direct and indirect costs have to be taken into account when weighing up whether to 'stand your ground' or 'cave in'. Situations where this could occur are on receipt of an instruction or demand from the owner or owner's representative, or when work to be done by others, and is required by the contractor, is not completed on schedule or satisfactorily.

Interrelationships

Consideration is given to the other party's point of view and from where the other party is coming. Other people's opinions are respected.

A cooperative environment is established where the probability of reaching a satisfactory resolution is increased. Dominance posturing is avoided. Aggrieved parties have grievances resolved to their satisfaction.

Regular non-dispute meetings are held to develop team spirit and to get to know people from the other party. A sense of mutual interest among stakeholders is cultivated.

Diplomacy and understanding are employed in all dealings. People act with appropriate moral and ethical standards; deceit and dishonesty (for example, covering up poor workmanship) are discouraged because these lead to project disharmony.

Some owners are reluctant to assist the contractor in solving design problems as they arise, because this is seen as the contractor's responsibility. In reality though, the cost and time extension become grounds of a claim and the owner ends up expending resources to counter the claim. These resources would be better applied to finding practical solutions to the original problems.

An alternative approach is that an extended project team, including both owner staff and contractor staff, might be established. 'Closed door', frank discussions (equivalent to 'toolbox' meetings) are held, where anyone is able to voice concerns and highlight any potential problems, which can then be mitigated in the best possible way from the project's viewpoint. The active involvement of people in this extended team and their openness would encourage relationship (both formal and informal) building and trust, which would also contribute to reduced problems.

Reviews

Regular design review and progress review meetings are held to discuss and resolve early any issues which could possibly lead to a dispute. Impending problems are given advance warning. The forum should be reasonably open for debate. A timetable is set for review meetings.

Areas of review would include:
- Management issues.
- Financial status of project.
- Current invoicing/payment situation.
- Technical design and construction aspects.

Both parties are kept well informed of the current status of the project, giving minimum cause for conflicts.

Communication, rapport

Good communication practices are established between the parties, such that differences and complaints can be identified early and dealt with. Dialogue is not confined to paper warfare. Everyone is kept 'in the loop'. Rapport is established between the contractor's and the owner's staff. Close coordination is established between the parties. There are no interface and coordination problems between the parties. Contact people in both parties are identified early. Lines of communication and authority levels are defined.

To eliminate potential confusion and misunderstanding, it is preferable to restrict the transmittal of information to a defined number of appointed recipients.

Example
One contractor organisation — Requests for Information, Contract Instructions, Shop Drawings and any design alterations are conveyed to a Project Coordinator, and then distributed to the persons involved. Day-to-day contractual issues such as program deviations, are dealt with directly by the Contract Administrator or Project Manager. Specialised functions such as relating to Safety and Quality Assurance have dedicated administrative persons, and are contacted directly. Contact details of appointed key personnel are made known to the other party.

Communication lines are open and facilitated to be open, rather than the parties not talking and working at crossed purposes. This includes a pre-construction meeting to discuss the expectations of both parties, and establish relationships between the owner's staff and contractor's staff. Open talking can change people's focus from being antagonistic to cooperative.

Example
On a recent project, the contractor invited the owner's engineer to be involved closely in the project's management and was very open with regard to the contractor's commercial position on the project. This was discussed at regular meetings. It also had a well-documented management process for presenting issues such as non-conformance. This open management style, shared by all project staff, was highly productive for the project.

Each party must be able to freely communicate its opinions, ideas and concerns to the other party. This can help clarify uncertainties, improve the overall project, and lead to the situation where both parties work together to find solutions that best meet the needs of both. The communication must be undertaken in such a fashion as not to be overtly passive or aggressive. Passive communication does not address the issues that lead to conflict; rather it tries to play down their importance, and invariably the needs of one or both parties go unmet, forcing them to take a more formal dispute approach. Aggressive communication can lead to hostility between the parties, stifling the opportunity for the mutual resolution of a problem and creating potential barriers to communication on future problems, and a willingness to freely try and solve problems.

The other party is kept informed of relevant information as it comes to hand, so that it has the greatest possible time for its own planning. Withholding vital information, or its late delivery, leads to distrust, ill feeling and potential disputes.

Resources are invested in formal and informal communication with adequate frequency. Communication is not allowed to break down; the parties remain on good terms. Regular meetings of stakeholders are useful in keeping all parties informed and aware of the current state of the project. Any grievances or reservations about what is occurring can be raised and dealt with before they become big issues. Familiarity with the project promotes unity, and understanding of what is trying to be achieved.

Before sending any correspondence which contains either a claim, instruction, request or any words that would cause the other party to take umbrage, communication either in a meeting, face-to-face, or via telephone is needed to arrive at an agreement suitable to both parties or to alert the other party to the correspondence. This additionally sets up an avenue for negotiating other, more serious matters.

All communications are confirmed in writing.

Sometimes directions provided by the owner or the contractor to the opposite's representative are not divulged to other staff with an interest in the matters, leading to perhaps uncoordinated or wrong actions on the part of these staff.

Personnel, teams

People are an integral part of running projects. Project teams are carefully organised with people who can work together. Appropriately qualified (technical knowledge, experience, skills) personnel are assigned to each task. Consultants and subcontractors with personnel who can work together harmoniously with others, are selected. Abrupt and abrasive characters create ill feeling. Some combinations of personalities just do not work. Personality clashes can develop, leading to antagonism.

The application of a little bit of psychology to a situation can be of great benefit. It is an advantage to be able to understand and foresee how other people may react. Project personnel need to be aware of any 'stress' or 'strain' in team members, and respond accordingly.

It is important to know the employees. If an employee has a fear of heights, it is silly to put that person where s/he is expected to work at height. From day one, that person will find ways to cause problems such that s/he never leaves the ground. Some employees will cause problems, inevitably leading to conflict and wasted time and money. A smarter approach is to build a team that is happy and productive.

Some people thrive on conflicts and disputes. Certain personalities are known to cause conflict on projects. People such as these should be avoided, or require special handling, and assistance may be needed. The owner's inspector, overzealous on trivial matters, and the contractor's project manager, who is claims motivated, are to be avoided. On some projects, battle lines are clearly drawn, with people from both parties solely looking after their own interests.

The level of adversarial attitude is very much dependant on the attitude of senior management. If the management is seen to condone or even encourage this attitude, then it will prosper. If, on the other hand, the management makes an effort to foster a good owner–contractor relationship, then the level of aggression will be diminished.

The legal nature of contracts can create a rigid environment that may stifle the development of a team approach; it can create a defensive attitude between contractor and owner.

Initially when people meet, they are polite and generally not intimidating. This character can change with time and circumstances.

All project personnel should get to know personally the people from the other party. Does everyone like each other or do personalities clash? There is a better chance of keeping problems in check if people like each other, and with a bit of 'give and take', problems can be resolved. This is not so easy if people don't like each other.

The responsible people in each party need to be identified by the other party — who is accountable, who is empowered to make decisions, who issues instructions, and to whom communications should be directed.

As the work unfolds, the strengths, weaknesses and overall ability of the various individuals implementing the contract will become evident. It is important that the selection of individuals to perform roles in the project remains flexible. Mismatches in individual calibre can lead to conflict.

Pareto

The countless smaller issues, which crop up during projects, are down-played and focus is placed on the larger substantive problems which both parties have a vested interest in resolving. Nevertheless, the smaller issues are not ignored. It is recognised that sometimes it may be more cost effective to forego the smaller issues.

Authority

People, from both parties, should be given the necessary authority and responsibility to deal with the day-to-day issues. Large organisations may only give the necessary authority to people high in the hierarchy, but it is suggested that people lower down, and dealing with the day-to-day issues, may be more appropriate.

Contracts administration

The parties should be conversant with the legal, commercial and technical aspects of the contract, and manage the contract provisions. But project personnel may place too much emphasis on managing the contract instead of managing the work. This can lead to substandard work management practices, while emphasising adversarial attitudes between the parties.

There is a need for competent contracts administration/management that has the support of senior organisational personnel. Contracts administration is sensitive to the calibre of persons involved. Poor contracts administration can lead to claims (costs and time), disputes, increased costs for both parties, and a deterioration in the relationship between the parties. Experienced and well-resourced administrators are recommended for both parties.

Both parties should adopt a positive attitude to the project and cooperate to avoid or minimise delay and disruption to the orderly performance of the project.

Large organisations tend to have elaborate contracts administration procedures, covering the practices of payment processing, claims administration, variation approval etc. Failure by either party to understand and appreciate the other's contracts administration procedures leads to dissatisfaction between the parties. There is a need to make contracts administration procedures transparent. However, against this, it could increase anxieties in the project staff who may feel that the other party is sitting there waiting to pounce on every mistake.

Effective procedures need to be in place to ensure effective contracts administration. These procedures include:

- Prompt attention to correspondence.
- Prompt assessment and processing of progress claims.
- Prompt responses to requests for information (RFIs).
- Keeping of good project records including that on project activities.
- Any time limitations within the contract are adhered to, in order to avoid claims being time-barred.

Good contracts administration, and in particular good project records, assist in the settlement of claims, and give each party the ability to defend its position should a dispute develop.

Certifiers and inspectors being independent, and not being paid solely by the owner, ensures that they take neither the owner's side nor the contractor's side on any issue. Being paid by one party introduces a conflict of interest, and possible partiality and unfairness. Certifiers must be realistic and be prepared to listen to the contractor's point of view for example in the event where the designer has specified requirements which are impossible to achieve (for example, 100% compaction on subgrades with extremely low bearing values).

Project *meetings* play an important role in ensuring the smooth progress of the project and the elimination of potential difficulties. The *initial project meeting*, held prior to work commencing (sometimes termed a 'pre-construction meeting'), attempts to establish the ground rules for the project, particularly administrative matters, and includes discussion on:

- Communication processes and protocols, including meetings, correspondence, records.
- The extent and method of coordination.
- The role of the owner's representative and certifier.
- Roles and responsibilities of the team members, subcontractors.
- Administration and site matters, including date of possession of the site, inspections, payments, certificates, time bars for notices, completion date.
- The adoption of a common drawing format and numbering system to allow easy drawing exchange between the parties.
- Procedures for handling requests for information (RFIs), claims and variations.
- The contractor's proposed method of work; work evaluation practices; quality assurance system.
- Provision of insurance.
- Security, guarantee, bonds.

- Appropriate safety procedures and training such that accidents are not the cause of delays.
- Schedules, progress monitoring and reporting — scope and frequency of reports.

An initial project meeting can provide an opportunity to air problems and differences, and to provide last minute information to avoid later surprises.

Ongoing meetings may be for the purpose of:
- Exchanging information.
- Reviewing performance and progress.
- Identifying and dealing with potential problems (industrial, technical and contractual) in a timely manner.
- Discussing variation orders, and claims.
- A first attempt at resolving any disputable matters.
- Decision making.

Example

Soon after the contractor commenced work, it sacked one of its site welders because of continuing poor workmanship and absenteeism. It turned out that this particular employee was a close relative of a high government official. Following this sacking, customs officers came onto the site and ordered the expatriate staff off the site as they were still on visitors permits, with which they had entered the country, and had not yet been issued with working permits and were therefore not legally eligible to work. Through the regular site meetings prior to the expulsion of the contractor's expatriate staff from the site, the owner's representative had been kept continually informed of the contractor's predicament. Through this regular contact, the owner's representative was able to quickly advise the owner that it would be in the owner's interest to arrange for appropriate permits to be issued as soon as possible, else project disruption and potential claims could result. Permits for the expatriate staff were later issued and a potentially disastrous situation soon diffused.

Meetings, inspections and correspondence should be carried out in a professional manner, with the rights of all parties maintained at all times. Minutes of meetings are kept, re-read before the meetings adjourn to ensure correctness, and circulated for action as soon as possible after the meetings.

Any indication from the contractor that it is being delayed or incurring extra costs as a result of actions of the owner or owner's representative, or by circumstances beyond its control, needs to be investigated. Any differences or difficulties that surface at meetings need to be resolved as soon as possible and not be allowed to deteriorate into slanging matches with further allegations and counter allegations at subsequent meetings.

Example

In one large public sector organisation, the senior project managers and directors spend the majority of their time involved in lengthy meetings, that achieve relatively little. Specialists are involved in discussions, that should be resolved in a separate forum. This lack of discipline bores other members of the meeting, and wastes their already scarce time. A lack of interest in the proceedings results, and the meetings adjourn with the more pressing issues not being addressed within the time allocated. Such disregard for the value of time and the blind application of procedure for its own sake is a constant source of frustration for project staff.

Many of the meetings result in a series of excuses from staff as to why actions requested of them at previous meetings have been carried out late, poorly or not at all. Each meeting results in a new set of promises ready to fuel a fresh collection of excuses at the next meeting.

The origin of these problems and subsequent disputes can be linked back to the fact that decisions are allocated to a committee rather than an individual, and due dates are sufficiently vague to avoid responsibility. Ambiguity of responsibilities results in inaction or poor performance.

2.6 CASE STUDIES

2.6.1 CASE STUDY — ROAD DEVIATION

This public sector project used an earthworks and drainage contractor and a pavement and drainage contractor to do the work. A partnering charter was included to help reduce and manage disputes. The project was only moderately adversarial in nature. All disputes, with the exception of one coming from a latent condition claim, were addressed at site level, and all issues were resolved within one year of completion of the construction. This was an excellent result which was felt, at least in part, to be due to the implementation of effective partnering. A small team of two engineers and three surveillance officers administered all contracts. Close contact was maintained at all times with the contractors, with monthly project meetings and an open door policy with all parties on site; this instilled a sense of ownership across all parties and developed good teamwork. While many issues arose during the project, most were resolved quickly. The structure adopted to review, and escalate issues within defined time frames allowed potential problems to be addressed at the appropriate level of responsibility in a timely fashion. This allowed the work to continue, and issues to be resolved at the workface, in the engineers' office or with the owner. The project was successful for all parties - the project was completed ahead of time, slightly over budget (mostly due to a latent condition claim) and with what appeared to be contented contractors and a contented owner.

The contract documentation was drawn from previously used and refined standard parts, that were familiar to both the owner and the contractors. To tender for the work, the tenderers had to be prequalified, to ensure that they were competent and capable of completing the work. The dispute resolution methods described in the general conditions of contract were not particularly good at resolving disputes quickly, but the implemented partnering procedures worked sufficiently well to negate this. Added to this was a mature owner, with many years experience with large-value contracts of a similar nature.

The project had formal partnering procedures developed early in the project, and a sense of trust and cooperation was developed.

Exercise
Summarise the practices that led to this project being undertaken in an adversarial/non-adversarial fashion.

2.6.2 CASE STUDY — HIGHWAY CONSTRUCTION

This public sector project used a joint venture contractor and a consultant contract administrator. A pre-construction risk assessment session was conducted to identify any areas of concern that should be monitored during the construction. This was a good idea, because this was the largest project to date for the owner and the owner's specification and other contract documents were still in a developing stage. The owner and the contractor had a very good working relationship and both were keen to ensure ongoing work was maintained between the two. (The close relationship between owner and contractor can be demonstrated by the fact that the contractor had a position within its site team seconded from the owner's organisation, to train and develop the owner's staff. This position was negotiated after the contract award.) It was refreshing to see this close relationship between the owner and the contractor. This close relationship did however have its drawbacks, in that at times the owner's contract administrator was short-circuited by the contractor going straight to the owner. While this was frustrating for the team of the owner's contract administrator, it facilitated quick dispute resolution, and in an overall sense was good for the project. A small contract administration team (three engineers, a quantity surveyor and two surveillance officers) was used by the owner. The relationship between the contractor and the owner's contract administrator was not particularly close, but this was possibly due to the close relationship the contractor had with the owner. This was advantageous in this set of circumstances, but may not be so in others.

This project was successful. It was the largest project of its kind for the owner. As well, the owner had no experience with the documentation on other projects. This did present some problems, but with a very competent head contractor keen to maintain good relationships with the owner, these problems were minimised. Disputes did arise, but were quickly dealt with due to the good lines of communication between all parties.

Trust and goodwill existed between the contractor and owner. And open communication was in place.

Exercise
Summarise the practices that led to this project being undertaken in an adversarial/non-adversarial fashion.

2.6.3 CASE STUDY — ROAD CONSTRUCTION

The contract administration procedures on site were driven by the contract documents without any real flexibility. There was no familiarity between the owner and contractor and hence no informal means for the contractor to state its claims. This led to a proliferation of paperwork in an attempt by the contractor to preserve all rights to claim extra time and/or costs and breach of contract in accordance with the conditions of contract. This in turn led to major claims for time and cost and resulted in many hours, weeks and months devoted to claims administration. Facilitation by an agreed third party was used to attempt to promote the resolution of disputes.

This project had relatively new contract documentation, while there was no close relationship between the contractor and the owner. This led to a formal approach to contract administration. The strict adherence to the contract procedures seemed to escalate small site issues into disputes and entrenched both sides into adversarial positions (not at the workface, but restricted to claims and dispute negotiation).

The project relied solely on the ability of the parties to generate trust and openness as the project progressed. Possibly even partnering would not have changed the path that this project took, as the strategy developed by the contractor was influenced by commercial pressures. The way disputes were handled could possibly have been helped by allowing for the timely escalation of issues important to the contractor.

Exercise
Summarise the practices that led to this project being undertaken in an adversarial/non-adversarial fashion.

2.6.4 CASE STUDY — HOSPITAL EXTENSIONS

This case study describes how an adversarial attitude developed as a consequence of the attitude of the people (from both parties) implementing a contract. It then details the lessons learnt (the hard way) and gives suggestions on how to enhance the likelihood of non-adversarial contract work.

The project was a hospital redevelopment. The contractor was a middle-sized company with its head office located some distance away. The design team was located also some way away and in a different city, and the owner's consultant project management team was located in yet another city. The work involved the construction of several buildings that housed complex hospital services. The work was particularly difficult because the new main building was built in the centre of the existing hospital site and required restrictive sequencing and handovers of areas to minimise interruption to the hospital's normal activities.

The seeds of the problems were sown during the design documentation phase of the project. The project was originally to be funded over several years as two stages. However due to a political commitment, the stages were combined two months before calling tenders. As a result, the contract documents included two separate sets of specifications, drawings not properly coordinated, and a hastily measured bill of quantities (BOQ). The hospital site was very old and the actual extent and location of existing underground services could not be determined accurately.

The contract was subsequently awarded to the lowest priced tenderer with a price well below the pre-tender estimate. This was acknowledged at the time as being 'worrying', but the view was taken that the cost saving was 'worth the risk'. It was later discovered that, when the contractor was finalising the subcontracts, the electrical subcontractor withdrew its tender price, because it had discovered an error. The next best price from another electrical subcontractor was quite a bit higher.

A mistake that the owner made was allowing the contractor to nominate an inexperienced and immature person as its site manager, with the assurance that this site manager would be appropriately supported by an experienced supervisor. In reality, the supervisor only came to site to attend meetings with the owner and when major problems occurred.

The owner became aware of the contractor's inadequate planning of the work when the site establishment was occurring. The contractor had not taken into consideration the restricted site boundaries and had not realised the consequences of sharing access roads with hospital vehicles and staff. The contractor was incurring additional costs for stacked site huts; long service connections to the huts; maintaining clean roads and footpaths; site fencing; safety hoarding and off-site storage.

A meeting was held with the preferred tenderer prior to awarding the contract. The discussion with respect to project background and tender clarifications was straightforward. However, when performance issues relating to site personnel experience, quality assurance, claims attitude and planning were raised, the tenderer would say anything to keep the owner happy. With hindsight, these performance questions, while legitimate, were ineffective in gaining the required commitment and assurances.

A confrontation occurred when the owner insisted on the submission of detailed construction programs in accordance with the contract. It was becoming quickly apparent that the contractor had done very little planning for what was a complex project.

In the first week of work, a backhoe dug up an electrical sub-main and this blacked out the hospital. The owner was, to say the least, not happy. It then became obvious that the subcontractors had no respect for the contractor's site manager, because of his age and lack of experience. The lack of experience was further illustrated by materials and work not being properly coordinated.

The owner tried to get the work organised by insisting on the submission of documents required under the contract (for example, fortnightly programs, shop drawing submission schedule, quality assurance programs, and equipment technical data). However, as the contractor had under-resourced the project, the site manager had to prepare this information and in so doing took his focus off the work. Consequently the work suffered again and the documentation provided was of poor quality.

The pressure of the dollar became evident with the contractor submitting or using: alternative equipment proposals; alternative construction materials; requests for additional site area; minimal quality assurance system; minimal site staff; and minimal site safety provisions. As well, there was excessive rubbish on site and rumours that subcontractors were getting pressured to lower their prices.

The lack of site staff experience resulted in a constant flood of requests for information (RFIs). The site manager took the view that if a design detail or section was not in the documents, he would ask for clarification before proceeding. Inevitably, with so many RFIs, the turnaround time was slow and in some cases RFIs were lost. This became a constant point of complaint from the contractor and a valid reason for not achieving some handover dates. This, in turn, caused the owner to become frustrated with the delays. The total number of RFIs exceeded several thousand by the end of the project.

In an effort to lower the subcontractors' costs, the contractor instructed the subcontractors to only price the BOQ, and that if there was any subsequent difference with the drawings or specification, it would be claimed as a variation. Consequently, the subcontractors would start work without having properly studied the drawings or read the specification. Throughout the work, the coordination of the documents and whether the BOQ correctly measured the work was constantly challenged. Claims were consistently on the high side and lacked detail, which prevented quick assessment. The total number of variations exceeded several hundred by the end of the project.

By this stage the resources of the owner's project management team were overstretched by the quantity of RFIs and variation claims. The team, in turn, also lost its focus on the project goals. It had a lump sum contract with the owner and additional resources were not allocated to deal with the extra work load.

The meetings with the design team consultants became very tense as the poor quality of the documents became increasingly evident. The project manager's requests to the consultants for a document review were ignored because the design fees (lump sum) would not pay for the time required. In turn, the frustration resulted in an aggressive attitude towards subcontractors when inspecting the work, and weak attempts at trying to argue against variation claims.

The site meetings with the contractor became very confrontational. The contractor became defensive when asked questions on progress, quality or shop drawings, and attacked on matters relating to the documents, slow RFI turnaround and slow assessment of variations.

The project had taken on an atmosphere of trench warfare and attrition. On a personal level the owner's project manager had no respect for or trust in the contractor's site manager, who in turn viewed the project manager as obstructionist, uncooperative and contract-focused. Personal discussions only occurred at the formal meetings and all other communication was done through letters and site instructions. The 'air' on site could have been 'cut with a knife'.

It was with this background that a strong adversarial attitude developed and remained until the end of the defects liability period. The contract documents became the weapons to drive the project.

There were no winners in this project and it took a great personal toll on all the people involved.

Exercise

1. The owner's project manager gives the following lessons learnt from this experience in how to make a project non-adversarial. Give your views and suggest other lessons.

   - If the owner decides on a change in the project scope during documentation, then it should ensure additional time and fees for the consultants to produce correct documents.
   - Complete documentation requires ensuring: an extensive site investigation and survey; one set of project specifications; and coordinated documents.

- The site manager requires appropriate experience and maturity. S/he is without doubt the most important person on site during the construction phase and will make or break the project.
- Assist the contractor in terms of: having a realistic construction period; minimising staged handovers of areas; providing as much land as possible for the construction site; separating hospital and contractor vehicle and staff movement; ensuring consultant fees include regular attendance on a remote site.
- The project manager must be understanding of the problems (and costs) being experienced by the contractor and assist wherever possible. This will include, at times, contributing to some project costs that could be viewed as being the contractor's sole responsibility (for example, the electrical cost during air conditioning commissioning).
- The contractor's site manager and the owner's project manager must develop a good working relationship through honesty, mutual trust and respect. This will then enhance effective communication and maintain a strong focus on the project goals.
- Use the contract as a last resort to resolve problems on responsibilities. Ask yourself if the position you're taking is fair and reasonable regardless of the contract's interpretation.
- Minimise changes requested by the owner during construction.
- Process progress payments as fast as possible.
- Assess, resolve and pay variations as fast as possible.
- Carry out the preferred tenderer meetings as follows:
  - Request the tenderer to make a presentation on how it will implement the work, including a basic program.
  - Discuss site constraints including: boundaries; access; consideration of hospital activities; site services.
  - Brief the tenderer on all project matters that may have an impact on its planning and costs.
  - Table contract administration procedures and a meetings schedule.
  - Carry out a separate interview of the proposed site manager, even though tenderers may not like this.
  - Get a commitment to the following ethos by both parties: honesty in all discussions; to be fair and reasonable when resolving problems; lodge only true variations and costs incurred; assist each other without question when required.
  - Agree to a dispute resolution procedure.
  - Ask the tenderer to declare any concerns regarding the work or contract that may affect its performance.
- The contract administration procedures implemented by the owner's representative can play a big part in making the contractor's life a lot easier. The following practices have been found to be welcomed by contractors:
  - Not insisting that every RFI be in writing. (Verbal questions will be answered.)
  - Quick responses to RFIs. (Verbal response followed by a written response if necessary.)

- Confirm all site instructions in writing.
- Allow the contractor to directly contact the consultants to resolve construction issues (subject to any variations being notified to the owner's representative before implementation).
- Ensure quick assessment and payment of progress payments.
- Allow direct discussions between the quantity surveyor and the contractor to resolve variations.
- Consultants are not to be 'picky' with respect to detail on shop drawings and review/return quickly.
- Consultants to attend on site when required by the contractor.
- Minimise paperwork and the lines of communication where possible.
- Give the site representative for the owner the authority to make quick decisions when required, to keep the work moving. The contractor to agree not to take advantage of this responsibility.

2. What form of monitoring would you envisage such that the owner's contractual position is not compromised or open to redress?
3. Does the success of achieving a non-adversarial attitude hinge on the success of the unwritten methods, rather than the contractual clause, for resolving disputes? If a dispute reaches the stage of implementing the relevant contract clause, have the parties already committed themselves to adversarial behaviour? That is, does the success of dispute resolution depend on the people rather than the words in the contract clause?
4. If a dispute is not resolved quickly, does it follow that an adversarial attitude will become the norm for the remaining contract work period?

2.6.5 CASE STUDY — PROFESSIONAL SERVICES

A contract applied to the provision of professional services. When submitting its proposal, the consultant specified services according to monthly manning schedules, which indicated the number of days each specialist would be engaged in the project.

After the consultant was notified of its selection, the consultant and owner met. At that meeting, the consultant told the owner's representative that the monthly manning schedules included in the proposal were only approximate, and that possibly they should be adjusted each month in order to take into account actual personnel allocation. The owner verbally agreed with the consultant on this matter, but no minutes were taken of the meeting.

Some months after that meeting, the consultant's accounting department noticed that the owner was only paying according to the manning schedules included within the proposal, and not the actual manpower usage (which was always higher). Each month a debt accumulated.

This state continued for many months up to a point where it became a serious concern. At that point, the consultant's project manager sent a strong letter to the owner indicating his concern about the owner's non-compliance with the verbal agreement. The owner replied that it had followed strictly the letter of the contract, and there was no documentation supporting the project manager's alleged verbal agreement. The consultant subsequently stopped providing service additional to that allowed for in the original contract, service that was necessary to properly do the work. The owner's opinion of the consultant's work dropped.

The result of this dispute was disastrous for the consultant. The consultant was banned for one year from tendering on other work funded by the same owner, and the owner carried forward a prejudicial view of the consultant.

Exercise
Suggest a better strategy that the consultant's project manager might have adopted on becoming aware that his total staffing expenses were not being met by the payments from the owner.

2.7 EXERCISES

Exercise 1
How can you demonstrate that, in the majority of cases, dispute avoidance is the most cost-effective approach to contracts management, and that no dispute should be the goal?

Exercise 2
Project staff tend to be well acquainted with the technical aspects of a project, including methods of work. However, their knowledge of legal matters, and in particular contractual matters, is often limited. How does such a scenario influence the way projects are managed?

Should a sound knowledge of contracts and contracts management be regarded as a key competency for all businesses?

Exercise 3
Some people enter into contracts with an adversarial intent because they fear that without this attitude, the other party will gain more than it gives. This may be the result of a previous bad experience or a simple belief that the nature of contracts is that you are out to get as much and give as little as possible. A person who has had his/her fingers burnt before could be forgiven for entering the next project with suspicion. How do you stop the cycle of bad experience, leading to poor attitude, leading to bad experience etc?

Exercise 4
Some project industries have a history of contractual disputation. It could then be argued that it would be an advantage to have experienced managers involved. However, would you expect more/less disputation amongst experienced or inexperienced managers?

Exercise 5
Many project industries have become fragmented through economic necessity, modern procurement methods and a reliance on the subcontract system. As well, there is a division of responsibilities between the various project participants (owner, designer, contractor, ...). How much does such fragmentation contribute to disputes?

Exercise 6
There is a view that the major factor in all disputes is money — the contractor is denied the opportunity to a fair profit, the contractor is greedy, the contractor underprices its tender etc. How useful is such a premise on which to base contracts management practices?

Exercise 7
In what way does uncertainty in projects influence adversarial attitudes? Would you expect projects with less uncertainty to be less adversarial in nature? By removing all uncertainty from projects, do you remove all adversarial thinking?

Exercise 8
There is a belief that the more competitive the industry, the higher the failure rate of projects; and the higher the chance of failure, the greater the tendency of the parties to be adversarial. Why might this be so? If it is so, how might the chance of failure be reduced, as a step towards reducing disputes?

Exercise 9
The presence of disputes can influence the attitude of the parties on future projects. Attitudes need to change in order to reduce the amount of disputation. How does industry get out of this circular logic?

Exercise 10
How does the practice of adversarial attitudes on one project influence organisation reputation and the potential for gaining future work? How is knowledge of attitudes on one project learnt about on future projects?

Exercise 11
One way suggested of reducing adversarial behaviour is to consider the other party's side of the argument. At the same time ask that party to look at the problem in the same manner that you would see it. In other words, 'put yourselves in each other's shoes'. How realistic is such a proposition?

Exercise 12
Comment on the following view as to why adversarial behaviour is exhibited in contractual relationships: 'Contracts themselves are not adversarial, but parties tend to respond with suspicion to the complicated formal and legal structure that forces people to conform and stifles individualism. This is heightened by the vagueness of the law that works on precedent and is not always predictable. These uncertainties add to the perception that parties in a contract must 'look after number one'.'

Exercise 13
How might total quality management (TQM) principles support long-term relationships?

Exercise 14
Some people view as commercial naivete, a contractor having a conscience, trying to develop a symbiotic relationship with an owner and adopting an ethical approach. What is your view?

Exercise 15
Dishonesty and greed can sometimes raise their heads. Are such traits ever present in some individuals and hence something that could be avoided if a long-term strategy was adopted with relationships?

Exercise 16
Would the elimination of a professional indemnity insurance requirement contribute to fostering a better relationship between owner and consultant? Discuss.

Exercise 17
There is a belief that the law often reverts to what is 'fair', and that those who impose unfair conditions on others simply because the contract permits this, will ultimately be reprimanded. Comment on this view.

Exercise 18
How can you demonstrate that the downstream benefits of providing good contract documents exceed the extra initial costs? At what point does it not become worthwhile to stop fixing up the contract documents any more because the return is less than the extra cost?

Exercise 19
What are the arguments put forward by those who advocate 'reinventing the wheel', with new contract documents for each project? How do such arguments weigh against using industry-accepted documents?

Exercise 20
It is commonly promoted that risks should be borne by the party better suited to bear them. What are some of the difficulties in translating project risks into obligations within the contract documents?

Exercise 21
How do you demonstrate that failure to carry out initial project investigations will lead to problems, considering that a project is a one-off entity?

Exercise 22
In the total life cycle cost of many facilities, the cost of design is very small and maybe only constitutes about 1% of the total cost. Cost cutting measures in design for short-term gain, at the expense of other costs, could therefore be considered ill-informed and bad practice. Why, then, do owners try to economise on design costs and initial background project investigations?

Exercise 23
There is a belief that contractors, engaged through the competitive tendering process, tend to be more adversarial in nature than those engaged by negotiation. Why is this so? Does it have anything to do with the contractor having different attitudes to the owner?

Exercise 24
A common practice is to ask tenderers to pay for the tender documentation, presumably as a way of the owner discouraging insincere inquiries. Sometimes the cost is many thousands of dollars. How is such a practice viewed by tenderers? Does it create animosity straight away, or is the practice supported by genuine tenderers?

Exercise 25
Is it a general truism that verbal contracts will lead to more/less disputes than written contracts, or is this a gross simplification of the issue? How strong is the need for scope clarification in writing?

Exercise 26
A common mistake made in dealing with others is to assume that you know another's needs and wants. How do you find out about another's needs and wants?

Exercise 27
For tenderers, when should a site inspection/view be held? Midway through the tender period? Other? How do you choose the timing such that it is a balance between giving the tenderers sufficient time to review the documents and leaving sufficient time to complete the tender?

Exercise 28
One question that seems to cause owners a lot of heartache is 'how much information do you need to give prospective tenderers'. A simple answer to this would be 'enough to allow them to make an informed decision about methods of completing the work, risk analysis, likely costs and what standards the work is to be carried out to'. Whilst this may seem a simple answer, delivering this information to prospective tenderers is not. What suggestions can you make on this matter?

Exercise 29
What do you suggest the reasons might be for tenderers informing owners of errors in the tender documents? To earn 'brownie points' in the hope of getting the work? Because it is ethical practice? Or other reasons?

Would it not be to the tenderer's advantage to say nothing if it finds an error, and then later claim extra money as a contract variation? What can you see are the negatives to such thinking?

Exercise 30
What actions do you have to take to work out what a tenderer's motives are?

Exercise 31
Frequently the time allowed by owners for tender preparation is quite short, and some would say too short. The counter argument to a long tender preparation time is that there are 'commercial realities' in the 'real world', but is this a valid argument? Discuss.

Exercise 32
Meetings and discussions consume a fair amount of time and effort on a project. How do you balance the cost to the parties with the advantages gained? How do the costs and benefits of meetings depend on the nature of the project and the contract?

Exercise 33
It is considered only natural for people to try and protect their livelihood. Why does the desire to protect one's livelihood lead to adversarial attitudes?

Exercise 34
What arguments are there against insisting that all communications on projects are committed to writing, signed and dated?

Exercise 35
What do you think of the project practice of each party 'placing all its cards on the table', including financial matters, that is complete transparency?

Exercise 36
In a disagreement, is compromise always the best option? What arguments are there against compromise?

Exercise 37
Is it so on a project that, without expediting, project activities will almost invariably fall behind schedule? Comment.

Exercise 38
Should an owner carry out regular audits on the contractor's quality assurance system? The reason this might be done is that by monitoring the contractor's management system, this may reduce the number of checks on technical matters. Or might this imply mistrust?

Exercise 39
Sometimes parties to a contract are working with different versions of the contract, with all attendant possible problems this might cause. How might such a situation be avoided?

Exercise 40
Sometimes work is commenced by the contractor before final agreement with the owner has been reached. From the contractor's point of view there is the danger that it may not be properly compensated for all the work done before final agreement is reached. What are the dangers from the owner's viewpoint?

In such a situation, how suitable is it to compensate the contractor based on its actual costs plus a reasonable overhead and profit?

Exercise 41
Consider the following scenario.

Many events cause delays on projects. Notices of likely delays are submitted by the contractor to the owner, as and when relevant events arise. Every possible delay event is covered by a notice. The owner responds to such notices by challenging the event notified as not likely to cause delay. The contractor reacts to such correspondence. Animosity develops between the contractor and the owner.

The owner assesses the contractor's time extension claims on a monthly basis and issues a formal notice of determination. Almost invariably there are differences between the time claimed and the time allowed. Claims are disallowed mainly on the ground that the event in question does not delay the critical activities.

The gap between the contractor's claimed time quantum and the allowed time quantum grows wider and wider as the project progresses.

How might this scenario be alternatively handled?

Exercise 42

Consider the following scenario.

As a project progresses, the paper war between the contractor and the owner escalates. Day-to-day administration degenerates into a trial by correspondence on many significant issues. The relationship between the contractor's project team and the owner's project team becomes polarised.

The contractor's claim strategy is simple — claim for everything. Not surprisingly, many claims are rejected. No serious attempt is made by either party to resolve the substantive question of the contractor's entitlement to an extension of time.

The contractor suffers loss on the project. The strategy is to recover as much loss as possible through claims.

The owner believes the contractor isn't doing what it contracted to do and is putting in unjustified claims for delays which do not affect the critical path. It appears to the owner that a number of the contractor's problems are caused by either its lack of management or coordination of the work, or that of its subcontractors.

How might this scenario be alternatively handled?

Exercise 43

Some people suggest that a good contracts administration practice is for the owner to adopt a strict acceptance or rejection of a claim. The intent is to encourage contractors to accurately report the situation, rather than embellishing the truth as the wronged party. Outrageous claims are rejected wholesale even though they may contain elements of fact. What are the merits of such an approach?

Exercise 44

For an extension of time claim for a delay caused by the owner, a contract required a written claim 'within 28 days after the delay occurs'. The contractor was delayed for approximately six months by the owner not supplying an item as agreed. When should the contractor submit its claim — after the first month or after the sixth month? That is, when does the delay occur — at the start of the 6-month period or at the end of the 6-month period? If the claim is made after one month, how will the contractor substantiate its claim, given that its total losses have not yet occurred? If the claim is made after six months, is this fair to the owner?

How could the wording in the contract be improved to remove the uncertain interpretation?

Exercise 45
Do all problems, that are left unresolved, invariably recur? Or is resolution-avoidance an acceptable strategy in some cases?

Exercise 46
In a potential dispute situation, how would you deal with the following types of behaviour?
- Loud and aggressive behaviour such as a bully or conniving person.
- A person who wants to lay the blame on someone.
- Someone who is resistant to your suggestions.
- A person who wants to avoid putting your solutions into practice.

Exercise 47
What role do active listening and empathy play in managing a dispute? Active listening refers to actually hearing what the other person or group is saying, not what you think they are saying. Empathy is trying to see the situation from another's point of view.

Exercise 48
Should communication be only between each party's project team head, or should communication be encouraged between all project team members? What are the dangers of having many lines of communication open? What are the restrictions of having only one line of communication open?

Exercise 49
Many organisations now commit to leadership and teamwork training programs prior to commencing projects, in an attempt to generate team spirit as well as instil quality and project goals. What is your view on such programs? Do such programs have to be experiential in nature?

Exercise 50
Should more than one person from each party have the authority to make decisions about the same issue? What are the advantages and disadvantages of having more than one person as a decision maker?

Exercise 51
Does it follow that, by placing emphasis on the management of the contract, rather than on the management of the work, the parties will tend to become adversaries, and it will lead to poor methods of management? Discuss.

Exercise 52
Does it follow that high dispute levels in an industry indicate ineffective contracts management practices? Discuss.

Exercise 53
Would individually allocated responsibility, by providing a single point of contact, reduce disputes from occurring and provide for their rapid resolution? Would such a change result in a better passage of information, and management? Would empowering project staff with responsibility and resources reduce the number of disputes and personnel problems, while increasing productivity?

CHAPTER 3

Trust, Goodwill and Cooperation

3.1 INTRODUCTION

Trust, goodwill and cooperation are some of the cornerstones of approaches such as partnering and alliances. However, in the absence of any partnering or alliance agreement, there are a number of practices that parties to a contract can adopt in order to engender trust, goodwill and cooperation, and hence develop a healthy working relationship. It could also be expected that within a partnering or alliance arrangement, similar practices would be attempted.

The reward is the avoidance of adversarial approaches to contracts management, approaches which are believed to, overall, cost the project time and additional administrative costs.

Long-term relationships and strategic alliances are based on trust, goodwill and cooperation. One-off and short-duration projects present a bigger challenge. Trust is best created at an early stage of the project.

It is the actions, of both parties, which can engender trust, goodwill and cooperation. Success only comes about with a lot of hard work from both parties. Trust, once lost, requires a lot more effort to rebuild than to maintain and enhance. It generally takes time for trust to be developed between parties, and the commitment must be maintained, because if any party breaches the trust, the chances of the other party trusting again are remote. Commitment may waver over time. Trust is based on honesty, and reliability, and these traits must be foremost in the actions undertaken by the parties.

The test for commitment can come from an examination of the motives of both parties. What are the vested interests of each party? This will decide their motivation and hence the amount of goodwill and cooperation that can be expected. The contractor will be enthusiastic if it is helping making a better than expected return. If the contractor feels that the owner is making demands on the contractor's good nature, or if the contractor feels genuine variation claims are being unfairly rejected or delayed, the contractor may adopt a more aggressive approach. The contractor's staff are unlikely to forget that their job descriptions require them to make the greatest possible profit, and do not require them to be amicable with the owner's staff. The owner will be aware of the contractor's position, and will initially be naturally suspicious of the contractor. The owner's enthusiasm for trust will depend on how it perceives the actions of the contractor.

The success or failure of basing a relationship on trust will depend ultimately on the personalities and professionalism of the project participants. Neither side will be able to ignore the risk involved in trusting the other, and will believe it prudent to test the other party. Only when both parties are confident that the other party is as committed to the arrangement as itself, and that the benefits to both parties can be seen, will the arrangement be able to work harmoniously.

It is possible to have trust, goodwill and cooperation within a contract environment. One of the most difficult things to do is to establish an association based on trust when money is involved. With unfair contracts or the intrusion of legal representatives, the original goal of a project may be lost with the parties to the contract communicating purely in a formal contractual fashion, without any trust, goodwill or cooperation, with the sole intent of protecting their own positions.

You can uphold the trust or honour of another country to the limits of its own interests.

Abraham Lincoln

Never shake hands and make an agreement with a man wearing sun-glasses. You can't see what his eyes are doing.

Anon.

Long's Note:
You can go wrong by being too sceptical as readily as being too trusting.

3.2 SUGGESTED PRACTICES

A survey of owners and contractors gave the following suggested practices to adopt to engender trust, goodwill and cooperation between contracting parties:

Organisational issues

Common contractor–owner position
- Give commitment at all levels from senior management down, without reservation.
- Be seen to operate with intentions of trust, goodwill and cooperation at all times; be proactive to instigate trust, goodwill and co-operation.
- Monitor the collective group.
- Develop interfaces between the contracting organisations on all levels; establish an overseeing 'board' linking the parties.
- Share responsibilities, share philosophy.
- Gain an insight to the mannerisms and thoughts of the other party.
- Show a genuine desire; a reason exists for practising trust.
- Form strategic alliances including with suppliers and trade contractors; develop long-term relationships; place value in long-term relationships.
- Establish an ethical code of mutual respect and trust.
- Restrain from utilising opportunities to take advantage of the other party.

- Remain free from suspicions of the other party; assume whatever is said is true until proven otherwise.
- Set the boundaries for the behaviour of the parties through something like a moral agreement.
- Define a responsibility chain of command, and levels of authority.

Owner's position
- Encourage innovation.

Contractor's position
- Involve the owner in the outcome of the project.

Goals

Common contractor–owner position
- Jointly identify, understand, appreciate and agree each other's goals and priorities; discuss differences in each other's goals and overlapping goals; address at the beginning of the project.
- Establish common goals; dedication to common goals; regularly review.
- Show mutual dedication; show interest in and commitment to achieving each party's goals.
- Seek win–win solutions.
- Understand that neither party benefits from the exploitation of the other.
- Understand the risk each is taking; where possible, share risks equitably.

Owner's position
- Understand that the contractor is entitled to a fair profit.

Contractor's position
- Understand time, cost and quality requirements of the owner.

Expectations and values

Common contractor–owner position
- Jointly identify each other's commitment to the outcome of the project.
- Jointly identify the intent of each party; avoid being cynical about the other's motives.
- Clarify, understand and give proper consideration to each other's expectations and values; understand each party's non-negotiable issues; remove misunderstandings of the other's expectations.
- Show respect.
- Give a commitment to each other's business success; symbiotic behaviour.
- Honour promises.
- Recognise the other party's right to fair and equitable treatment.
- Reassure the other party that its concerns will be addressed.

Sharing of knowledge and expertise

Common contractor–owner position
- Recognise benefits to both parties through alternative or improved methods.
- Seek input from all parties at the conceptual stage.
- Share information.
- Accept a reduced control.
- Help the other party; reciprocal help.
- Promote a sense of a joint effort; combine effort.
- Conduct joint award ceremonies to recognise and reinforce cooperative effort; recognition for anything done that goes beyond the minimum required.
- Pass on third party praise.

Contract and contract documents

Common contractor–owner position
- Jointly work through the contract to ensure an understanding of the rights and obligations of both parties.
- Have thorough documentation; clarify any grey areas; ensure no design deficiencies.
- Interpret the specification as intended, not as written; alternative products may be more effective.
- Honour the contract.

Owner's position
- Negotiate a contract rather than owner seeking tenders; contract jointly developed.
- Use fair contracts.
- Clearly delineate scope and services required (responsibilities of each party); clearly define contractual obligations leaving no areas of doubt over responsibilities.

Planning and control

Common contractor–owner position
- Agree on a realistic schedule/program/timetable and budget; unrealistic schedules require tampering with in an effort to maintain a completion date.

Owner's position
- Provide available information to the contractor to base its planning on.
- Avoid changes to plan and design.
- Confirm that the work is at expected quality.
- Verify that the end-product is as desired before the project is at a stage where it has progressed too far to make changes.

Contractor's position
- In planning, identify long lead items, note work to be completed by others, estimate manning levels, included milestone dates and plan the sequence of project events.

- Avoid program clashes.
- Demonstrate a willingness to work around any problems caused by delays and changes, rather than issuing 'delay', 'acceleration costs', 'extension of time' and 'damages' notices as permitted under most contracts.
- Establish quality assurance procedures; produce a quality product; put in place quality controls to increase the owner's comfort; don't hide nonconformances; reduced prospect of 'surprises'.
- Monitor progress; regular review of progress, problems, concerns etc.
- Carry out promises to do work by a specific date, or inform other party of a new date, and the reason for the change.
- Meet or exceed delivery commitments.

Contracts administration

Common contractor–owner position
- Display honesty, adopt an 'open book' approach, nothing to conceal.
- Clearly define the 'rules of play'; fully understand and accept right from the start.
- Take responsibility for own problems; don't pass the problem to someone else or pretend it doesn't exist.
- Manage the project as if it was being managed by one entity.
- Carry out own work efficiently and on program, so as not to affect the other party.
- Record everything potentially contentious, not as ammunition against the other party, but rather to have the facts clear and unclouded by time.

Owner's position
- Establish key performance indicators (KPIs); view effort in the light of the prevailing project conditions.

Contractor's position
- With a contract that involves a design component, present the design to the owner early.

Project finance and payment

Common contractor–owner position
- Incorporate an incentive scheme for all; profit share in relation to risk carried.
- Take risk in proportion to the potential gain.
- Use open book accounting/invoicing; allows viewing and traceability.

People

Common contractor–owner position
- Select appropriate personnel; success of the approach will depend on who are the project personnel.
- Retain key personnel for the life of the project to provide continuity and stability to the established trust and goodwill relationship that may exist.

- Avoid personality clashes; be aware of the ability of people, particularly leaders, to work together.
- Implement ongoing training as team players; ongoing training to change old adversarial habits and mind sets to being cooperative.
- Encourage liaison between project participants of both parties, to understand one another.
- Respect others; do not criticise others behind their backs.
- Develop cohesion and team spirit; rapport between personnel.
- Improve people's communication skills.
- Ensure ready availability of people to talk to their opposite numbers.
- Ensure reliability in performance and competency of staff.
- Develop motivation based on reward.

Contractor's position
- Select the project team with the required level of expertise and credentials; employ competent personnel who can gain the respect and confidence of the other party; employing people with a 'can do' attitude will help reduce the level of anxiety about possible failure.
- Professionally administer industrial relations.
- Establish a good track record in health and safety.
- Organise social gatherings/interaction outside of work hours to foster friendship, and establish a talking relationship.

Resources

Common contractor–owner position
- Reduce the duplication of equipment needs, administration, drafting requirements etc.
- Share amenities and facilities.
- Commit sufficient resources to back up commitments.
- Be aware of the size, capacity and current commitments of the other party, to ensure delivery.
- Jointly address value analysis/constructability issues; brainstorm to achieve cost and time savings.

Communication

Common contractor–owner position
- Ensure continuous communication.
- Establish clear, open, frank and honest communication.
- Don't hide things; openly talk about factors affecting the final outcome of the project.
- Ensure two-way communication.
- Have regular constructive meetings involving all stakeholders (and occasionally senior management); provide an open forum for input from all; minute meetings.
- Have openness of one party's meetings to the other party.

- Encourage both formal and informal communication/dialogue at all levels in the organisations; define a communication structure and team members' roles.
- Provide requested information in a timely fashion.
- Do not instruct the other party's employees, suppliers and/or contractors on any issues.
- Listen carefully to discussion carried on at the worker level; dispel rumours and mis-interpretations openly and quickly.

Contractor's position
- Advise the other party of the real status of the project; tell the other party what's happening; share reporting.
- Convey commitment to achieve a quality product, convey the method, convey inter-pretations.

Disputes and conflicts

Common contractor–owner position
- Agree and clearly define procedures for resolution should disputes arise.
- Remove issues that may cause disputes.
- Avoid defensive case building.
- Bring problems out into the open for resolution before they become a major issue; openly discuss; treat problems as something to be solved, rather than looking for some-one to blame.
- Admit mistakes without trying to shift the blame.
- Encourage genuine discussion and negotiation, without firstly quoting from the con-tract, or putting a position in writing; 'off the record' type dialogue before submitting any contractual claims.
- Avoid temper tantrums, erratic and unhelpful abuse in both direct or written replies; on sensitive issues a cooling off period should precede a response; respond in a rational and objective way.
- Agree the price of variations.
- Empower lower level employees to resolve disputes at the level at which they hap-pen.
- Resolve disputes and conflicts with minimal disruption.
- Implement timely resolution; resolve quickly (same day).
- Resolve disputes to prevent any future occurrences.
- Be prepared to compromise.
- Seek win–win solutions.
- Have ongoing training in conflict and dispute avoidance.
- Avoid third parties unless agreed; keep lawyers out of the project.

Owner's position
- Discuss, before rejecting a claim or making a decision which affects the contractor's progress or budget; be sympathetic and use discretion when determining claims; use judgment when invoking 'time bars' for claims — potentially inflammatory; make fair and rational decisions; deal with claims on their merits.

Contractor's position
- Firstly discuss the claim, fully explaining its basis; claims that appear in writing without prior discussion may offend; back up with support documentation; do not over-value variations or claims.

3.3 COMMITMENT TO TRUST, GOODWILL AND COOPERATION

How can you be sure that the other party is also committed to trust, goodwill and cooperation?

It is not possible to be absolutely sure of the other's commitment to trust, goodwill and cooperation. Whatever mechanisms are in place for a successful liaison, a future dispute of significant enough proportions could instantly jeopardise the entire arrangement. The risk associated with showing trust, goodwill and cooperation needs to be managed (not ignored, as some people might suggest). As well, a lot can be done to improve the possibility of trust, goodwill and cooperation being present.

Commitment to trust, goodwill and cooperation can be abused by one party. The betrayal or abuse might become apparent as the project progresses. The decision can then be made to resolve any differences or imbalances, or to abandon the commitment to trust, goodwill and cooperation, and revert to a combative style. For example, a contractor could attempt to be open, and providing the contractor ensures that it has left itself sufficient time to pursue the issues through the contract, this gives the contractor a chance to gauge the owner's reaction; if the owner 'throws' the contract at the contractor, the contractor then knows where it stands.

Indicators of commitment

It may be difficult to be sure that the other party is committed to trust, goodwill and cooperation, since in practice the opposite of these qualities might be more readily observable. It is also possibly not in the spirit of the arrangement to monitor too closely the actions of the other party, since this is itself an admission of distrust.

Indications of commitment/non commitment may be evident in the manner in which a party conducts itself, fulfils its obligations to the contract, and reacts to problems or apparent failures, while individual's body language can also be an indicator. 'Actions speak louder than words.' For example, strong indicators of intentions of trust, goodwill and cooperation may be found in the following.

Common contractor–owner position
- A willingness to cooperate in the 'fleshing out' of the details of the project and demonstrated flexibility in arriving at common goals for the project.
- No disputes on the project; looks for speedy solution to conflicts or disputes.
- Honest and truthful reporting and communications; high level of communication.
- Timely submission of matters that have contractual time bars.
- Active participation of senior officers; commitment at all levels of an organisation.
- Contactable at any time.
- Excuses not continually made.

- Resources (people and equipment) the project appropriately; people with appropriate skills and experience.
- Other party allowed time to deliver on its promises.

Owner's position
- Few demands for minor work scope variances.
- A willingness of the owner to cooperate in the completion of critical path items by pushing through approvals at a faster rate then it is required to do so by the contract.
- Contractor given assistance in achieving its completion by maximising access to the task and assisting in the task completion.
- Required information passed on quickly; design drawings, technical queries or other submitted project documentation reviewed quickly; being constructive and objective when reviewing drawings or other project documentation submitted; being realistic and practical when reviewing claims for extra costs; taking a positive stance on industrial relations issues which affect the contractor on the site; and by following the correct lines of communication set up by the contractor.

Contractor's position
- Project schedule and completion dates maintained.
- Recovery on certain cost variations not pursued.
- Work executed within specification criteria; no 'short cuts'.

Principles of trust, goodwill and cooperation are not common in project industries, and hence monitoring commitment without appearing to be questioning these principles is difficult.

To ensure commitment, both parties must understand that success is reliant on both parties. Trust, goodwill and cooperation must be mutual.

The willingness of the parties to show trust, goodwill and cooperation will become evident when a problem of significant proportion arises. Consideration must be given to and allowance made for the initial human reaction. In the cooler light of day, the rational mind will hopefully come through, and so will the commitment towards the goals, if it really exists. The unreasonable initial reaction of people cannot be considered conclusive as an indication that there is no commitment towards the goals.

Indicators of non-commitment

Indicators of a breakdown in trust, goodwill and cooperation might be, for example some of the following.

Common contractor–owner position
- The flow of information, resources and cooperation becomes one way; all information is not exchanged; unwillingness to communicate; withholding information which should be shared.
- A lower flow of knowledge which reduces the ammunition that can be used against a party at a later date; thus each party ends up in the dark as to the other's motivation and position, with an associated inability or unwillingness to make mutual decisions.

- Not responding properly, openly and promptly to communications; defensive or evasive answers to questions; leading questions possibly as part of case building.
- A change in tone of the correspondence.
- A change in the attitude of staff.
- Obstructive attitude; unwillingness to work together to solve problems; too busy to discuss issues.
- A tendency to retreat to the contractual 'bunker'.
- Financially, the other party wants it all its way.
- An unwillingness to indulge in informal communications and resorts only to formal means of communication.
- A tendency to use threats, be they verbal or of a punitive nature.
- Abuse of the other party's trust; takes advantage of the other's openness.
- Not honest about own shortcomings.
- Not being honest with excuses for failures; shifting blame.
- Non-genuine promises; promises not kept; no effort made to keep promises.
- Third party reports of 'bad-mouthing' by the other party.
- One party continually looking over the shoulder of the other.

Owner's position
- Refusal to put verbal instructions in writing.
- Refusal to discuss issues relating to time; refusal of extensions of time.
- Approval of drawings delayed; unnecessarily pedantic about technical issues, time and quality.
- Justifiable claims rejected.

Contractor's position
- Ambit claims; frivolous claims; claims 'out of the blue'.
- Not giving early notice of failures; waiting to see if something fortuitous turns up in the hope that it will turn out all right.

Not showing any of the above signs could be interpreted with reasonable safety that the other party is committed to trust, goodwill and cooperation.

If commitment is not there, further work or training may be required to achieve this. The potential benefits that come from trust, goodwill and cooperation should help maintain the level of commitment.

Monitoring, and performance measurement can remove misunderstandings.

Pre-contract

Prior to the project, a party's potential commitment to trust, goodwill and cooperation might be gauged through:
- Reference checks from past and present organisations which have had contractual dealings with the other party; the other party may have either kept quiet or might not have considered something important enough to mention.
- Meeting and discussing practices that lead to trust, goodwill and cooperation, and observing the reaction.

At the time of contract award, everybody is happy and harmonious, and generally have just agreed to work together to achieve the end goal. It is not until the problems and queries start to develop that it is possible to know if the commitment to trust, goodwill and cooperation, made at contract award, will be followed through at the time it is really required, or will the other party simply adopt a contractual position. It would be hoped that professional morals and ethics would ensure that any non-contractual commitments made, such as working together for both parties' mutual benefit and communicating, would be honoured.

3.4 EXERCISES

Exercise 1
It is often said that: 'Each party needs to realise that the ultimate goal of the project is the important issue and if energies are devoted in this direction, it will be to both parties' best interests.' How realistic is such a statement?

Exercise 2
Would the road to project success be easier if an initial project workshop was convened to establish guidelines for behaviour (trust, goodwill and cooperation)?

Could these guidelines be modified over time to take into account the changing project scope and expectations of the parties?

Exercise 3
One contractor notes that after being involved in projects with guidelines involving trust, goodwill and cooperation, he was adamant that no venture (whether it be a subcontract, purchase order — complete with conditions of sale — or a joint venture agreement) involving any type of financial gain should be entered into without guidelines of some sort. In business, the contractor argues, respect is gained, but trust is not. Goodwill is earned and does not develop without consistent effort. How coloured a view of the contracting world is this?

Exercise 4
Good project performance gives the owner confidence, not only on the current project, but for future work as well. If a project is on, or ahead of, schedule it engenders goodwill and ensures the project is run without adverse influence from the owner. The quality aspects of the project also fall into this category. If the quality is as per the specification, it saves costs at the end of the project by not having a lengthy defects listing. Owners also seem to remember projects by defects lists and it does not engender goodwill to have adverse quality issues on the project. Why is it that issues such as defects and claims can strongly influence owners' perceptions of a contractor?

Exercise 5
Contracting is very much a people business. Every day project participants negotiate, plan and socialise with the other project participants — the owner's staff, subcontractors' staff and the contractor's staff. Body language influences perceptions. When you meet people, you usually know from the first meeting if you will get on with them or

if you will place your trust in them, or if they are being truthful. The more experiences you have with those people the better you understand what they mean from what they say. How influential are such perceptions in the development of trust, cooperation and goodwill?

Exercise 6
A more formal way of gauging trust is to set up a test case with little risk and see how it pans out before you commit fully to cooperation. This is usually forcing (nicely) the person into making a commitment, and then testing them on the commitment s/he has made. It may be, for example, a simple ruling on a specification; then at a more public meeting with the person, bring up the ruling s/he has made and see if s/he still wants to own it. Does the fact that you are using such a test say something already about your relationship?

Exercise 7
Would you expect trust, goodwill and cooperation between the parties to be more easily established under a fixed price contract or a prime cost contract? Consider with respect to the risk sharing between the parties to such contracts.

Exercise 8
How would you view the situation where, prior to the work starting when each party is discussing its expectations and goals, it is found that one party to the contract does not appear to be gaining anything from the contract? What might this do to considerations of trust, goodwill and cooperation?

Exercise 9
A supplier has underpriced a component. Is it better for you to pay the extra or should you let the supplier bear the expense? Do you ignore the supplier's problems? How do you balance the contractual and commercial considerations with a willingness to help the other party? Does the monetary value of the component affect your thinking? Depending on the course of action you take, what happens on the next project involving you and the same supplier?

Exercise 10
Is it possible to feel comfortable with someone, yet not trust that person?
    How do you demonstrate openness? What else to openness is there besides making statistics available, demonstrating audit trails, and making financial results available? Does openness involve reducing the uncertainty the other party may feel about the relationship?
    How do you demonstrate honesty?

Exercise 11
One of the main reasons for trust, goodwill and cooperation between contractor and owner is in the building of a long-term relationship. Without a long-term relationship, the contractor has no guarantee of future work, and the owner has to be vigilant with new contractors. Should one party always be attempting to engender trust, goodwill and

cooperation in every relationship regardless as to whether it is or is not reciprocated by the other party?

Would you expect a relationship to improve if one party is committed to the ideals of trust, regardless of the other party's commitment?

Exercise 12

Many people have happy stories to tell about their partnering experiences. However there are many others who will tell you that partnering works fine until problems arise, when goodwill and cooperation go out the window. Why does trust, goodwill and cooperation have trouble coexisting with problems?

Exercise 13

In maintaining your own commitment to trust, goodwill and cooperation, will it usually be reciprocated?

Do you trust someone who says 'trust me'?

'Once a company/person is outside its border of interest, then you cannot be sure of its commitment to trust.' Comment on this view.

Exercise 14

Care needs to be taken to carefully judge what the intentions of the other party really are. At the project start, does it come down to reading body language, and a 'gut' feeling of whether to run with trust or not? What else may be involved in your decision to run with trust or not? How important is the initial meeting between the parties?

CHAPTER 4

Formal Cooperation

4.1 INTRODUCTION

A poor economic climate makes it hard for contractors to make a profit. Low margins from tight competitive bidding often result in an adversarial relationship developing between the parties. Each party tries to exploit the other for its own gain. There is an owner's insistence on high quality at low cost, while needing the services of contractors because many owners have downsized to their basic core competencies. From such a business and management environment (partly driven by faddish thinking), there has been a recent re-invention of owners and contractors working together. New faddish names have been thought up to describe this working together (Carmichael 1996, 1998).

Viewed from a disputes angle, the costs associated with high levels of commercial disputes have resulted in project outcomes becoming more dependent on the relationships between the parties involved rather than the actual form of delivery.

Relationships

Project outcomes are strongly influenced by the relationships between the contracting parties. Relationships that have degraded to the point of mistrust and lack of respect, lead to destructive adversarial thinking. Adversarial thinking can also be present right from the start of a project, when each party assembles its own stand-alone team, and formulates its own goals without regard to the other party's needs and expectations; communication is poor and the paths of the parties diverge.

Relationships may be developed:
- Long term between owners and contractors, and between contractors and subcontractors and suppliers, over a number of projects.
- One-off, project specific, either starting at project concept stage or later in the project.

Much of the application of cooperative type relationships in project industries has been of the second type. Non-project based industries look more towards the first type.

A downside to working together is that contractors and consultants, tied to a particular owner in the long term, may be excluded from rival owner work, for fear of a possible conflict of interest, either real or imagined. Long-term relationships may also introduce complacency, and competitiveness may be lost.

Some contractors and owners are looking for mutual goals, and win–win solutions to problems, rather than the I-win-you-lose solutions. This thinking has resulted in the recent development of approaches such as partnering and alliances.

Naming

Rather than attempting to work together under existing banners, new names have been introduced. A cynic might say that the new names have been introduced to create work for management and legal consultants, to trick owners into thinking that there is a panacea for their problems. Others might say that new names are necessary as a form of revitalisation. The point remains that there is nothing new embodied under the new naming, only a shuffling and repackaging of existing ideas.

Partnering is a formalised attempt at getting the parties to a contract to work together and in trust, cooperation and goodwill, to improve communication between the parties, and to resolve disputes quickly. An *alliance* (equivalently *relationship contracting, cooperative contracting, alliance contracting or alliancing*) has a similar endeavour. It, not so much represents a strictly defined approach but rather, represents more a way of thinking about how parties will work together on a project. Each application of alliances could be expected to be different depending on the project circumstances.

Whether partnering or alliances, the owner's organisation and contractor's organisation remain as distinct entities, even though some business functions may be integrated for mutual benefit, their strengths may be pooled and risks shared.

Selling the ideas

Many arguments put forward by proponents of partnering and alliances, compared with straight contracting, are hard to sustain — arguments relating, for example, to greater owner involvement, or greater project flexibility. In each case, the proponents of partnering and alliances are comparing the worst case scenario in straight contracting with the best case scenario in partnering and alliances. In fact there are examples of straight contracting that perform to everyone's satisfaction and beyond, while there are examples of partnering and alliances that have been dismal failures. Straight contracting can also embody all the elements of partnering and alliances if the parties so wish, without even mentioning the words partnering and alliances. Clearly the situation, mode of implementation etc. are important in the success or otherwise of any project, and that anecdotal evidence should not be used to dogmatically argue that something is better than something else. As well, people shouldn't jump on the *faddish* wagon of a new project delivery name that offers a much sought after panacea, but rather should examine the fundamental practices on which it is based. Adopting the fundamental practices should reap the same rewards, irrespective of the encompassing grand-sounding name. The appendices address faddish thinking.

Attitude and culture

For partnering and alliances to work, it requires a change of attitude and culture for both parties. Changing human values on trust and goodwill take a long time, some people even suggesting a full generation is needed before a complete change occurs. The cultural change required should not be underestimated.

4.2 ALLIANCES

Two forms of alliances are mentioned here:
- Strategic alliances, involving long-term relationships with service, material or product suppliers.
- Project alliances, especially set up to deliver a project.

However, the term 'strategic alliance' may be used (inappropriately) by some writers to refer to project alliances, perhaps because it sounds more impressive.

4.2.1 STRATEGIC ALLIANCES

A strategic alliance represents a long-term commitment for the purpose of both parties achieving their goals.

An organisation may enter into strategic alliances with, say, suppliers for the provision of products used extensively by that organisation. Examples might include stationery, chemicals, instrumentation, consumables, office furniture, film and associated products, and hardware and building supplies. Such suppliers represent preferred suppliers to the organisation and would be chosen through conventional tendering practices.

No contractual relationship might exist between the supplier and the organisation, and so the organisation is free to use other suppliers if it so wishes. However the preferred supplier arrangement provides the organisation with the opportunity of purchasing from suppliers which are attuned to the needs of the organisation, not only in terms of highest quality and best price, but in terms of efficiencies to be made through the rationalisation of purchasing and accounts payable administrative systems. Such practices result in operating cost savings to both the preferred supplier and the organisation.

Example — cable, insulator supply

Cable suppliers agreed to a long-term relationship or strategic alliance in order to reduce an electricity distribution organisation's costs of obtaining, storing, delivering, preparing for use, installing and using cable. The strategic alliance was a cooperative arrangement between the electricity distribution organisation and the supplier, and did not mean a relationship where any liability was created, shared or transferred.

Contractors supplying and delivering electrical insulators for overhead lines aimed to achieve a predictable and reliable quality of supply for maximum mutual benefit, leading to minimum stock holding by the electricity distribution organisation, resulting in an overall system cost reduction.

Each supplier had the electricity distribution organisation's interests foremost. They established relationships to assist the electricity distribution organisation.

4.2.2 PROJECT ALLIANCES

Project alliances are an attempt to get the parties of a project to work together rather than in an adversarial fashion.

Roles and responsibilities of the parties are defined in an alliance agreement. A 'virtual organisation', rather than a legal business entity, is created, and this is looked over by an executive comprising representatives from all parties. The parties share any project cost savings or overruns. The performance of the parties is judged according to set key performance indicators (KPIs).

The alliance agreement exists in conjunction with contracts between the owner and contractor, consultants, ... The contract conditions are selected with a view to some equitable risk sharing of obligations, rather than being, say, one-sided or onerous. The contract conditions are selected in the spirit of cooperative behaviour between the parties, and may even come about through consensus of the parties.

Human resource management issues, such as establishing working relationships between the parties and developing a cooperative project culture may also be tried.

Project alliances may be referred to as *alliance contracting, relationship contracting, cooperative contracting*, or *alliancing*.

4.3 PARTNERING

4.3.1 OUTLINE

Partnering is a formalised attempt at getting parties to a contract to commit to work together and in trust, and to improve communication between the parties. Apart from the formalism, the approach represents nothing new. Every benefit claimed for partnering is achievable without partnering, and everything embodied in or packaged as partnering existed before the term 'partnering' was coined. Some would say it is an attempt to return to olden-day values, or at least to what people perceive were olden-day values, of trust and co-operation, of putting the handshake back into business. Its intent, as a management tool, is to try to eliminate some of the adversarial behaviour of contracting parties, and to improve owner–contractor (and generally all stakeholder) relationships. Parties to a contract can be less protective of their legal positions, freeing unproductive time and effort that might be expended in anticipating or dealing with disputes. It provides a visible focus for commitment.

The formal notions of partnering are said to have evolved from American construction practices in the late 1980s/early 1990s. It was originally promoted as a way of minimising disputes and of bringing trust back into the relationship between contracting parties.

Some people speak of partnering as a 'moral contract' (but note its (legal) contractual force is not clear). It is not a legal partnership, joint venture or other legal entity.

Partnering describes a management process ... to overcome the traditional adversarial and litigious nature of the construction industry. Partnering uses structured procedures involving all project participants to: define mutual goals, improve communications and develop formal problem solving and dispute avoidance strategies within the boundaries of a project.

(NSW Government, 1993)

In project industries, partnering tends to be undertaken on a project-by-project basis, rather than in a long-term sense.

The intent of partnering is praiseworthy. It is perhaps a sad indictment of industry that such frameworks are necessary to achieve what is possible without partnering, but which industry may never contemplate without partnering. The simple concept of repackaging existing ideas has captured many disciples, and many success stories are told.

Features

- Jointly agreed *mission statement*, related to contract administration details and procedures, mutual project goals, business goals and dispute resolution. Non-overlapping interests of the parties to the contract still exist. The process may include all project stakeholders, not just owner and contractor.
- The production of a *charter*, not necessarily of contractual force, yet followed out of integrity and honour; this is in addition to a contract. An ethical code of mutual respect and trust. A set of guidelines that define everyone's needs and wants, define lines of communication, identify possible areas of conflict and a means to deal with disputes, spell out intellectual property matters etc.

  There is a view that the partnering charter might have applicability in interpreting the contract should there be ambiguity in the contract. In that sense, the charter takes on contractual force. The position on this is unclear. It also has implications for acting in good faith and fair dealing.
- A required *commitment* from senior management and/or a partnering champion in each contracting party's organisation. Commitment at all managerial levels.
- *Training* and indoctrination of project personnel from both parties (including subcontractors and suppliers), aimed at changing their mind sets from one of being adversarial to one of being co-operative. Development of a team atmosphere to facilitate the project's progress and prevent disputes.

  An initial workshop ('partnering seminar'/'team building workshop') where project personnel meet each other. People are able to advance their ideas on the project, how it could be better run, what changes would be of benefit to them and the opportunity to question and provide feedback on such things as project documentation.

  On-going training and workshops throughout the project.
- Consideration of each other's *expectations*, interests, needs, constraints and risks. A sharing of risks if feasible. Trust in the other, and teamwork.
- An attempt at resolving *issues* quickly either at the level at which they occur (empowerment at lower levels) or by more senior personnel, rather than letting them develop into disputes. Potential disputes are 'nipped in the bud'.

  A time frame is placed on issue resolution, and if an issue is not resolved within this time frame, it is escalated to a higher level, for a commercially sensible solution. Parties are forced to address issues, and stand-offs are prevented, in the same way that disputes review boards operate.

- Timeliness in *communications*. Open and honest communication. Defined communication structure.
- Regular partnering *meetings*, and *performance questionnaires*, to keep everybody informed of project and partnering progress, and to evaluate how well the mutual goals are being met.

Conditions of tendering

Partnering can be initiated by either the contractor or owner.

The benefits of partnering appear best with long-term relationships between the parties. However, there is the fear that a long-term close-knit relationship with a contractor may become inefficient, and possibly corrupt, without the benefit of opening up the work to competition for every project.

Because of probity issues in the public sector, a partnering relationship will usually not be established on a long-term basis but rather not until after the contract is awarded, following a transparent competitive tendering process, on any given project. This is in the interests of accountability of public money to protect the public interest. It is accordingly difficult for the public sector to enter into anything other than project-specific partnering. However there is nothing to prevent contractors and subcontractors, and contractors and design consultants from entering into long-term relationships.

A common comment on the practice of partnering is that it is introduced too late in the project, and not sufficiently close to the beginning of the supply chain to gain most benefit.

In the conditions of tendering, tenderers are advised of the intention to include cooperative arrangements between all project participants — contractor, subcontractors, suppliers and consultants — and that such cooperative arrangements are voluntary. Partnering seminars/team building workshops are flagged. Partnering and its expectations are outlined. There may be an attempt to disclaim any collateral agreement, representations, the implication of acting in good faith or fair dealing, and any fiduciary relationship. Communications are to be conducted on a 'without prejudice' basis.

Benefits

All the benefits, associated with parties to a contract working together and in trust with good communications, should follow. These include reduced time resolving and preparing for disputes, improved timeliness, cost and quality, and reduced stress and contract administration costs (for example, no defensive case building, and less paperwork). In spite of publications to the contrary, actual benefits cannot be measured directly because all projects are one-off with many variables; favourable benefits have been inferred.

As with most management fads, initial enthusiasm could be expected to produce benefits and this situation may last until the fad becomes usual practice, whereupon a new fad is needed to raise productivity again.

The hype

The champions of partnering link it with everything good and wholesome in the world, but these things follow from the parties' cooperation and trust, from adopting good project and contracts management practices, and from improved communication, and not from partnering per se.

Many of the benefits claimed relate to an ideal world; but in an ideal world partnering would not be necessary. Most of the arguments supporting partnering cannot be rigorously defended, primarily because of the one-off nature of projects.

Proponents talk of synergy between the contracting parties such that the combined result is greater than the results of each party working independently.

It is linked with total quality management (TQM), value management/analysis/ engineering (VM), improvements in project timeliness, cost and quality, equity, commitment, respect, psychology, sociology, leadership, empowerment, organisational redesign, organisational culture, and nearly everything else. It has become a gravy train for 'facilitators'.

The downside

- Increased staff and management time up-front is hopefully repaid by benefits down the track. There is a common perception that the cost (usually shared) and time in setting up partnering may not be repaid, particularly where partnering requires a significant change in mind set. Contractors working on small profit levels will have this profit eaten into by the cost of partnering.
- The approach requires commitment from both parties and an ignoring of the risk involved in trusting another. In many project industries, the existing culture works against trusting the other party. Changing to a trust culture cannot be done in the short term. One-off projects work against developing long-term trust.
- Experience seems to be that when both parties genuinely apply the principles of partnering, the outcomes are successful. However, the approach can be easily abused by one party taking advantage of the other's trust. This effectively nullifies partnering. For many, the experience of using partnering has not matched their initial expectations.
- Partnering, when used, appears to be favoured on a project-by-project basis, rather than in developing long-term relationships.
- Partnering in the public sector generally cannot be applied until the successful tenderer is notified that it has won the work. The tenderer then has to transform from being a competitive animal (in order to win the work), into being a non-competitive animal (for the partnering process). (The private sector has more scope for awarding work in a non-competitive manner.) Winning the work competitively may mean small profit for the contractor, which may consequently show more commitment to its own financial viability than to partnering.
- Partnering might only be applied at the management level with little filtering down to lower levels, and may not involve all project stakeholders.
- One party may only pay lip service to partnering. Underneath, it may be business as usual.

- Partnering would appear to work best where there is equity between the parties. Where one party dominates the other (through size, power, financial strength, ...) partnering will have difficulty working. Conventional contractor–subcontractor relations and many owner–contractor relations tend to be unequal.
- Generally the owner and contractor have different goals. Each satisfying its own goals, may work against satisfying the other's goals.
- There is the potential for project participants to become immersed in the mechanics of partnering, to the point where the project loses focus.
- Partnering was initially thought to not create any legally enforceable rights or duties; the contract provided the legal relationship, while the partnering charter was a moral agreement and established the moral obligations and working relationship between the parties. However the legal situation may not be as black and white as this. ... *the distinction between moral and legal rights is blurred, and thus the introduction of a partnering agreement often does not add to the clarity of the relationship, but in fact may detract from it* (Jones, 1996).
- Partnering encourages free and open communication. In claims, ... *the parties must be careful not to allow their correspondences to become too casual. A number of adverse legal consequences attach to such a state of affairs* (Jones, 1996). Reference here is to complying with contractual notice provisions, and also making representations. ... *if one party makes a representation which is relied upon by the other party, the representor may either lose its rights to pursue some claims, or, in fact become liable to a claim itself* (Jones, 1996).
- Dispute outcomes are not always of the win–win kind as most writers hope. When a dispute arises, the partnering relationship becomes strained, and the relationship might not be strong enough to withstand a dispute. *In this eventuality, the fact that a project was partnered may in fact exacerbate the dispute* and results from a lack of *clear upfront specification of responsibilities*, the *so called 'fuzzy-edge disease'.* (Jones, 1996).
- *The agreements and representations associated with the partnering process, even if not intended to be legally binding, may impact upon the contractual risk allocation in the following ways:*
  - *the implication of a contractual duty of good faith.*
  - *the creation of fiduciary obligations.*
  - *misleading and deceptive conduct.*
  - *promissory estoppel and waiver.*
  - *confidentiality and 'without prejudice' discussions* (Jones, 1996).

... partnering is rather reminiscent of the nursery rhyme:
There was a little girl,
Who had a little curl,
Right in the middle of her forehead.
And when she was good,
She was very very good,
And when she was bad, she was horrid.

Jones, 1996

4.3.2 CASE STUDY — SIGNAL INSTALLATION

This case study shows how, for a standard contract, partnering was introduced as a perceived method to overcome the inadequacies of both the owner and the contractor, but was then later discarded when it appeared not to fit within the owner's desired contractual outcome.

Situation

Paulan (contractor) entered into a contract with a rail corporation (owner) for a resignalling and remote control installation project over a 12-month duration. The contract was awarded to Paulan on the basis that technology aspects would be provided by a third party (Charleco), but responsibility lay with Paulan.

During tender discussions and negotiations, which stretched over a six month period, all technical and commercial aspects were extensively discussed by all parties.

Throughout the discussion, Paulan was totally reliant on Charleco, having little or no expertise in the detailed application of this particular technology. This was not an unreasonable position, as Paulan had previously undertaken similar work with Charleco with satisfactory results.

At all times during the discussions, Charleco advised that the system offered was proven technology, used in similar applications elsewhere, and provided an off-the-shelf solution requiring essentially no development.

At the end of the tender discussions, both the rail corporation and Paulan were comfortable with the contract.

The contract work commenced, and all appeared to be straightforward and proceeding basically to program. Approximately nine months into the work, it became evident that Charleco did not fully understand the complexity of the work, it was not in control, and an enormous amount of design and development work still had to be undertaken to complete the scope as understood and now required by the rail corporation.

This situation persisted for several months with both sides establishing their contractual positions based on their respective understanding of the scope.

Approximately twelve months into the work, the rail corporation realised neither party was blameless, with significant weak points within the specification causing problems, and decided that a more cooperative approach would be beneficial towards completing the work.

To this end, the rail corporation, in agreement with Paulan, proposed that 'partnering' be introduced. The main people from the rail corporation, Paulan and Charleco initially met on neutral ground, chaired and guided by a facilitator, to air their real and perceived respective opinions about the state of the project and everybody's performances to date. Once these opinions had been aired and all parties agreed to work under partnering to achieve project completion, a number of meetings were held with the appropriately responsible staff from both the owner and the contractor to formally lay out the issues and agree solutions that could be supported by all parties.

In the early days of partnering, both the owner's staff and the contractor's staff worked in harmony, using people at the appropriate levels to mutually solve problems. In time however, the owner's senior management realised the enormous task that still needed to be completed, and decided that partnering would not best serve the owner commercially, and retreated to a direct contractual position.

The project remained in this situation with the duration extending to over several years. The applicable contractual penalties were applied by the owner, and senior management of both the owner and the contractor attempted to reach a negotiated settlement.

Exercise

1. The contract documents were documents traditionally used by the owner, and did not consider partnering. The contractor believed the conditions of contract to be quite onerous. Should contract documents be modified to cater for partnering? If so, what modifications would you suggest? If not, why not?

2. From the contractor's perspective, the owner had failed to meet significant obligations under the contract, particularly in relation to design documentation approvals and the raising of additional requirements perceived necessary to meet the intent of the specification. As usual in such a situation, conflicts developed with communication exchanges addressing contractual issues, and this did nothing to progress the work. How else can the parties work around differences of opinion without going straight to the contract clauses to demonstrate that the other party is at fault?

3. In an effort to make progress in spite of the contractual issues, the rail corporation, in conjunction with Paulan sought to bring harmony and cooperation back into the project with the introduction of partnering. Although there was 'dedication to common goals and an understanding of each other's individual expectations and values', it could not be said that the 'relationship was based on trust'. How do you get trust into a relationship?

4. Whilst true partnering recognises the contract fully and makes provision for dispute resolution, including litigation and arbitration as a last resort if necessary, all the parties concerned were unable to be completely frank and open about their own interests, and this caused partnering to fail. How do you get openness and frankness into a relationship?

5. At the initial partnering meetings, the owner's staff and contractor's staff jointly agreed goals for completion of the project, and the implementation and management procedures to meet the goals. These meetings were held in the form of workshops and at this level, there was good cooperation between the parties.

   The workshops allowed detail discussion between the hands-on design staff, and allowed agreements to be reached and implemented during the workshop sessions, resulting in significant progress being made outside the workshops. This aspect of partnering was generally successful and saved considerable time that might otherwise have been absorbed during the normal to-and-fro of the design and approval processes of a project. This approach was in accordance with the partnering features of resolving issues at the level at which they occur, and timeliness in communications.

So, the basic features of partnering were followed in the setup and initial implementation stages. Indeed, other than agreeing and adopting a formalised partnering
charter, the appropriate systems inherent in a partnering relationship were established. Would a formal partnering charter have helped?

6. In parallel with the partnering workshops and meetings, the owner's senior management and the contractor's senior management continued meetings aimed at reaching
   a commercial settlement within the contract. These meetings had the effect of both
   parties not being able to ignore the risks involved in trusting one another. Is this an
   inherent downside in partnering arrangements? Perhaps if a commercial settlement
   had been approached under the partnering arrangement, trust between the parties may
   have been developed. What is your view?

7. Partnering helped the contractor to complete its work. However, it became apparent
   to the owner that partnering was not to its commercial advantage, because it required
   significant commitments in time and effort from its staff for activities it believed
   should be done by the contractor, and it felt that any failure by its staff to meet these
   commitments would ultimately provide the contractor with some contractual time
   advantage not due under the contract. At this time the owner started to distance itself
   from the association and to retreat to a purely contractual position. This lack of commitment from one party when not convenient commercially, caused the breakdown
   of the partnering relationship. How can you prevent such situations from occurring?

8. Whilst, in this particular case, partnering was not implemented successfully, it is felt
   by the contractor that there were obvious benefits to the project, during the period
   when it was implemented. All subsequent tenders for the rail corporation included the
   requirement for a partnering commitment from the successful contractor. Would it
   therefore follow that the rail corporation recognised its contract documents contained
   weaknesses that could give rise to disputes that might best be solved by cooperation
   or partnering?

4.3.3 CASE STUDY — ONE PERSON'S EXPERIENCES

On three recent projects that a contractor's project engineer has worked, one involved
partnering. Experiences with working relationships have varied from 'good friends' to
'worst enemies'.

The first project (out of the three projects) did not have partnering but it could be said
that it was a good model of what partnering is all about. A very good working relationship existed between the owner, the owner's staff and the contractor's staff. As a result,
the resolution of issues proceeded smoothly.

The second project was the complete opposite. Prior to the arrival on site of the engineer, a bad relationship existed between the contractor's staff and the owner's staff. The
engineer's attempts to use some of the methods from the previous project were unsuccessful because the project had been running for nearly a year and had gone downhill
from day one. The end result was a lot of unresolved issues which could have been
solved very easily had there been a good working relationship. It was believed that the

problems related to the delivery method — design, document and construct (DD&C). The documentation was unclear and possibly a result of the owner not making its intentions clear enough. Also, there were a few personality clashes which added to the problem. A formal partnering process may have made both parties more willing to work together to find the appropriate solutions to the issues that arose.

The third project had formal partnering. The engineer became involved in the project three months after it had started, and missed out on the partnering workshop held at the start of the project. The workshop was held to introduce the project personnel to each other. A charter, which outlined the aims of the project and how issues were to be resolved, was developed. Part of the partnering process was to conduct monthly meetings where representatives from both parties attended. The meetings gave each person the opportunity to openly rate the various aspects of the project and discuss the areas that needed improving. The partnering process encouraged an open dialogue which resulted in the prompt resolution of the many issues that arose.

From the contractor's point of view, the majority of recent contracts won have involved partnering. As a result, the directors have decided partnering is a good thing and have embraced its ideals. The company has been successful in winning continuous work with a particular owner, and has maintained the same project staff to further improve the existing good working relationship.

Exercise

1. Many people state that every benefit claimed from partnering is achievable without partnering. The project outcome and relationships experienced in the first project mentioned support this statement. All parties at the time agreed that open lines of communication and trust were required to ensure a successful project. In this particular situation, formal partnering was not required. On the other hand, for the second project, maybe partnering would have saved the project from the way it ended up. There were no open lines of communication and no trust between the parties. On the third project, all went fine and partnering worked. Partnering helped things along. Is there any way other than by anecdotal evidence, such as above, to support or not support partnering?

2. For partnering to be successful, commitment is believed to be required from senior management. The enthusiasm shown by senior management ensures the project staff act swiftly to resolve any issues as soon as they arise. Can partnering work without senior management commitment?

3. A downside of partnering, in particular the increased staff and management time, can be a problem. Generally the partnering meetings on the third project ran for an hour and involved approximately sixteen to twenty people. The meetings involved rating a series of items and discussing any points of interest. The upside was that the forum provided the opportunity to discuss 'without prejudice' any issues that might have been affecting any party from performing to expectation. The engineer believes that the cost and time spent can be repaid, especially if issues are resolved quickly, instead of dragging beyond project completion. Can a formal benefit–cost analysis on this be carried out, or will it always remain one of people's perceptions of benefits versus costs?

4. Partnering is said to work best with long-term relationships between the parties. Unfortunately, in many cases, partnering attempts tend to be short term, running for the duration of the project only. Is this primarily due to the competitive bidding system?

4.3.4 CASE STUDY — MINE DEVELOPMENT AND OPERATION

The project (as part of a larger mining project) to construct mining infrastructure and develop and operate the mine throughout its life used a partnering agreement. The project required a large injection of capital by the owner to acquire the ore reserves, the relevant freehold land, community infrastructure requirements, and to complete the construction of a 'state of the art' mine. The contractor completed all of the fixed infrastructure and built the mine for a guaranteed maximum price (GMP, an upper limiting amount). The contractor also provided all of the mobile equipment required for the project and was rewarded with a tonnage based production service fee.

Partnering was seen as an opportunity to enhance the otherwise average economics of the project, through getting increased trust between the owner and the contractor. Once it became apparent that the partnering concept for the project had merit, teams from both organisations formalised the arrangement.

A partnering agreement was developed to reflect the corporate aspirations of both parties, with commitment to the concept being driven from upper/senior management down.

The owner–contractor joint mission was to be a world class mine operator. Common goals included excellence in safety, environmental practice, product reliability, low cost production and to be a preferred workplace for all involved in the project. These goals were to be achieved through the highest level of commitment and consultation with customers, shareholders, staff and all stakeholders.

The corporate intent of both parties was reflected in their guiding principles which included: an emphasis on safety, quality, the environment and productivity; maximising returns to both parties; world's best practice; and sharing all risks and returns.

The design-and-construction (D&C) contract for the mine was prepared by a small group consisting of project managers and legal representatives from both organisations. A D&C delivery meant that the owner specified what outcome was required, leaving the responsibility up to the contractor to determine how these performance requirements would be accomplished. Standard general conditions of contract were used as a starting point for contract development. However, it required a lot of manipulation to appropriately reflect the intent of the partnering agreement, mutual commercial goals and the basic contractual obligations of both parties.

An operating agreement was also devised to cater for the life of the mine that followed the commissioning of the ore preparation plant. The contractor was responsible for the operation and maintenance of the mine, whilst only passing on a proportion of the direct operating costs to the owner. This agreement provided strong incentives for the owner to provide adequate capital to produce a facility that could operate efficiently, and for the contractor to wisely spend capital provided by the owner. If both these occurred, then both parties should enjoy the benefits throughout the mine's operating life.

The construction was essentially completed on time and under budget. Safety performance was not as good as hoped, as tough zero lost time injury (LTI) targets were not accomplished. Safety performance improved over time. Production met and even exceeded production targets.

Exercise

1. Partnering started with the notion that the contractor was honourable, trustworthy and reliable. It is believed that this underlying principle assisted in avoiding disputes that would ordinarily come from suspicion, a lack of trust and a confrontationist attitude, on both sides of the fence. It is believed that partnering, in this case, successfully reduced the number of disputes, related delays and associated costs. How could this belief be demonstrated?

2. In this case, the owner and the contractor were able to establish a long-running and relatively successful relationship. Issues such as trust and mutual benefit were able to be discussed openly. Wasted costs and time associated with a usual tender process and usual contract administration were avoided. How do you demonstrate that savings were made in usual tendering and administration costs, compared with what replaced them?

3. As the work did not go out to tender, it could be argued that the lack of competitive tendering forces provided flexibility for the contractor to inflate the price. How do you demonstrate that the partnering relationship gave the owner a fair price?

4. As the work did not go out to tender, it could be argued that other innovative designs and methods of construction were not able to be reviewed and considered. How do you demonstrate that the partnering relationship gave the owner the best design and method of construction?

5. As the work did not go out to tender, it could be argued that the tender cost was not passed on to the owner. However, is it reasonable to assume that costs accrued from past tender failures were partially built into the contractor's price? If so, is the saving in the tendering cost of significance, because after all the contractor may only win 1 in 5 or 1 in 10 tenders?

6. In the D&C delivery the owner specified what outcome was required, leaving it up to the contractor to determine how these specified performance requirements would be accomplished. It could be argued that if the contractor is given such a large amount of flexibility, large reductions in the quality of the construction may result in an attempt to maximise its profits. However, in this instance, any shortfalls in quality or reliability of the finished mine, as a result of shortcuts, poor workmanship or inferior technology, would impact on the mine's output of clean ore, thus directly affecting the contractor's future profit margins. This meant that whilst the contractor had a large amount of flexibility to be innovative and fast-track the project to maximise its profits, it also had to be particularly mindful of completing the mine infrastructure to the highest possible standard in a timely and cost efficient manner. If not,

the contractor would potentially delay and even jeopardise future profits during the operating phase of the mine. In what way does this indicate that partnering was not needed on this project?

7. The mine development was based on a guaranteed maximum price by the contractor. However it provided flexibility for the contractor to schedule the construction, and to unbalance its progress claims (front-end loading was possible) to provide it with adequate cash flow, and to assist financial security. Front-end loading, of course, makes it hard for the owner to clearly define its maximum financial commitments (cash flow) prior to commencement of the project, thereby affecting its budgeting. To what extent would you expect front-end loading to differ between a project without partnering, and a project with partnering?

8. Would you think that a partnering agreement would work best with a traditional delivery or D&C delivery? Lump sum, schedule of rates, cost plus or guaranteed maximum price? Why?

9. How do you measure the success of a partnering agreement?

10. The strength of the relationship between the two parties grew and all key issues were resolved, without difficulty, at a technical rather than contractual level. The operating agreement was also designed to reflect the true spirit on partnering as both parties appeared to share in the gain and pain of market forces. Is the essence of a good and lasting partnering agreement whether disagreements don't arise, or whether they are positively, quickly and equitably dealt with by both parties? Does a partnering agreement need to put pressure on the project participants to work out any problems, so as to ensure future success?

11. Improved or good safety performance was not achieved. Are partnering agreements supposed to produce significantly improved contractor safety performances? Or is it the case that a workforce still requires constant reminding, training and management to maintain standards such as safety, which appear readily sacrificed in the pursuit of more glamorous and lucrative goals such as production targets? Should key issues and targets, such as safety, be tied into the contractor's performance measures (KPIs) so as to impact on the hip pocket of the workforce and the contractor?

12. It is often argued that partnering lowers contractor costs. How are these lower costs achieved? Is it: by removing duplication of people, implementing shared risks, increased productivity due to work practice changes, an increased familiarity with the work environment and its processes, other?

13. Risks were believed to be equitably shared and appropriate work practices adopted, as both owner and contractor worked together to take advantage of each other's experience and practical, technical and administrative strengths. How can you demonstrate equitable sharing of risk?

14. In what way does this case study show how partnering can provide a fundamentally better way of doing business? What influence does the participants' attitude and trust in each other influence the chance of success? Can partnering overcome differences in culture and business outlook more readily than conventional forms of delivery? In what way does this case suggest that partnering can work in other environments?

15. Rank the following key elements, to this partnering agreement, that you believe contributed most to its success:
    • Commitment: Came from upper management down and included clear policy statements.
    • Trust: A good working relationship with clear and concise communication within the team was developed.
    • Interests: Each party's interests and goals were considered and satisfied.
    • Goals: Both parties worked together to develop and define mutual goals.
    • Implementation: Strategies for the implementation of stated practices were developed together.
    • Evaluation: Performance, relationships and operational techniques and processes were continually evaluated.
    • Responsiveness: Timely communication and decision making were achieved.

16. Amongst the owner's staff there was a distinct lack of experience in dealing with partnering agreements. Consequently, it was a huge leap of faith taken by the owner that fortunately paid off. How do you assess the risk associated with 'faith'?

17. Contracts management and dispute resolution short courses were attended by the vast majority of the owner's staff. Is such training more, or less relevant with partnering compared to having no partnering?

18. Team building sessions were not conducted because of the nature of the contract, and because the small number of staff on the owner's team were nearly all heavily involved in the development of the partnering agreement and subsequent contract. Consequently, the owner's staff had a large amount of contact with the contractor's staff prior to work commencement. How might missing out on formally scheduled team-building sessions have affected the interaction and communication on the project?

19. Penalty clauses were avoided and incentive clauses were primarily focused on bonuses, with profit sharing related to over-budget production. How could safety targets and cost/tonne targets have been tied into the contractor's KPIs and profits? How could shared goals have been established with the contractor such that some of the contractor's profit was locked in to assist the owner in meeting its goals? Is it possible to maximise the returns to both parties, or is it a 'zero sum game'?

20. The conditions of contract were modified to reflect the partnering agreement. Should the contract and the partnering agreement be separate or interrelated? What contrac-

tual force does the partnering agreement have in this case? What contractual force do the early communications between the two parties over partnering have in this case?

21. The partnering agreement was believed to be successful. What faults in its development and/or execution can you identify?

4.3.5 CASE STUDY — MINING EQUIPMENT

Introduction

The case study relates to aspects of a partnering agreement as it was applied to an equipment supply and maintenance contract. It was a long-term arrangement.

The partnering agreement

The parties to the agreement are referred to here as the miner, the maintenance provider and the original equipment manufacturer (OEM).

The miner operated a large truck and shovel operation.

The maintenance provider was the local agent for an international earthmoving equipment manufacturer. At the miner's operations, it supplied and maintained the majority of the in-pit trucking and dozer equipment. The performance of the equipment had a major impact on the operational costs of the miner.

In 199A the miner proposed a tripartite partnering agreement between the miner, the maintenance provider and the OEM, with a view to exploring ways of mutually reducing costs and improving the performance of, specifically, the OEM's haul truck fleet. A tripartite steering committee was established and subsequently a charter was drafted setting out the goals, expected outcomes and measurements for each partner. Key areas of haul truck potential improvement were identified and project teams formed to develop solutions. These teams were very successful in improving both truck performance and logistics between the miner, the maintenance provider and the OEM.

The charter was presented to the miner's contracts committee for ratification in 199C. Although the partnering process was accepted in principle, the charter was not signed as it was considered that the benefits arising from the agreement were too biased towards the miner.

Though the joint cooperation continued between the parties, it was decided in 199D, to again try and re-establish the partnering process. This time it was approached at a higher corporate level, thereby focusing on business opportunities, and not just technical improvements. A workshop was set up which established a clear understanding of each party's business strategy and the preferred style of the relationship. From these talks a vision of the partnering relationship was drafted.

Subsequent workshops, using outside facilitators, established the direction and partnering strategy, and developed specific action plans for establishing the relationship.

The partnering relationship was established in 199E. It allowed for the OEM to be the sole supplier of this type of equipment. The maintenance provider undertook the total maintenance and warranty of the equipment. This partnering relationship required the close integration of the operations of the miner with the maintenance services provided by the maintenance provider.

Some years on, quiet concern was expressed by some people involved in the agreement, as to whether it was achieving the desired outcomes or whether it was successful. Others involved in its set-up were quick to defend it, but the rhetoric on its benefits was broad and subjective.

Comment

The intent, of what was trying to be achieved in the agreement between the miner, the maintenance provider and the OEM, was at getting the parties to work together and in trust, and to improve communication between the parties.

Some of the features that were missing in the partnering agreement were:

• Commitment of senior management.
• Dispute resolution procedures.

The initial three years of the partnering development were too closely tied to technical developments and the technical staff of middle and lower management. Focus was at a low level of the operation, and consequently the finalising of the agreement floundered. It was only when the senior management became fully involved and the focus moved to the business level that the impasse was overcome. Once the agreement was struck, the commitment of senior management waned.

There was not a process in place that enabled disputes to be openly discussed and quickly resolved. There were annoyances on the maintenance provider's side that got a better airing in the social sphere than the workplace. This could partly be because the miner was not always properly listening to what the maintenance provider had to say.

The benefits from the partnering agreement were calculated before the agreement was entered into. These benefits were based on assumptions about future economic conditions, future equipment purchases and future production increases. Unfortunately these assumptions only remained valid for a few years. The local economic conditions decreased to a level that no one could possibly have predicted even twelve months before.

The sole purchasing section of the agreement assured the OEM the possibility of more sales, but later production analysis determined that these trucks were possibly no longer the most economic types. The next truck order might be larger trucks with electric drives, a type of truck that was not available in the OEM's product range. Mining survives on low cost per tonne. If a business case was shown for the larger trucks, then no further purchases from the OEM would result. The OEM and maintenance provider would still benefit from the sales of spares for the existing fleet, but at a reduced level.

Lastly, production was always expected to increase. In the long term this was still expected to happen because it was tied in with a strategy of maximising returns of existing reserves. In the short term though it was unlikely to occur because of market oversupply and depressed commodity prices.

Even had all the assumptions proved to be correct, it still would have been difficult to clearly establish what was the amount of gain to be made from partnering. With so many improvement initiatives being undertaken at any one time, the originators of these programs were quick to claim that any production gains were a direct result of their improvement work. There were also other influences on production such as the negative effects of high rainfall, and the positive benefits that come with a prolonged drought.

Exercise

1. In the 'Standard Handbook of Heavy Construction' (J. J. O'Brien, J. A. Havers and F. W. Stubbs, McGraw-Hill, 3rd ed., 1996, ch. A9), the authors point out some successes achieved in partnering. However, it goes on to say 'Unfortunately, the owner and the contractor are still bound by a low-bid, fixed price contract managed by employees being judged in their performance evaluations on how much they make for the contractor or how much money is saved by the employee of the owner.' This statement identifies one of the problems of the case study, that is that employees on all sides are still evaluated on their performance to drive down their company's costs and increase their company's revenue. No matter what charter has been signed or agreement made, most employees of the partnering firms have this fundamental belief that a large part of their company's success and their performance evaluations will be tied to these issues. How do you address this issue?

2. A second issue related to the above is the aspect of risk and reward. In the case study, rewards were in the form of additional purchases and an extension of the contract maintenance work. The contract was regularly reviewed against a set performance criterion. With a good performance review, the contract would be extended by a further three months giving the maintenance provider the ability to apply long-term planning. With poor performance, the maintenance provider may have had its contract period shortened by up to three months, thereby reducing tenure. How do you address this issue?

3. With respect to the performance criteria, had the maintenance provider been removed from site, there was not a local organisation with the experience or capacity to take over. The more likely result of poor performance would be that discussions would have been called to analyse and resolve the situation. There was too much capital invested in the trucks to use unproven maintenance providers. In any case, performance criteria for a complex operation can be subjective and can be easily challenged. How do you address this issue?

4. An alternative form of reward driver may have been to distribute a proportion of the miner's profit to the maintenance provider and OEM. This would have more closely aligned their goals, as each party would be interested in the overall performance of the mining operation. A downside to this though is the risk involved from the profit being heavily influenced by outside factors such as changes in commodity supply and demand. So would such a reward, which can be influenced by factors beyond the parties' control, work?

5. One partnering issue is the win–win scenario. The case study did not provide a win–win scenario for all parties. The maintenance provider felt that the miner was not

sharing any cost savings being generated. As soon as a saving was identified, the miner absorbed the benefit by setting a new baseline for performance.

The miner also spoke about the warmth and fuzziness of the relationship, but did not act to make the other partners inclusive. True openness in communication was lacking.

How do you address these issues?

6. There is a feeling that over the long term, a partnering relationship will lose any cost competitiveness that originally existed. You may not be happy with the balance of the relationship. The drive to be cost competitive will wane. Even if you have a good relationship, over a period, complacency can still easily set in. Like any relationship, you need to make sure that the desire to continue and improve is nurtured. What is your view?

7. There were many early believers in the concept of partnering. Now such people are not so convinced. While still believing that the features of partnering (trust, communication, working together etc.) are worth embodying in any project, there are some drawbacks and some issues to be considered in the application of a partnering agreement:
    *   Partnering is never a solution for poor management.
    *   Some people are naturally adversarial in their nature and will never change their mind set.
    *   Partnering is not for everyone.
    *   Partnering can breed complacency.
    *   Partnering may sometimes just be a part of a cyclic process where the other half of the process is adversarial relationships.
    *   Partnering is more suited to stable business environments. Environments where there is a large step change or continual change over a long period may not be suitable for maintaining long-term stable relationships.
    What is your position?

4.4 ALLIANCE CONTRACTING

4.4.1 OUTLINE

The fundamental practices of alliances, alliance partnering, alliance contracting, cooperative contracting or relationship contracting are applicable to all delivery methods, with single or multiple contracts, and with multiple parties if applicable.

Alliance contracting shares most of the values espoused by partnering. (In fact, some people use the terms alliance and partnering interchangeably.) The partnering charter is usually a separate document from the contract itself, while an alliance contract incorporates the expressions of cooperation. However hybrids will be found in practice.

Amended standard conditions of contract may be suitable for alliance contracts, though legal consultants would suggest that totally new conditions of contract are necessary, and new on a project-by-project basis expressing the particular alliance members' individual wishes. Developing new conditions of contract, of course, brings with it its own problems and suspicions. As well, there are non-contractual legal issues such as fiduciary duties, acting in good faith and fair dealing, and misleading and deceptive conduct which may be present.

The relationship between contracting parties and its strength are seen as paramount to a project's successful outcome. Alliance contracting gives a basis for establishing and managing such relationships. Conventional contracting in recent years is perceived as having shortcomings, often leading to adversarial relationships, and cost and time overruns.

Central to alliance contracting are:

- Behaviour and culture modification (by both parties) away from more common adversarial thinking.
- Closer alignment of owner and contractor goals; a focus on project results, with success measured by key performance indicators (KPIs); emphasis on 'value' rather than lowest price.
- Removal of barriers between the contractor and owner; removal of elements that cause adversarial and confrontational relationships.
- Strong people relationships.
- An integrated project team of senior staff from all parties, without a conventional owner–designer/contractor/subcontractor hierarchy, with responsibility (individual and collective) and accountability for all key decisions.
- Mutual respect and trust.
- Transparency and information sharing; open and honest communication.
- Fairness in the contract; a sharing of risks, and a sharing of rewards (gainsharing/painsharing); an equitable (commensurate) risk/reward balance; an appropriate sharing of cost under-runs and cost over-runs.
- Collectively, the contractor and owner working together should achieve something better than if they work towards their individual goals.
- Cooperation for the benefit of the project rather than focusing on penalising contractual non-conformance; a 'no blame' approach — success or failure is a joint responsibility.
- A focus on finding solutions to problems, rather than fighting over problems.
- Continual attention to building the relationship between the owner and the contractor, during projects and from one project to another.
- Altered values of the contracting parties by reflecting on and commitment to the above points leading to a win–win outcome; altered values of legal advisers in drafting conditions of contract such that the conditions are not confrontational or adversarial in nature, allocating risk to the party best able to manage it; altered values of third party advisers regarding opinions on proposed commercial arrangements, proposed contractual arrangements, and reviewing and reporting progress.

The core values and guiding principles are summarised in Table 4.1.

The approach mirrors the way contracting would like to be carried out by both parties in an ideal world. It could be said to be no more than 'commonsense', but also could be said to challenge existing mind sets within industry.

The approach works alongside the framework of existing delivery methods, but assumes that the best method is chosen for each project's particular characteristics and requirements.

Core values	Guiding principles
Commitment	Total commitment to the achievement of the project goals — actively promoted by the chief executives of all parties.
Trust	To work together in a spirit of good faith, openness, cooperation and no blame.
Respect	The interests of the project take priority over the interests of any of the parties.
Innovation	To couple breakthrough thinking with intelligent risk taking to achieve ... good project outcomes.
Fairness	To integrate staff from all parties on a 'best for job' basis.
Enthusiasm	To engender enthusiasm for professional duties and the project's social activities.

Table 4.1. Core values and guiding principles of [alliance] contracting (ACA, 1999).

The approach works alongside existing good practice in the management of the technical side of projects (quality assurance, planning and control, value engineering, meetings, integrated phases etc.).

The approach works alongside existing good owner actions (prequalification of tenderers, good scope definition, choice of appropriate contract payment type and delivery method etc.).

On any project, if changes are to be made, they should ideally be made early in a project. Alternatives should be examined early on. The 'cost to change' curve for a project goes from nothing at the start of a project to a maximum at the end of a project. ACA (1999) argues that the 'cost to change' curve is flatter where the contracting parties are working together, as opposed to an adversarial relationship.

The intent is to make project industries more globally competitive.

Key performance indicators

Key performance indicators, measuring the success of a project, might include:
- Costs — project, operating, life cycle.
- Milestone completion dates.
- Environmental performance.
- Occupational health and safety.
- Productivity.
- Standards.

A framework

There is involvement of short-listed tenderers in:
* Agreement on the delivery vehicle (chosen to suit the project's characteristics and requirements), project scope, parties' goals, commercial arrangements (sharing of rewards and losses), organisation structure and decision making processes.
* Developing a clear statement of the obligations of each party.
* The recognition of all project risks, and an allocation of the responsibility for managing them; a risk allocation schedule included as part of a contractor's tender.
* Commitment to cooperation first, followed by the contract documents (drafted through negotiation) second.

And involvement of the successful tenderer in:
* The training of personnel from both parties towards more cooperative behaviour and culture.
* Development of a project team involving all project stakeholders.
* Workshopping as an integrated team, led by a suitable facilitator.

Benefits

Benefits are said to be in project completion time, project completion cost (and life cycle cost), risk management, enhanced business relationships, satisfaction of people involved, increased flexibility to suit changing project requirements, and enhanced standards.

Risk treatment

Not uncommon practice is for owners to use contracts and delivery methods that ask the contractor to carry risk that is beyond the contractor's ability to manage. In a competitive contracting environment, contractors may be reluctant to include something in their tender prices to cater for such risks, for fear of being non-competitive. Owners then have transferred risks without any apparent cost. However when the contractor starts to lose money, the project environment may turn adversarial, leading to disputes, and time and cost overruns and poor performance on the project.

However, this strategy often fails, creating an adversarial climate, a high level of commercial disputation, time and cost overruns and overall poor performance ...

Faced with a risk transfer strategy, it is often not in the contractor's interest to be flexible. However, given the adversarial nature of relationships, it may be in the contractor's interest to allow a problem to unfold rather than deal with it positively. At its worst, the contractor's interests may be best served by pursuing strategies aimed at increasing the overall cost to the [owner].

Contracts fail if [owners] attempt to transfer all project risk to the contractor, or if the contractor seeks higher returns without accepting a greater proportion of project risk.

(ACA, 1999)

It is argued that better project outcomes would be achieved with a sharing of risk between the parties, such that those which can manage the risk are asked to take responsibility for that risk.

Dispute resolution

Some form of alternative dispute resolution such as expert determination would be recommended. However if the relationship between the contracting parties has reached the point where an outsider is needed to resolve a dispute, it may mean that the alliance contract has outlived its usefulness and should be terminated.

The downside

Owners have a genuine cynicism towards offering transparency, if there is potential for it to be abused. Many years of 'hard dollar contracting', where the contractor was solely interested in its own goals, can be difficult to put to one side.

Alliance contracting ideals tend to be promoted at senior levels of contracting organisations, but little seems to filter down to the people at project level, where performance is judged by bringing the project in to schedule and under budget, irrespective of the owner's goals.

Project staff may not be appropriate and able to adopt cooperative contracting values. Behaviour modification from more common adversarial type thinking may not be possible. Staff selection, training and performance monitoring issues would need to be addressed. Subcontractors may have no interest in any relationship between the owner and the contractor; contractors would have to be made more responsible for their subcontractors.

Project work is won on the basis that a working relationship between the contracting parties will be established, but may be put to one side when commercial pressures arise.

The project may be based on a sharing of rewards/losses, but parties become adversarial in circumstances involving losses.

4.4.2 CASE EXAMPLES

Offshore oil platform (ACA, 1999)

Project description
The project required the development of an oil field, located offshore Western Australia, ... in relatively shallow water depth ...

The physical nature of the deposit created reservoir engineering problems which had to be resolved, meaning that the field was always a commercially marginal and high-risk deposit. The engineering solutions adopted to recover and process the oil had to be delivered within tight time and budget constraints to ensure that the returns satisfied [the owner's] acceptance criteria.

Project duration — 26.5 months. Project budget A$480M.

Project process — contract implementation

... An alliance should be thought of as a virtual corporation — a separate body with its own identity and culture, supported by the participants.

A contract is set up to provide that each party's profit and corporation overhead is at risk, based on the overall project result, not just that party's portion — hence rewards flow from joint, rather than individual effort.

To achieve this ... requires a trusting relationship between all parties, with a clear understanding of each other's expectations and values. A successful outcome from alliancing requires ongoing learning and alignment of management, staff and individuals, so that the process becomes the evolution of a culture of collaboration, mutual respect, integrity, innovation and 'no blame', with the focus on results.

At the core of alliancing is the courage to embrace a new way of doing business — of breaking with past methods when appropriate, of shedding corporate and individual baggage and of generating new possibilities. It's about changing the mind set.

Why an alliance?

The concept of alliancing came to the attention of [the owner's] senior management as an arrangement where owners, both large and small, could deliver large projects with only a small management team. Furthermore, overseas research proved that alliance teams could be 'super' enthusiastic; organisations using alliancing were forecasting extraordinary savings in time and cost; and seamless integration is achievable between [owners] and contractors with efficiencies flowing from the breaking down of traditional barriers.

The alliance team was selected on the basis of assessment against criteria established by [the owner and] included technical competence, acceptance of a business assessment, acceptance by the CEO of each participant to a total commitment to the alliancing concepts and agreement to put 100% of gross margin (that is, profit and corporate overhead) at risk in return for gain sharing arrangements.

After selecting the alliance team, a technical study period of three months was commissioned to analyse numerous alternative solutions, identify the optimum platform configuration and determine whether that solution was able to meet the [owner's] criteria for project sanction.

A project office was established ... and an alliance board was also established with representatives of each party entitled to exercise one vote, with all decisions requiring the consent of all parties. The [owner] and contractors were therefore equal in decision making and influence on the alliance.

The board appointed a project director to mould the various groups into an alliance operation. In this instance the person chosen was a representative of [the main contractor].

[The owner] formed a special purpose company, ..., with the same board as the ... alliance, as the vehicle for major procurement of services and materials as well as provision of offices and general support and administrative services.

Direct procurement in the name of [the owner] was particularly useful as it preserved all warranties for the end user and achieved significant cost savings.

Developing the 'alliance' culture
The alliance sought to develop its own unique culture, requiring individuals to become result focused and willing to challenge conventional standards. The main principles underlying this culture were to:
- Work together in a spirit of openness, cooperation and 'no blame'.
- Use innovative methods to bring the ... field on stream at the lowest possible cost whilst meeting the design basis, operating standards and schedule.
- Disclose to each other cost and technical information.
- Bring full commitment to effective interfacing between the parties.
- Strive for continuous improvement.
- Integrate staff from one party into another where it best suited project needs.

An external consultant ... was appointed for the duration of the project to facilitate the alliance [goals] and coach the team members in developing and maintaining the culture.

Gas field development (ACA, 1999)

Project description
[The project developed] a gas condensate field located offshore Western Australia
...
Project period — 22 months. Project cost — A$270M.

Project process — contract implementation
[A] joint venture was selected by the [owner] after submitting an expression of interest along with other potential consortia.

Three contracting parties, including the owner/operator, formed an 'alliance' which contracted with the 'owner'.

A project office was established and staffed by personnel from [the owner and the two joint venture partners]. An alliance board was selected with two representatives from each of the three parties to the alliance with the chair being taken in turn by each party.

The alliance board selected the project manager and section managers on the basis of skill appropriate for the roles. In the event, each party provided managers to the project team.

The next step was the development of a design and budget. This was done in an open manner with the three parties being full partners to all negotiations over design options and price. With the concept design and budget price developed, an alliance contract was agreed and a profit and loss sharing formula determined.

The best technical, project control, safety, quality and procedures available from three parties were adopted by the integrated project team.

A rigorous reporting system was adopted and regular detailed progress review meetings, between the project team, the alliance board and, at less regular times, with the owner's stakeholders, ensured all the issues were aired.

In addition, target performance criteria were established for all the issues mentioned and a personal performance incentive system was instituted to allow all project team members to share the contract incentive payment if all the performance targets were met.

Selection of people for the integrated team was on a 'best qualified for the job' basis, as was the need for a 'team player' track record. This ensured a very high quality team. Risks were managed by those best qualified/experienced to do so, and this helped address the major problems effectively.

The thrust of an alliance is to align the goals of the [owner] and the contractor — to get them on the same side. This is done by structuring the contract so that the contractor's reward increases as the total cost reduces and [its] reward reduces as the total [project] costs increase.

A difficulty which can arise is in agreeing the target cost. The contractor would like a higher target cost, the [owner] a lower target cost. [On this project] the difference was 10%. The solution was to accept both figures with a reduced bonus/penalty sharing of 1:6 between the two figures.

Figure 4.1 shows the agreed risk and reward arrangement. Point 1 on the graph was the [owner's] estimate, point 2 was the contractor's estimate. For costs below point 2 earnings were shared equally. For costs above point 1 losses were shared equally.

In fact the final actual direct costs after a number of major changes were very close to the [owner's] target.

Why an alliance?
The preferred option in contract style was an open cost, risk/reward sharing alliance — with target cost, schedule and operating costs and availability targets agreed at the outset.

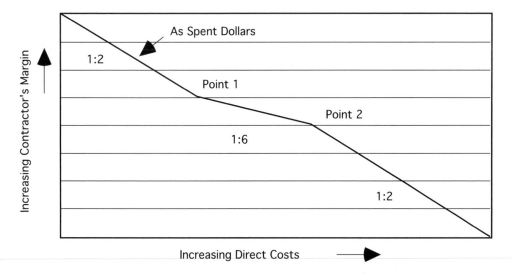

Figure 4.1. East Spar Alliance Risk/Reward — As Spent Dollars.

This style was chosen primarily due to the inability to define the project's scope at the outset, the very tight window of opportunity to meet the delivery date of gas ..., and thus a necessity for flexibility in the contract to address expected significant changes — an expectation which was realised, with the open, risk sharing format ensuring a successful, minimum conflict result.

Major benefits of the alliance contracting style
- Saved time and met schedule.
- Flexibility allowed critical path work to proceed while offshore problem areas were addressed.
- The [owner]–contractor relationship was very amicable and technical problems and issues were addressed without the normal commercial problems preventing the best solution being found quickly.
- Working relations between team members were focused on doing the job without letting contractual problems interfere — a 'whole of life' approach to the job was thus possible.
- Overall cost was minimised since all stakeholders had an incentive to improve in order to share in the outcome.
- The traditional inertia on technical issues was removed and significant innovation was achieved due to the incentive to improve outcome.
- A 'fair' margin was negotiated in the original contract and then some additional margin was achieved.
- The ability to develop long-term relations has been beneficial to negotiating future work between the parties.
- There was no owner's representative team which resulted in cost savings and removal of a source of conflict.
- Project decisions were taken by mutual agreement between all parties and thus implementation was able to proceed far more quickly.
- A 'whole of life' cycle approach to job decisions, equipment selection etc. was successfully implemented due to the cooperative team approach and the contract provisions to adjust the target if an issue justified it. An example was a $1M increase in target to use common equipment to the existing gas treatment compressors.

Factors critical to the project success
Openness, cooperation and the sharing of risk/reward were all critical factors to the success of this project.

The commercial structure on [the project] was the key to providing the 'natural incentive' to ensure that openness and cooperation actually occurred. The attractive risk/reward regime was the core of this natural incentive since it was the driver of the aligned commercial interests.

The ... project presented very significant risks to each party, but also presented the opportunity for extra reward and a fair based margin as well.

Selection of project personnel, a strong facilitated training program for people to work in the new environment of openness, and a major effort by the alliance board and senior project team members to develop good personal relations, allowed problems to be addressed with the 'best interest of the project' always at the centre of decision making.

A distinguishing feature of the project was the great satisfaction experienced by all participants. A general expression was how refreshing and stimulating it was to work in an atmosphere where everyone was constructive and trying to solve problems.

Unexpected latent soil conditions required a complete change in concept which was accommodated smoothly with innovative engineering. Under a conventional contract there would have been long delays and cost overruns.

The stark difference between the constructive, cooperative atmosphere of an alliance with all parties having a common goal and the legalistic, adversarial atmosphere of a conventional contract was very apparent.

4.5 EXERCISES

Exercise 1
Some people view attempts to develop long-term working relationships between organisations as protectionist and barriers to new players. What is your view?

Exercise 2
What issues are there in the management of strategic alliances with suppliers and liaising with suppliers?

Exercise 3
Why does industry feel that there is a need to formally introduce trust, cooperation and communication in projects? Why is it thought to be not present already? Why can't we trust and cooperate with others, as a matter of course, without the need for something like partnering?

Why does industry feel the need to follow the latest management fad?

Why does industry feel the need for a banner under which it must operate? What is wrong with just doing 'good practice'?

Exercise 4
There are examples of failures of partnering where the contractor has taken advantage of the trust shown by the owner; where defensive case building was still carried out 'just in case'. Are such practices unavoidable in partnering, despite the best intentions of both parties? Does it show something about human nature?

Exercise 5
What is the legal standing of partnering charters in terms of being part of the contract or aiding the interpretation of the contract?

Exercise 6
Partnering presents probity and integrity challenges to any public sector body contemplating working with private sector companies. Such challenges tend to prevent long-

term partnering, and reduce partnering to one of a project-by-project nature. Comment.

Exercise 7
If the law required contracting parties to act in good faith and fair dealing (perhaps an implied duty), would that do away with the necessity for partnering? Is good faith synonymous with the cornerstone of partnering? Good faith here relates to the spirit of the agreement rather than to the letter.

Exercise 8
In partnering, the parties are encouraged to work together to solve problems. Working together could be expected to have a greater creative output. What happens to intellectual property developed by both parties working together?

Similarly, what happens with admissions or disclosures (made in the interests of the project) by one party to the other that it made a mistake? By so doing, does that party expose itself to some liability, and also compromise insurance policies? Does using the term 'without prejudice' overcome this problem?

Exercise 9
Does partnering represent a Utopia? Is it unrealistic, in that it ignores basic human behaviour?

What happens if trust between the parties breaks down part way through a project?

Exercise 10
Is it feasible for tenderers to submit, as part of their tenders, what partnering arrangement each envisages? Does this then make partnering part of the contract if the tender is accepted? Or is it better for the owner to take the lead role in raising the issue of partnering?

Exercise 11
In partnering workshops and application, would you expect a one-to-one relationship between people and levels of a contractor's organisation, and the owner's organisation? Discuss.

Exercise 12
Suggestions, made by some writers, for mutual goals that the owner and contractor may have are: completing the project under budget and ahead of schedule; minimising disputes; improving safety; successful implementation of quality systems; timely resolution of issues; etc. Why are these 'mutual goals'? Do the contractor and owner really have mutual goals?

Exercise 13
Is there anything new in partnering, or is it a repackaging of old ideas? If it is a repackaging, why do you think industry has seen a need for repackaging?

Exercise 14
Many published statements about partnering are difficult to substantiate. Much is anecdotal in origin. Why do reporters not feel the need to be objective?

Exercise 15
One recent partnering model applies a modifying factor to the management fee component. Here, all the parties have an opportunity to rate the success or otherwise of each other's team and their respective abilities to interact as a team. The higher the rating, the greater the fee modifier and vice versa. How practical is such an approach?

Exercise 16
Performance questionnaires may be completed regularly (for example, monthly) throughout a project's life by the contracting parties. 'Open and honest communication' may be one of the performance indicators used. What might be a suitable measure of 'open and honest communication'?

Exercise 17
[Alliance] contracting is founded on the principle that there is a mutual benefit to the [owner] and the contractor to deliver the project at the lowest cost — when costs increase both the contractor and the [owner] are worse off (ACA, 1999). How does such a statement reconcile with: (1) the often stated belief that many contractors tender low, but make their profit out of variations; and (2) the practice of loophole engineering?

There can be no blame — success or failure is a joint responsibility (ACA, 1999). What happens if failure is the result of poor practices by the owner, designer or contractor? Are the other parties expected to say 'never mind'?

Exercise 18
Common contractual practices over recent years have tended to be adversarial in nature. Alliance contracting requires behaviour modification. How successful would you expect this behaviour modification to be — in the short term; in the long term? What needs to be done before people will change their practices? Or are the behaviour changes being asked for just too much? Can people actually change their behaviour? Assuming alliance contracting was adopted universally, what would happen to the 'hard dollar' contractors that couldn't adapt?

Exercise 19
One common problem is that owners draw up contracts with a view to cooperative contracting, but inadvertently create an adversarial climate. What are the fundamental things needed in the contract documents, and also needed to be left out of the contract documents, in order to encourage cooperation from the contractor?

4.6 APPENDIX: MANAGEMENT FADS

Based on D. G. Carmichael, 'Management Fads', Journal of Project and Construction Management, Vol. 3, No. 1, pp. 115–125, October 1996.

Introduction

The writer has for a number of years touted in public the view of the undesirability of much of what is appearing in the popular management literature and which is not only not advancing the state-of-the-art of management but also is believed to be regressing the state-of-the-art. The term fads is perhaps the closest description of the developments.

Recently others have gone public on the same issue, for example Shapiro (1995) and Hilmer and Donaldson (1996). All plea for a return to objectivity and the removal of dogma, unnecessary jargon, unfounded beliefs and formulas. '... prevailing fads have the potential to lead managers down false trails ... that have been falsified by hype and oversimplification' (Hilmer and Donaldson, 1996).

Hilmer and Donaldson (1996) express a concern shared by the writer in terms of what is happening to the practice of and writing on management, namely 'the substitution of dogma — platitudes, homilies and fads — for careful, sustained professional management.' To this should be added the subjugation of rational objective and logical model building in management to opinion, anecdote and subjectivity. '... otherwise sceptical and pragmatic executives [seem] to have put their reasoning powers on hold and [are] being lulled into believing the messages.'

Managers are looking for quick fixes that will solve all their problems. Accordingly, as marketers would suggest, products have evolved to satisfy that desire. Yet 'such short-cuts and simplifications are effectively undermining — trivialising and denigrating ...' the practice of management. '... the quick fix and fad mentality [is] corrupting the professional practice of management.'

Hilmer and Donaldson argue that 'the fads are symptoms of a more fundamental problem — the loss of respect for management and for professional managers.' The writer would contend that it is symptomatic of other matters, namely — that every second person is called (often inappropriately) a manager, that such people have inadequate education in management fundamentals before being allowed to practice as 'managers', that the state-of-the-art of management is crude by comparison with the bodies of knowledge of other professions, that management terminology is imprecisely defined, and that the proponents and users of these fads have inadequate training in objective model building.

'It is time for management to be redeemed.' But, in the writer's view, only if the above symptoms are addressed.

It is recognised that many organisations operate in 'turbulent markets' and these are 'crazy times'. Hilmer and Donaldson argue that successful companies 'rather than flit from idea to idea or seek a range of crazy solutions, they are dealing with competition and complexity by calmly, purposefully, and professionally developing and adapting sound and seemingly simple ideas. It is excellence in the doing, not the cleverness of their ideas nor nifty jargon, that sets them apart.'

Management and rigour

Disciplines like engineering and the sciences have developed over many hundreds of years. They have established bodies of knowledge developed largely through the so-called 'scientific method'.

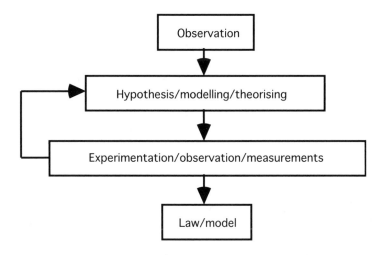

Figure 4.2. Essence of the 'scientific method'.

The 'scientific method' is based on a process of observing behaviour, postulating or hypothesising a model or theory of behaviour, and conducting experiments, observations or measurements. Should the experimentation support the hypothesis, a law, model or theory is established. Should the experimentation not support the hypothesis, the original hypothesis/model/theory is modified or replaced with another and the process repeats itself (Figure 4.2).

The essence of any sound and rigorous research in engineering or science is based on this 'scientific method'.

It would seem reasonable (at least to an engineer or scientist) therefore that a good place to start the development of a body of knowledge in a discipline such as management would be to use the 'scientific method'. This appears to have been the case in management studies in the first half of this century. But many recent 'contributions' to the management literature appear to have forgotten about rigour. 'While we see management as parallel to other professions, we recognise that it is still in its infancy. There are thus lessons to be learned by management in the evolution of other professions that sought to develop clear language and to act on the basis of high ethical standards, logic, and proven ideas' (Hilmer and Donaldson, 1996).

The following comments refer to much of the more recently published literature on management, to what is sometimes referred to as management fads. It is noted that there are ongoing rigorous contributions to management but that these receive little publicity; the non-rigorous fads grab all the headlines.

The management of people, in particular has become a highly emotional subject. Opinions run rampant. Few recent contributions in the popular management literature bring a logical and rational approach to bear in developing and organising factual as well as subjective information in presentations which clearly and unemotionally portray what is true or false and what is likely or unlikely. The 'scientific method' requires careful observation and equally careful inference or deduction from these observations, and is fundamental in establishing the exact nature of behaviour. Management fads tend to be presented in subjective ways with a lot of hype.

Modern fad 'theories' of management tend to be based on opinion and anecdote. There is little justification given that the 'theories' have any generality. The 'scientific method' is absent. Axing of a fad is based on similar dubious justification.

Donaldson (Hilmer and Donaldson, 1996) puts it another way: '... good ideas are those that are properly supported by facts and whose logic can withstand the tough criticisms of peers. By these standards, much of what is touted as a pathway to better management is at best unproven and at worst misleading and dangerous.' '... novelty seeking seems to be displacing truth seeking in much of the research on management.' 'If these ideas were well supported and logical one might hail this flow as the beginnings of a vibrant science of management. Unfortunately, this is rarely the case.'

Can it be argued that management is progressing at all with these fad theories? Do any of the plethora of theories provide insight into management and advance the state-of-the-art? The answer is felt to be 'no' or 'almost no'. Most of the fads have at least a few good ideas, albeit in many cases not original ideas. It is these ideas which are worth considering, in conjunction with objectively developed models, when practising management.

The success stories of fads are many, but mostly appear in the originating publication. The success stories are peculiar to particular individuals, work type, organisation type etc. From this it is argued that success can be generalised. However this subjective, anecdotal approach by its nature lacks objectivity and repeatability across different situations. It fails by any measure of the 'scientific approach' to establishing laws.

It goes without saying that applying the approach/practice without regard to the situation is silly.

The proposed faddish practices can be viewed as being based on models. These models tend to be simple (for example, in terms of a few basic steps to follow), but almost all lack rigour and general applicability. The appeal of using simplified behavioural models is understandable, particularly when the complexity of the system being modelled is considered.

Panaceas and hype

'It is evident that since the advent of commercial and industrial management early last century, managers have tended to look for quick fixes for their problems. And, they have been willing to pay ready cash for "instant pudding" ... managers have always sought seven or eight simple steps to solve all problems, and in doing so have "leaped from fad to fad"' (Mitchell, 1995).

Yet in the 1915 words of Taylor: 'No system of management, no single expedient within the control of any man or any set of men can insure prosperity to either workmen or employers' (Mitchell, 1995).

'Much of this "management in three easy steps" literature is too glib to solve the real problems facing corporations' (Hilmer and Donaldson, 1996).

The popular management literature is full of panaceas, with new panaceas replacing others that have lost favour, on a seemingly endless conveyor belt. 'Today, everyone is a re-engineer. Before that they were disciples of flat structures, and before that, of total quality. Along the way they made detours into empowerment, gainsharing, niche marketing, and culture' (Hilmer and Donaldson, 1996).

Managers, organisations and individuals struggle to deal with the programs and mantras just in time for them to be overtaken by some new fad. The net effect is consuming of time, sometimes detrimental to organisations, but always lucrative to management consultants, writers of popular management books, and presenters of popular management seminars.

Managers have difficulty coping with this continuous supply of wonder drugs. They all contain pieces which can help management, but all need careful application and consideration. Like much in management, their effectiveness and application are situation dependent. Cookbook recipes don't exist.

It is, of course, tempting to adopt these management panaceas, because the potential returns, should they work, are enormous. They are sold in terms of opportunity lost should they not be adopted. But the sellers — management consultants, authors and seminar speakers — have a conflict of interest.

Interestingly, even though one panacea may not work, people are ready to grab the next quick-fix panacea that comes along. Never is it 'once bitten, twice shy'. The latest fad is the magic which will do the trick — it doesn't have flaws like the previous one.

How can people be encouraged to admit that the practices are fads and still let them save face? Many organisations are currently adopting one or more of the practices. How then do you get people in such organisations to admit their naivety? When these people are confronted, it seems that all the practices are fads except the particular ones being trialed in their organisations.

From an industry viewpoint, how do you convince people to stop looking for panaceas and to start thinking about the fundamentals of management? How do you convince people that situation-dependent solutions have to be structured for each organisation — there is no general panacea, there are no instant answers? This requires courage from management, as well as a general elevation of the education of managers, a general elevation of the teaching of management beyond the restrictive case study approaches, and a general elevation of the theory of management.

Some fads, of course, don't warrant serious attention, but when sold with fervour, some unsuspecting naive souls convert to the new religion. Their saving grace is that, given the current crude state-of-the-art of management, who can tell the difference between a good and not-so-good manager? After all the disciple is applying the 'latest theories'.

In the same vein, is it because the theory of management is so crude compared to technical disciplines, that anyone can call themself a management guru and espouse the latest theory or fad?

The case for adopting any fad is put forcefully and simply. To not adopt is suggested is unfashionable and politically incorrect. The 'message' put by the gurus always sounds appealing. Masses uncritically accept the 'word'. The masses must be right and so therefore everyone should adopt the fad.

The basis of the fad is never questioned by the masses and also never given by the promoter. For example, can management be reduced or simplified to a handful of steps? Is not the behaviour of an organisation and its people a bit deeper than this?

Successful companies would appear to adopt management thinking far more thoroughly and more deeply than the simplified views expressed in the fads.

The paradox is that, otherwise quite clever people so readily accept the 'word'. There is no other discipline that so readily accepts fads and quick fixes. Does it reflect the fact that management is an extremely complicated field, yet there are continual pressures on managers to be seen to be doing something. (Thinking, of course, is something that can't be seen.)

'A manager who is using the latest technique supported by an eminent expert or who is following the widely applauded prescriptions of a best-selling book can hardly be criticised, while those who ignore the latest trend risk being judged old-fashioned and unprofessional. Moreover, because many management problems are complex and persistent, executives often become frustrated. Therefore, the assertion that management is being overcomplicated, that there is a need and way to cut through the web, finds a ready audience' (Hilmer and Donaldson, 1996).

The fads, put forward as eminently reasonable practices to follow, are hard to avoid. They receive a disproportionate amount of publicity, as well as a disproportionate prominence in book stores. Are publishers and the media partly to blame for the emphasis on this faddish literature? As well, many managers know of no other practices and hence a discussion on management invariably revolves around this popular management literature; a discussion on management fundamentals is not possible because many managers do not have an adequate educational background.

Recycling

Many fad management theories have been promoted. The theories have shifted emphasis from people to process to function to end and back again. Many are a case of shifting the deck chairs around. Old approaches, variations on old approaches and combinations of old approaches are being renamed. New management gurus emerge as the repackagers of the old.

The number of proposed theories together with many people's dissatisfaction with them has led to the expression 'the management theory jungle' to describe the body of proposed theories.

It is interesting that the fads appear to contain few ideas that didn't exist prior to their heraldic releases. In many cases they are mutton dressed up as lamb. New trendy terms are introduced to replace existing mundane terms. Existing ideas are packaged and assembled in different combinations; the number of permutations and hence the number of management fads accordingly appears unbounded.

They trade on the knowledge that managers are ill-informed of all the historical development work in management fundamentals. A cynic might also comment that these fad espousers are similarly ill-informed of any historical work in management fundamentals.

The application of fads

Commonly it is argued that the techniques of these management theories are adopted by industry, but that the approaches don't work because the philosophy of the approaches has been ignored. Presumably an organisation's culture has to be altered with every new theory introduced into the workplace — a lucrative pastime, dare one say gravy train, for management consultants?

Whilst it may be possible to cite the prevalence of certain patterns in particular work types and/or environments, and this can provide support for one particular model, the applicability of the same model in other situations will vary markedly because of the many variables that impact on the individual, group and organisation. The validity of any assumptions needs to be questioned when attempting to use them in areas outside the conditions within which they were framed.

One of the by-products of these fads is that they perturb the steady state, making people work differently and perhaps more effectively than before. After a while, however, another steady state returns which requires a further fad to perturb it. Are these fads, then, cunning management tools designed to lift worker productivity through perturbation rather than through anything intrinsic in the fad itself? Perhaps something like Figure 4.3 applies.

There is an argument that organisations introduce a management fad as a way of enthusing the work force. After a while the transient effect disappears and the organisation settles into a steady state, possibly of perceived lower productivity. In order to raise the productivity, another transient is introduced in the form of another management fad whose effect after a while also disappears. Thus the logic is to never let the organisation reach a study state, but rather to keep it in a continual transient state, of perceived higher productivity.

Similar observations are made on the introduction of incentives; these have to be continuously changed else the workforce regards them as the norm. Nine-day fortnights, flexible hours etc. are well received initially but their effect disappears with time. Consider the good old tea/coffee break.

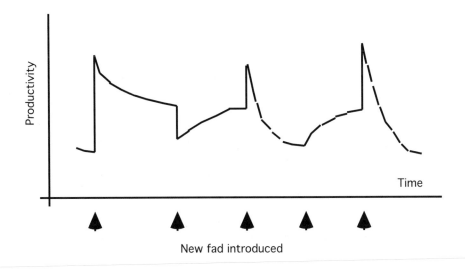

Figure 4.3. Speculated effect of fads?

There is a sort of Hawthorne effect operating here.

Many people would say that if the practice increases productivity, decreases costs, improves customer relations, ... then go with it whether it is rigorous or not. Such a pragmatic argument is hard to resist. However there can be no real assurance that these increases/decreases will occur. There may in fact be the opposite effect.

In many instances, the introduction and implementation of an approach/practice is counter to productivity. This can occur with blind application and without proper understanding. Interestingly, there appears to be no attempt to predict, or is there the ability to predict, the influence of changes on productivity. Changes seem to be implemented in the belief that they will work. Sometimes it is a case of change for change sake — 'We have to have change!' Why?

'... the desire to oversimplify and to reduce complexity is leading to an excessive reliance on simplistic prescriptions as the essential tools of management. In our view, there are no gimmicks that explain sustained success. The best football teams run, pass, block, and tackle with awesome skill based on years of individual and team practice. The best writers spend countless hours crafting and recrafting their manuscripts so that ideas and feelings shine through. Management is no different. The best managers are relentless in the pursuit of facts and reality and in their commitment, energy, and skill in leading groups of people to achieve well-defined and considered goals' (Hilmer and Donaldson, 1996).

Promoters of these fads (though they don't call them fads) often advocate the discarding of existing practices, thoughts and experiences, to be replaced by the 'new'. The development of management knowledge up to this point in time accounts for nothing. This is in stark contrast to the way other disciplines, such as engineering and science, progress — there is considered to be value in studying the practices and thoughts of those that have gone before. A compulsory part of any engineering or science research is a literature/background survey together with a critical appraisal.

The fads, either directly or by implication, imply that existing management practices are failures and a radical transformation is necessary. A doomsday picture is painted unless managers change their ways. The fads are called 'modern management' in complete ignorance of all historical and ongoing rational work in the theory of management.

Almost without exception, ideas that are being promoted in the fads are being, or have been practised elsewhere, though generally under less glorious titles and without the fanfare.

Simple prescriptions seem to encourage poor implementation. There also seems to be an expectation on the part of the user that results will be achieved quickly. When the results don't materialise quickly, a new solution is sought. It could be expected that good management practices will evolve and adapt to the situation at hand and take some time before the full returns are obtained. Both managers and investors want instant returns.

There would also appear to be a cost to not only the organisation but also society from companies continually trying quick fixes instead of getting back to fundamentals and aiming for longer term returns.

Back to fundamentals

Often the approach/practice is presented as a set of recipes in a cookbook without regard to the fundamentals. Often the cookbook is applied without understanding the fundamentals. Therein lies a major criticism of most of the fads — they are short on fundamentals but big on hype. It is argued that managers would be better advised to study fundamentals in people management etc. including that relating to organisations, groups, teams, conflict and similar, and work from there. Against this, such fundamentals-oriented approach lacks impact and razzle dazzle — both necessary to perturb an existing organisation and hopefully improve the bottom line. In contrast, a fundamentals approach could be expected to incrementally improve productivity through improved understanding of how an organisation and business works.

In spite of this, if a manager still wishes to adopt one of the fads, it should be through full recognition of the underlying fundamentals. The temptation of trying to fit a square block in a round hole may then be avoided.

This century has seen the development of an interest in the study of management, the making of decisions, how activities are coordinated, people handled and performance evaluated. Each firm is different yet there are basic essentials of management which are common to all — so-called generic management principles.

Interestingly, despite this emergence of the study of management and a considerable amount of time devoted to researching the matter, there is no generally accepted correct way of managing and no universally agreed theory of management. The fundamentals of management remain to be discovered and the practice of management is more of an art than a science. There is usually no one right answer to any given management problem and several management theories may all be applicable. In many cases it is even difficult to define the problem.

Hilmer and Donaldson concur: 'There are few remedies through instant cures; rather progress comes from the disciplined application of fundamental concepts guided by values and reasoned analysis.'

The worry with fad theories is that they tend to be situation-specific. What happens when the situation changes? Only a knowledge of fundamentals will prepare a person for diverse situations.

Increases in productivity, decreases in costs, improvement in customer relations, ... are not assured using the simplified and generality-lacking models present in fads. More certainty is attached to the outcome if a fundamentals approach is adopted, because of the better models.

Examples

Some recent examples of faddish thinking include:
- The horizontal organisation where hierarchies are removed/ pyramids are flattened.
- The empowerment of workers.
- Satisfying and delighting customers totally through customer focus and listening to customers.
- Extolling the virtues of mission statements, visions and strategic planning.
- Continuous improvement.

- Shift paradigms.
- Becoming a learning organisation.
- Total quality management.
- Re-engineering corporations.
- Partnering.
- Benchmarking and best practice.
- Self-directed work teams or similar.
- Value management.
- Risk management.

Hilmer and Donaldson (1996) identify five 'trails that have been falsified by hype and oversimplification, namely:

- Flatten the structure: Until all hierarchy is abolished, unproductive bureaucracy must be exposed and eliminated.
- The action approach: Don't become immobilised by planning, reflection, or analysis. Allow people to follow their natural instincts and intuition. Focus on visions, not numbers.
- Techniques for all: As problems arise, find the right technique, apply it quickly, and then move on to the next one. There is no situation that can't be fixed by methods such as portfolio planning, value-based planning, niche strategies, total quality management, benchmarking, re-engineering, or gainsharing.
- The corporate clan: Model the organisation along the lines of a happy family rather than a hierarchy. Create a corporate culture that guides and encourages. Abolish rule books and procedures manuals and rely on culture to define what is good or bad, right or wrong.
- The directors to direct: Fix the board of directors to better scrutinise management actions and decisions. Make sure that the board chair and a clear majority of directors are tough-minded, independent nonexecutives who are prepared to call the shots and ensure that management keeps the shareholders' interests paramount at all times.

While there are grains of truth in each of these trails, there are also mistaken assumptions and dangerous implications.'

Managerial background and expertise

The field of management is very crude when compared, for example, with established engineering disciplines of age. Management models tend to be largely verbal models with all their attendant drawbacks. Many management contributors are incapable of modelling otherwise because of their backgrounds.

The reputation of management suffers from people undertaking the task without being aware of all that is necessary to perform the task successfully. People without the necessary skills and qualifications practising management damage the discipline. At present anyone can hang up a shingle or advertise that s/he is a manager. Accreditation is one suggested way of rectifying this situation.

Likewise it is suggested that managers using these quick fix approaches damage the profession. The management profession will not gain credibility unless knowledge is applied in an objective and reasoned fashion. The body of knowledge used should not rely on dogma or unproven models.

Is it feasible that a manager can work in discipline areas without having technical expertise in these areas? Or is some technical expertise, albeit perhaps only an overview, a prerequisite to being a successful manager?

It is argued that some case study based learning and some experiential learning promoted by many management educators are some of the main contributors to preventing a critical, analytical approach to management. Some system and structure has to be attempted, rather than purely entertaining the pupils and making everyone 'feel warm inside'.

Closure

Let's give the fads, bandwagons and hype away. Let's go back to fundamentals, look at the underlying system and structure of management, and progress from there. Let's develop management thinking through the 'scientific method', rather than by opinion and anecdote. Productivity, worker satisfaction, profits etc. must improve.

4.7 APPENDIX: GURUS OF FADDISH MANAGEMENT[*]

Based on D. G. Carmichael, 'Gurus of Faddish Management', Journal of Project and Construction Management, Vol. 4, No. 2, pp. 77–84, 1998; also appeared in Proceedings of the International Conference on Construction Process Re-Engineering (ed. K. Karim, M. Marosszeky, S. Mohamed, S. Tucker, D. G. Carmichael, and K. Hampson), ACCI, UNSW, Sydney, pp. 365–371, 1999.

Abstract

It appears that in every aspect of the management of individuals and organisations, there are management gurus at work. Gurus are the pushers of fads. No matter how challenging current day management is and the need for a rational approach, the gurus hold power over organisational thinking. New ideas sweep the board, only to vanish in a puff of mediocrity when fashions alter. Faddish management has become an industry in its own right. This paper attempts to deglamourise the phenomenon and to put current practices of management into a clearer perspective. In the main, organisational society has adopted an anti-rigour stance. The move to anti-rigour in management has flourished to such an extent that there is concern for the whole future of rigour in management.

Introduction

A dictionary definition of a guru is that of a leader or chief theoretician of a movement, especially a spiritual or religious cult. The term derives from Hindi/Sanskrit, but is used here not referring to any particular religion, rather to the influencing of management thinking along religious fervour lines.

---

[*] Quotations in this appendix are all from Ford (1982).

It appears that in every aspect of management of individuals and organisations there is faddish management and gurus at work. Comments have been made on this, for example, by Shapiro (1995), Hilmer and Donaldson (1996), and Carmichael (1997). There is also the humour of Adams (1996, 1997). 'New ideas sweep the board, only to vanish in a puff of mediocrity when fashions alter' (Ford, 1982). No matter how challenging current day management is and the need for a rational approach, the experts/gurus hold some power over organisational thinking. 'Yet if we are to make sense of the future, it is time the phenomenon was deglamourized and put permanently into a clearer perspective.'

Faddish management has become an industry in its own right. 'If you have the right long words, no matter if there is no meaning behind them. If you have the grandiose aim, no matter if there is no integrity. If you are obsessed with power, no matter if you lack wisdom or worldliness. You have now become a [guru], and your craft is ... Nonscience' (Ford, 1982). The pronunciation of Nonscience is part of the pun. Although the term is used by Ford in the world of science and its experts, it holds equally well to the world of management and its gurus/experts.

'The mushrooming growth-industry of Nonscience (pronounce it how you will) is the real enemy. Its monstrous power structure is nothing more than a confidence trick. The public, who are too poorly educated these days to tell the difference, have been encouraged to accept a range of beliefs that all serve to keep [gurus] in their unchallengeable position.'

People, with a healthy criticism for faddish views, are sparse indeed.

'Once a viewpoint is fashionable, there is no end to the slavish support it attracts.'

'Millions of pounds can be diverted into repetitive investigations by innumerable institutes, all chasing the same trendy topic whilst other, equally deserving subjects go by default.'

As an example, re-engineering is currently a trendy subject for publishing and attracting the attention of people away from more fruitful areas. Witness the number of recent publications in this area. Have any contributed anything to the state-of-the-art? The writer believes not. And of course there are those publications that add re-engineering to their titles in order to gain ready acceptance by the bandwagon community.

Fads come and go. It appears important for people to be fashionable. 'Stepping out of fashion these days is a serious offence.'

'... public opinion is moulded by [gurus] who make proclamations that ordinary people are in no position to challenge.'

'If you have a heart attack, there are plenty of spokesmen who will tell you to take as much exercise as you can, to build up those damaged cardiac muscles. There are just as many who will advise you to rest as much as possible, to give the muscles a chance to recover. This principle of contradiction is found everywhere.' Management advice excels in this principle. For example, why flatten organisational structures unnecessarily (Carmichael, 1996)?

Management knowledge, on the surface, may appear sophisticated but really, little is known. But no management guru admits this. 'It is a cardinal rule of the [guru] code that you always simulate total confidence.' Few admit to there being areas of ignorance, or incomplete logic in the argument. The gurus know astonishingly little about management fundamentals. Management is an area of ignorance.

With the entry of gurus, independent decision making by managers has taken a back seat to the views put forward by the gurus.

'[Gurus] are unassailable and superior individuals who use a language of their own to cloak their inner whims in a spurious aura of authority. As a result, everyone from the man-in-the-street right down to the politician takes very seriously what they say. All [gurus] delight in changing the world we know and love: not necessarily for the better, perhaps even for the hell of it.'

The management literature has seen a procession of fads over recent years, for example re-engineering, TQM, benchmarking, partnering, self directed work teams, value management, flat organisational structures, and so on, and so on. And gurus have emerged to champion the causes.

What fad is going to be introduced next? Which of the existing will be cast aside? What name will be given to old ideas revamped and shuffled? The only certainty appears that fads will come and go like the seasons.

Why do gurus hold so much power over the thinking of practising managers and many academics? They possess magic and mystery which is worshipped by their followers. Are the same processes involved here as in religion? Are the fads opiates for the masses?

Education

The gurus trade on the poor education and demoralisation of managers. From the gurus' perspective it is important that such managers remain uneducated and demoralised, and future generations of new managers are educated less and demoralised more, in order to keep managers and gurus in their respective places.

'But it is important that the effects of the deterioration are disguised. [Gurus] have evolved two methods of doing this. The first is the progressive lowering of standards, so that people gain spurious qualifications ... At the other end of the spectrum, the job-description business has continuously up-graded vacancies so that they sound progressively more prestigious as time goes by.' Everyone today is a 'manager' of some sorts, yet is extremely poorly educated for the role. Few have any knowledge of past developments in a rigorous management body of knowledge.

And there appears no attempt to seriously upgrade the credentials of managers, even acknowledging the plethora of management postnominals on business cards.

'As it is, we are increasingly ruled by men and women who know little about anything apart from their own capacity for power. The very people who have the greatest ability to threaten us all are uniquely narrow-minded, and almost culturally illiterate. We do not need a race of moronic super-specialized [gurus] who know nothing about almost everything, but wise and insightful [managers] who try to understand what little we know so far.'

Rigour

In the main, organisational society has adopted an anti-rigour stance. The move to anti-rigour in management has flourished to such an extent that there is a concern by some, including the writer, for the whole future of rigour in management.

Today's trend towards guru-led faddish practices in management threatens the serious evolution of a rigorous management body of knowledge. It leaves '... a burden of pollution future generations will resent.'

Gurus and people appear selective in the arguments they apply to support their stances. Anecdotes and opinion are rife. Few attempts appear to be made to present a balanced and all-inclusive rigorous picture of management scenarios. All issues are not put in proper perspective. Significant issues seem to be glossed over.

Language

'The contorted language of the [guru], designed to keep outsiders at bay rather than communicate, is a hallmark of Nonscience.' Re-engineering is an example of this. Re-engineering is no more than Work Study rejigged, or if fundamentals are preferred, is no more than systematic problem solving. Why solve a problem in a systematic fashion when you can dress your work up in the clothes of re-engineering? And why think about improving your processes if you can't call the practice something glamorous? Which companies waited for the fad of benchmarking to be introduced before they compared their performance with competitors? Which organisations would have thought about improving their practices without having TQM as a label? Who would have thought of trust, cooperation and goodwill without partnering?

'Yet their language embodies a heady illiteracy.' For example, verbs are built out of nouns, such as 'to re-engineer', 'to scope', ... Obscurantism blinds the followers.

'The simplest term can instinctively be made complex by the [guru] mind. One elementary means is the rule of superloquation: "never use one word where three will do".'

For example (courtesy of ABC Radio — 2BL, after P. Broughton), faking it with work content language, using a Systematic Buzz Phrase Projector. The system employs these thirty buzz words:

0. integrated	0. management	0. options
1. total	1. organisational	1. flexibility
2. systematised	2. monitored	2. capability
3. parallel	3. reciprocal	3. mobility
4. functional	4. digital	4. programming
5. responsive	5. logistical	5. concept
6. optional	6. transitional	6. time-phase
7. synchronised	7. incremental	7. projection
8. compatible	8. third-generation	8. hardware
9. balanced	9. policy	9. contingency

The procedure is simple. Think of any three digit number, then select the corresponding buzz word from each column. For instance, number 736 produces 'synchronised reciprocal time-phase', a phrase you can drop into any report with a ring of authority. 'Nobody will have the remotest idea what you're talking about. But the important thing is that no one is about to admit it', says Broughton.

The language of many gurus is not intended for communication but rather to impress. Gobbledegook, expertise, obscurantism or metalanguage might all be terms applied to the language used. Simple, straightforward situations are described in verbose, unsystematic, circular and contorted fashion.

'A few choice paragraphs of elongated and contorted pseudo-prose can keep the most dazzling mind at bay.' A relatively straightforward management situation can be made to seem exceedingly complicated, necessitating an exceedingly complicated solution. The guru is able to convince readers/listeners that s/he knows more than them and hence must be right.

Cause and effect

People will do whatever a guru says, no matter how preposterous, simply because the guru says. A non-guru is ignored, even if more rational.

Gurus enjoy a god-like status.

A cynic might conjecture that gurus don't solve problems, only perpetuate them. By solving problems, a guru is put out of work. Arriving at solutions would imply a contribution to the management body of knowledge is being made. Experiences and data are selectively assembled to support the latest theory.

'A [guru] is always right, primarily because he says so.'

Management discourses should strive for the truth, for understanding and simplification. The gurus almost get the third part right but end up with trivialisation instead, and in so doing cloud the picture.

For example, the use of acronyms such as LEAD and TEAM are popular. Leadership is reduced to four actions beginning with the letters L, E, A, and D. Team building is reduced to four actions beginning with the letters T, E, A, and M. And so on for other acronyms. Have the proponents of these acronym approaches discovered some fundamental laws of nature?

Changes proposed by gurus are for the most parts entirely pointless. Old ways may be familiar to people or the ideas may already be known. Change creates work for the gurus by adding to the problems of others. Change is put forward as part of progress, but is arbitrary, inconvenient and costly. Change should mean improvement, betterment and moving forward, but not so to the gurus. To the guru '... progress means changing anything that hasn't been altered round for a while. "If it moves, shoot it," used to be the motto of the great white hunter of the twenties. To today's [gurus] the wording has changed, but the result is almost as bloodthirsty. "If it stands still, change it," they say.'

'At the core of their unique mental processes lies an unremitting failure to identify cause and effect. If there is one single factor behind the success of Nonscience (I only said "if", there isn't) then it must be this blindness to one of nature's most fundamental truths.'

False cause and effect links are established, conveniently ignoring other possible contributing factors. Some statistical correlation is possible between many variables. Why are gurus selective in the variables chosen? This leads to spurious cause–effect relationships that do not assist managers, because they don't tell the whole story. And based on such relationships, managers are asked to change their practices, perhaps to something worse.

'There have been many examples of inferred causality which proved to be baseless.' For example, the statement that hierarchical organisations are inefficient (Carmichael, 1996).

Fashion

An examination of the popular management literature will show a changing emphasis on management issues. Why is it that these issues, so urgent at the time, suddenly are no longer popular? What happened to such important issues?

Could the rise and fall in popularity of the issues be attributable to fashion? 'A spectacular summit of interest, followed by eclipse — that is the fate of any topic when a [guru] gets hold of it. The student of Nonscience knows this as Fashionism. In choosing which subject to study or which aspect of society to alter next, the [guru] must always be guided by the constraints of Fashionism, and his choice need never be tainted by considerations such as "humanitarianism" or "relevance"' , 'or worthwhileness'.

'Do not be persuaded by [guru] arguments that these trends are a reflection of real priorities, or that they reflect genuine merit. Fashionism does not act like that.'

Interestingly, management issues can exist and be tried over the years, but then one particular or several gurus attract all the attention. This is not because of novelty, painstaking research effort by the guru, improved understanding by the guru, fastidiousness of the guru, new ideas etc. Why then, does one particular guru attract all the hysteria, exposure, worship and accolades. Equally, why does the hysteria shift to another guru on another issue, a short time later? In fact the original issue and guru commonly then become discredited. There is a backlash. The fall of the idea and guru are as sharp as their rise to stardom, like popular music going up and down the music charts.

The difficulty remains for observers, such as the writer, of this phenomenon, of trying to predict what will be the next fashion in management. How do fashions in clothing come about? Which particular songs, out of all those recorded, will make the Top 50 chart? What will be the popular colour of motor cars next year? How high will hemlines be this summer? Will flared-cuff trousers, peddle-pushers and hotpants make a return disguised as different products?

'Fashionism has the power, not only to take empty-headed topics to the peak of popularity, but also to influence the corporate criteria of a whole community until people begin to doubt the evidence of their own senses, and to lose touch with the most basic levels of reality.'

The writer is unaware of any management issue that has climbed steadily up the chart of popularity, when it did not exist in the first place. Why are not issues more carefully scrutinised by the public before acceptance? The reason perhaps is the lack of management education by the public and also the guru.

Interestingly when an issue becomes fashionable, at the height of the craze, there may be thousands of people publishing and consulting and making hay while the sun shines. But of these, only a handful appear to be actually making a contribution to the state-of-the-art. The rest are simply jumping on the bandwagon, perhaps hoping that some of the aura will rub off on them, but at the very least making a living out of it. Will the thousands of 'groupies' take on a similar mantle of eminence as a guru?

There can, unfortunately in this climate, be pressure placed on people to come up with something new or original at the expense of professional honesty and ethics and the scientific method. Data can be invented, unreliable data can be used, experiences selectively sampled and so on. Some people appear to start with the results and work backwards. Given the complicated nature of management, it can be difficult to establish what is fake and what is not. Certainly such practices do not advance the state-of-the-art, and it can take years to undo the damage.

Closure

'The spreading acne of Nonscience has taken the new and wild-eyed race of [gurus] into every walk of life, driving the bewildered victims like withered leaves before a plastic brush in Autumn. It is surprising how seriously they are taken; but there is a self-centred headiness, almost an ebullient mania of oppressive righteousness, about the [guru] of today. He may have done nothing more than [something minor], but he intimidates all outsiders with the piercing eye, the jutting chin, the uncompromising manner which demolishes all opposition.'

'[Gurus] always tell you things, they never ask. Every [guru] pushes aside criticism or probing, he merely asserts. [Gurus] fly from ideas like midges from fire-smoke; they congregate around piles of data instead. Above all, they adore their power to rise above everyone else in the scramble for prestige, without the slightest wisdom, or worldliness, or even common-sense, behind them.'

'Perhaps the cult of the [guru] is a movement designed to occupy our waking hours and prevent us from becoming worried by too many major decisions.'

CHAPTER 5

Negotiation

5.1 INTRODUCTION

It is generally better to deal by speech than by letter; and by the mediation of a third than by a man's self. Letters are good, when it may serve for a man's justification afterwards to produce his own letter; or where it may be danger to be interrupted, or heard by pieces. To deal in person is good, when a man's face breedeth regard, as commonly with inferiors; or in tender cases, where a man's eye upon the countenance of him with whom he speaketh may give him a direction how far to go; and generally, where a man will reserve to himself liberty either to disavow or to expound. In choice of instruments, it is better to choose men of plainer sort, that are like to do that that is committed to them, and to report back again faithfully the success, than those that are cunning to contrive out of other men's business somewhat to grace themselves, and will help the matter in report for satisfaction sake. Use also such persons as affect the business wherein they are employed; for that quickeneth much; and such as are fit for the matter; as bold men for expostulation, fair-spoken men for persuasion, crafty men for inquiry and observation, froward and absurd men for business that doth not well bear out itself. Use also such as have been lucky, and prevailed before in things wherein you have employed them; for that breeds confidence, and they will strive to maintain their prescription.

It is better to sound a person with whom one deals afar off, than to fall upon the point at first; except you meant to surprise him by some short question. It is better dealing with men in appetite, than with those that are where they would be. If a man deal with another upon conditions, the start or first performance is all; which a man cannot reasonably demand, except either the nature of the thing be such, which must go before; or else a man can persuade the other party that he shall still need him in some other thing; or else that he be counted the honest man. All practice is to discover, or to work. Men discover themselves in trust, in passion, at unawares, and of necessity, when they would have somewhat done and cannot find an apt pretext. If you would work any man, you must either know his nature and fashions, and so lea him; or his ends, and so persuade him; or his weakness and disadvantages, and so awe him; or those that have interest in him, and so govern him. In dealings with cunning persons, we must ever consider their ends, to interpret their speeches; and it is good to say little to them, and that which they least look for. In all negotiations of difficulty, a man may not look to sow and reap at once; but must prepare business, and so ripen it by degrees.

Sir Francis Bacon, Of Negotiating, from 'Essays on Counsels, Civil and Moral'

Negotiation is common in all forms of interaction between people and is part of people's general problem solving processes. At the firm or project level it is most noticeable in settling contractual problems and disputes, and in the industrial relations area in discussions between labour and management. At the personal and family levels, negotiation is always present. Most matters are negotiable. The emphasis, however, given in this book is on negotiation as a means of resolving contractual differences and disputes.

> Helga's Rule:
> *Say no, then negotiate.*

Negotiation can involve 'dickering', 'horse trading' or 'wrangling' over a price. However in its purest sense, negotiation is a process that attempts to maximise the interests of both parties or attempts to produce a settlement or agreement, or assists in finding a resolution to a problem.

Negotiation is a process of attempting to reach and possibly reaching an agreement between two or more parties. It may, but does not have to, involve compromise.

The *negotiation process in summary form* entails:

- Each party having an initial position.
- Each party evaluating the position of the other party.
- Each party adjusting its initial position in the light of this evaluation.
- Repeating these steps with adjusted initial positions until a resolution is reached or the negotiation is abandoned.

  The final *resolution* may involve each party coming half way, one party not shifting from its initial position with the other party capitulating, or something between these two extremes, or something lateral to the original stances. The final resolution may or may not be perceived as being equitable. Along the way parties' goals may change.

Negotiation will only proceed as long as both parties believe that an acceptable outcome is possible. As a *fallback*, both parties have the *status quo* should the negotiation not be proceeding to an acceptable outcome. This may or may not be acceptable in itself and may drive the negotiation on, or stop the negotiation from proceeding in an unacceptable direction.

The fact that a negotiation only proceeds as long as both parties believe there is an acceptable outcome implies that for a final resolution to be reached different to the initial positions, then both parties have to believe that they have won something. One party negotiating on the basis that it is to *win* and the other party is to *lose* may lead to the termination of the negotiation by the other party. Certainly if a lasting resolution (rather than a legally binding agreement) is a desirable outcome of a negotiation, then both parties have to believe that they have won something. Both parties winning something, in turn, implies *concessions*. Good negotiating trades concessions rather than giving them away.

There are no rules to negotiation but rather culturally accepted styles. Parties have a large degree of flexibility as to how negotiations proceed. Successful negotiation, nevertheless, depends on an *understanding of human behaviour*, preparation, personal skills relating to self-discipline, objectivity and decision making and many more matters.

A study of the negotiation process gives an awareness of such matters and improves the chances of a more favourable resolution.

Where negotiation involves an adversary, matters of *ethics, truth* and *morality* may take a back seat in such cases. Openness is not necessarily considered a desirable trait in negotiators. The analogy between a poker player and a negotiator is sometimes drawn to indicate what may be necessary in order to be successful.

Judging the success in negotiation

Whilst each participant has different aims and goals, success in negotiation might be established relative to criteria such as:
• The outcome is better than the alternatives available.
• Parties' needs of satisfaction have been maximised. This could be broken into need types:
  – Substantive.
  – Emotional/psychological.
  – Procedural.
• The outcome has been reached in such a way, that parties are prepared to associate again with each other.
• All parties feel they have negotiated rather than being compelled to accept something.

5.2 PHASES OF NEGOTIATION

5.2.1 DESCRIPTIONS

A number of references on negotiation describe the negotiation process in terms of phases. For example,
Preparation
Wants
Propose
Bargain
Agree
Follow Up (Rose, 1987)
or
Introduction
Differentiation
Integration
Settlement (Richards and Walsh, 1990)
or
Introductory (Forming)
Differentiation (Storming)
Integration (Norming)
Settlement
Not all phases occur in all negotiations and it may not be a simple matter of proceeding through the phases one at a time from the first phase to the last phase without itera-

tions and feedback between phases. There may also be no clear distinction between one phase ending and another starting. For the present purposes, the following phase descriptions have been adopted:

- *Preparation*
- *Familiarisation*
- *Opening*
- *Proposal*
- *Bargaining*
- *Agreement*

These are followed up by debriefing, review and similar activities.

Two-party phasing

Note that in any negotiation, it may not always be the case that both parties are in the same phase at the same time. Part of the skill of the negotiator is to understand which phase each party is in and adopt an approach to match.

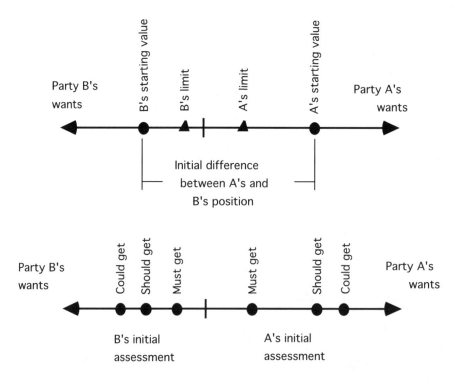

Figure 5.1. Example range charts.

Withdrawal

Should the negotiation be proceeding in a manner not considered desirable by either of the parties, they may withdraw at any of the phases.

Range chart

A device used in negotiation is a range chart (Figure 5.1). A range chart is a line on which both parties' wants are indicated or marked. The line extending in one direction is in favour of one party, while the line extending in the other direction is in favour of the second party. Common ground for the negotiation is in the middle of the line. It assumes all wants are commensurable. The scale of the line is not important.

5.2.2 PREPARATION PHASE

A poorly prepared negotiator will always do worse than a well prepared one.

Rose, 1987

In all matters, success depends on preparation.

Confucius

Cavanaugh's Discovery:
Never, ever, enter a battle of wits half-armed.

Sobel's Law:
There is no substitute for genuine lack of preparation.

The Preparation phase involves obtaining information on most matters to do with the negotiation including:
- The cost and other characteristics of the item or matter that is central to the negotiation.
- Acceptable alternatives.
- The facts which are disputed/not disputed.
- Distinguishing between matters of fact, legal or rule interpretations, matters of expert professional opinion, and matters of personal or only partly-qualified opinion.

And for each party:
- The wants and needs together with priorities. (This will mean acquiring as much information that is available or can be accessed of the other party, and a client if negotiating on behalf of the client.)
- Testing of facts and assumptions.
- The likely negotiating style and tactics to be used by the other party and the planned style and tactics to be used in return.
- Options available to both parties including the possible preservation of the status quo.

- Possible deadlines for each party and the consequences of exceeding these deadlines.
- The goals.
- Items that can be conceded and their cost.
  Assumptions need to be checked.

ASSUME makes an ASS out of U and ME.

Wethern's Law of Suspended Judgment:
Assumption is the mother of all screw-ups.

Examples of assumptions going wrong are:

Example
... when a builder was offered some earth fill free of charge he assumed that both the fill and the cartage were free. The other party assumed that the builder would pay cartage. You guessed it! A dispute arose out of these innocent assumptions.

Rose, 1987

Example
A formwork contractor quoted a price per square metre, that is on a schedule of rates basis. (Formwork, in this case was plywood supported by timbers and props. The formwork was for a concrete slab.) On completion of the work, the owner measured the number of square metres of formwork used, multiplied it by the contractor's quoted rate, and accordingly paid the contractor. The contractor queried the magnitude of the payment, and proceeded to measure the area of the slab in square metres.

Because the slab was supported on walls below, the area of formwork was less than the area of the slab. A misunderstanding had occurred over the price quoted by the contractor and accepted by the owner.

Negotiating with people from different cultures and countries makes the checking of assumptions imperative, else misunderstandings may eventuate. Assumptions don't necessarily translate between different societies.

A party's *wants* and *needs* may be difficult to isolate where the party is, say, a large organisation with many departments or divisions. The other party's wants and needs can be explored, for example, by taking a devil's advocate role and related strategies used by lawyers preparing a court case.

In the Preparation phase, the range chart shows the limits between the worst result which could be agreed, to the best that could practically be attained, with specific points which could be conceded or gained shown in between. The range chart is based on an assessment of the 'must get', 'should get' and 'could get' categories for each party.

This information, together with other information gleaned during the negotiation, is essential to a well-ordered negotiation stance. It will determine tactics, and indicate the possible tactics of the other party. Although information obtained during the negotiation is useful, the negotiation itself should not be used as an information gathering exercise and, as such, an excuse for not doing the necessary preparatory work.

Further work carried out in the Preparation phase is:
* The choice of *location*.
* The choice of *timing* and *time*.
* The development of an *agenda*.
* Ensuring that the other party attending has the necessary *authority*.
* Choice of *attire*.
* Selection of *style* and *tactics* to be employed.

More is said later on location and timing and time. The agenda, itself negotiable, can shape the negotiation by including or omitting items or by the order in which the items are listed. There is always, of course, the hidden agenda which may be deliberate or a result of the parties not recognising the real issues but perhaps only recognising the symptoms or perceived issues.

5.2.3 FAMILIARISATION PHASE

This phase precedes the negotiation proper. It is common to go through *introductions* and some *small talk* before beginning the negotiation in earnest. It can act like an *ice breaker*.

This time can be profitably spent *developing a rapport or relationship* between the members of each party. One approach to this building of rapport is to talk of common interests (and definitely not talking of divergent interests, at least initially) and to show concern for the other party's problems.

There is always the difficulty of switching from being nice to starting the negotiation proper, but in effect the negotiation has already begun.

Some cultures insist on a long Familiarisation phase, while others regard anything other than a handshake as a delaying tactic.

Example
Observe the behaviour of Australians.

Typically a greeting is something to the effect of 'How are you?' or 'How are you going?' ('How ya goin?') even though the asker has no real interest in the other's health; in fact the asker becomes irritable if the other replies anything other than 'Well, thanks. And yourself?', and the asker becomes particularly irritable if the reply lists a series of illnesses or misfortunes. The obvious question then is, why do Australians ask of another's wellbeing when they have no real interest in it. The process has become ritualised to the point that many people reply 'Well, thank you.' even though the other person has not asked 'How are you?'; there is an expectation that this question was asked, even though something else may have been said.

This ritual can be witnessed ad nauseam on talkback radio. Almost without exception every call starts by saying to the radio announcer 'How are you?'; the announcer replies 'Well', 'Not too bad' or something similar, and this is often followed by the same 'How are you?' question. Neither the caller nor the announcer have an interest in the wellbeing of the other.

The terms 'Gidday' (Good day) or 'Hello' seem to be reserved for situations where a conversation may not be wanted. They only prompt similar replies.

The initial greeting over, Australians commonly enter into small talk. This often revolves around some pointless comment on the weather, inquiries after a spouse's and children's health, how the local football team did on the weekend, or other inconsequential matter. If the two parties know that they have something in common, for example an interest in a particular sport, inconsequential remarks may be made about this.

At some stage, generally only minutes after meeting, the discussion becomes serious, sometimes prompted by 'Let's talk about ... now' or similar abrupt introduction.

5.2.4 OPENING PHASE

The Opening phase is one in which the parties *express* their *wants* and an attempt is made to establish the limits of the other party's position.

Example
Often initial wants are inflated, though if excessively inflated, they may have a negative impact on the negotiation. Inflated initial wants are used on the assumption that they may be reduced as forms of concessions during the negotiation. This is the familiar bargaining tactic used by street vendors — the seller proposes an initial high price; the buyer offers an initial low price, and then haggling ensues until some middle ground is reached. Should the seller's initial price be too high, the buyer may walk away without purchasing anything, in the belief that no amount of haggling will reduce the price to something acceptable. A similar situation of no sale may result if the buyer's initial price is too low; perhaps the seller feigns that s/he has been insulted.

Example
An employee when asking for a rise in pay may couch the initial request within a larger package of wants.

Wants from both parties will generally include what each:
- Must get.
- Should get.
- Could get.

The wants are prioritised, though it may be difficult to work out the *priorities* of the other party. The body language used by the other party may give some indication. There may also be some unstated wants from each party, done as a deliberate tactic.

You need to find out: Why, What, When, Where, How and at What Price?

When stating Wants or listening to the other side state [its], there are two important rules:

Be General About Specifics: *e.g. We want the major part of this order delivered by Friday, 25th.*

Be Specific About Generalities: *e.g. We must have a meal allowance which is equitable.*

This is done intentionally to allow room for movement. In the first case we are specific about the date but prepared to talk about what constitutes 'the major part'. In the second example we demand a meal allowance but are prepared to discuss the amount.

Being specific about specifics leads to brinkmanship and being general about generalities appears vague (Rose, 1987).

Unstated wants, like priorities of wants, need to be drawn out of the other party. Knowing the unstated wants, which might be more fundamental than the stated wants, increases the chances of finding a solution.

It is unclear as to which is more advantageous, whether a party states its wants first or second. Going first seizes the initiative in the competitive environment but may intimidate the other party. When going second, there is always the possibility that the wants will be modified in the light of what has just been heard.

The assumptions used in the Preparation phase are reconsidered in the light of information obtained in this phase.

5.2.5 PROPOSAL PHASE

The wants, and their priorities, of the other party are continued to be probed in order to establish where that party is prepared to move ground and where it is isn't, and whether alternative wants can be established. Proposals, at first tentative and non-committal and sometimes hypothetical, are introduced and are based on *areas of common ground or potentially common ground.*

Proposals commonly take the 'if/then' form and invariably are conditional. For example, 'If you do (something specific) then I'll consider (something vague).'

The condition is expressed first followed by a trade for something less. The proposal may involve grouping a number of items in order to embed unpalatable wants amongst palatable ones.

It is important to make one proposal only and then stop and wait for a reaction. There is a tendency to make another proposal if met with a straight 'No' or the tactic of Silence.

... If the result of your proposal is a flat 'No' or silence, ask the other party for a proposal in return. Do not make another proposal yourself or you will find yourself in an Aunt Sally game, where you set up proposal after proposal and the other party shoots them down.

The counter proposal can be linear (that is, contain the same elements), or parallel (containing different elements) (Rose, 1987).

Proposals from the other party help to elucidate its wants and priorities. Any new wants need to be clarified openly. One party may still be in the Opening phase while the other has moved on.

The Proposal phase should also help to elucidate the real needs of the other party through critically examining the other party's proposals and its response to proposals put to it.

5.2.6 BARGAINING PHASE

Bargaining involves the exchange of concessions. The phase is different in emphasis to the Proposal phase. In Bargaining you move from the mode: 'If you ... then I'll consider ...' to 'If you will ... then I will'.

... Care needs to be taken in the Bargaining phase. It is important to state your condition first then be specific about what he will get in return.

It is important that you have room to move. In this regard the way concessions are made and the increments give important information about your bottom line (Rose, 1987).

The point most good negotiators emphasise is that concessions should never be given away free, but rather should be traded. And, the increments in concessions should be made smaller and smaller in order to signal a limit or resistance point to the other party (even though an actual limit may be much greater/smaller).

There is always a possible situation of coming across a person who is in the negotiation for the *game* or the *sport*. Some people like to haggle even though the end result may be predictable and could be reached much more quickly.

Other seemingly illogical behaviour may occur with *people's utility* and *value of money*. One dollar to a rich person is different to one dollar to a poor person. A cheque and cash for the same amount are perceived differently. Big ticket items and savings on them are treated differently to small ticket items and attached savings; for example saving one dollar on a one thousand dollar item is perceived differently to saving one dollar on a ten dollar item.

The essential element of good bargaining is to trade that which is cheap (relatively) for you to give and valuable for the other party to receive ... in return for that which is cheap for the other person to give and valuable for you to receive (Rose, 1987).

The presence of *deadlines* may act as a spur to making concessions.

... all major concessions occur on or after the deadline. ... even Parliament passes more legislation in the last weeks of sitting than any other time (Rose, 1987).

Example

Some deadlines which have been used to advantage in negotiations include plane departure times and car park closing times.

A situation occurred where people in the host country stalled the negotiation until close to the departure time of the foreigner. As the departure flight time approached, the foreigner reluctantly agreed to a less than favourable settlement, for fear of missing the flight and returning to his country empty handed.

Interstate plane flights are not as drastic as international flights, but in both cases a recommended practice is to not divulge flight times to the other party, on the basis that the divulging party may be more inclined to come to some settlement as its flight time approaches.

Car park closing times have influenced at least one mediation. The mediation had been going all day. The car park building closed at midnight. Agreement was reached late in the evening.

A similar outcome, in another mediation, was reached because the parties' solicitors had evening flights to catch, and it being Friday night with no one wishing to spend the weekend away from home.

Much *bluffing* and *'sham'* bargaining can go on in the Bargaining phase.

The Attorney's Rule, or The Rule of Law:
If the facts are against you, argue the law.
If the law is against you, argue the facts.
If the facts and the law are against you, yell like hell.

Swipple's Rule of Order:
He who shouts loudest has the floor.

Deadlocks or *impasses* may prevent the Bargaining phase from continuing just as they may also prevent the Proposal phase from continuing. Such deadlocks may be real or staged by one of the parties. In the latter form it can be a useful tactic.

Fear of deadlock or a stage managed deadlock can cause the inexperienced negotiator to give away concessions instead of trading them. Deadlock is also used by a party who is interested in preserving the status quo and, therefore, wants to stall settlement for as long as possible.

Deadlock must be put in perspective. There is little point in conceding yourself into a win–lose (you lose) position just to break the deadlock. The worst that can happen is a lose–lose situation if the deadlock continues. (Rose, 1987)

As well as going back to a discussion on wants, the following are *suggested ways of breaking deadlocks:*
- *Change negotiators or levels (i.e. summit meeting of [senior executives]).*
- *Split the difference.*
- *Round out numbers.*
- *Go for a lateral solution.*
- *Concentrate on real wants.*
- *Postpone difficult items.*
- *Have a break.*
- *Go 'off record'.*
- *Try a mediator ...*
- *Point out the disadvantages of deadlock to both parties and try mutual problem solving procedure* (Rose, 1987).

Dr Nordstrom's First Rule of Debate:
It is difficult to win an argument when your opponent is unencumbered with a knowledge of the facts.

5.2.7 AGREEMENT PHASE

As a good strategy it is recommended that only agreement be given on the *final package*. Agreeing on an item-by-item basis may lead to some exposure, unless this is intended.

Detailing or *summarising* the final package might raise unpleasant matters covered earlier in the negotiation, but it is essential that both parties know exactly what is being agreed to and that each party doesn't walk away from the negotiation with a different view of the final package. Definitely where a binding agreement is expected at the end of the negotiation, then the agreed items have to be summarised. In other cases, not summarising the agreed items more than likely will lead to another negotiation episode in the future.

Common practice is for the agreement to be drafted at the close of the negotiation, by either party in consultation with the other, in rough form and then tidied up shortly after while the negotiation is still fresh in the minds of both parties.

'In principle' agreements, whereby the negotiator doesn't have the final authority to make the agreement but is only able to make a later recommendation to others that they accept the agreement, are undesirable but in many cases unavoidable. Negotiation works best when both parties have authority to commit their sides to an agreement. Good negotiation practice is to request in the Preparation phase that only someone with authority attend the negotiation.

In writing up an agreement it is good practice to ensure that there are no loopholes that permit the other party to not perform or delay performing its obligations under the agreement.

5.2.8 POST NEGOTIATION

The negotiation completed, there are a number of follow-up issues. There is the reporting of the results of the negotiation to relevant personnel. And there is a process of reviewing the performance of both parties, and feeding back this review data to improve performance in future negotiations.

5.3 THE NEGOTIATOR

5.3.1 TRAITS

> *A complete negotiator should have a quick mind, but unlimited patience, know how to dissemble without being a liar, inspire trust without trusting others, be modest but assertive, charm others without succumbing to their charm, and possess plenty of money and a beautiful spouse.* Hawkins and Hudson (1986, p. 62) In other words, a great negotiator is the perfect communicator, lucky, charming and attractive. Everyone else needs as much practice as they can get.

The *personal traits* of the negotiators on both sides are strong influencers of the direction and outcome of any negotiation.

Some suggested traits required of negotiators are (original source unknown):

Cognitive abilities
  clear rapid thinking
  verbal and quantitative intelligence
  creative — conceptual fluidity
  perceptual acuity
  memory

Emotional characteristics
  poise, self-restraint
  maturity, ego strength
  enthusiasm
  toleration of ambiguity
  tolerance — empathy
  dominance
  humour
  consistency (but not rigidity)

Communication characteristics
  patience
  tact
  fluency
  empathy

5.3.2 STYLE

Style refers to the way negotiators go about their work. Styles range from being over-bearing, aggressive and dominating, that is competitive in the win–lose mould, through to being collaborative, that is in the win–win mould.

Certain types of negotiation call for a certain style and it is the good negotiator who is adaptable. Some negotiators are not adaptable and retain the same style from one situation to the next.

Superimposed on this competitive–collaborative spectrum are the human traits related to emotion, anger, humour, vacillation, weakness etc.

Style is influenced by many factors including *past experiences*, the negotiators power relative to the other negotiator, the degree of familiarity with the subject, and the degree of authority. Style can be thought of as having a component related to the person (*personal style*) and a component related to the situation (*situational style*). Certain situations seem to bring out a certain stereotype of style in people; they unconsciously adopt a particular style as if it were an expected cultural behaviour.

Different negotiating styles might be classified along the following lines:

Competing: Distributive (win–lose) bargaining
Satisfying own party's needs is important; satisfying the other's needs isn't important.

Collaboration: Integrative (win–win)
Satisfying both parties' needs is important.

Compromising
Satisfying both parties' needs is moderately important.

Avoiding
There is indifference about satisfying either party's needs: no action is likely.

Accommodating
Yielding by one party because it doesn't matter, though it matters to the other party.

While it can be seen that there are several different types of negotiating styles, the two most commonly attempted are *collaborative* and *competitive*. Frequently the collaborative style degenerates to one of compromise.

Competitive implies one side wins and one side loses. Dominant strategies in this mode include manipulation, forcing, and withholding information.

Collaborative implies both sides win; the dominant concern here is to maximise joint outcomes. Dominant strategies include cooperation, sharing information, and mutual problem solving.

The *competitive style* is characterised by:
1. *There is no on-going relationship expected (one-off negotiations tend to be competitive).*
2. *Where the relationship is so strong or perceived to be a monopoly relationship, then the style chosen might be either [competitive] or [collaborative]. This is based on the assumption that the use of the [competitive] style will not fracture the relationship. Domestic negotiations, some business deals and some industrial relations negotiations fall into the competitive category even though there is an on-going relationship because it is perceived that the parties have no choice but to deal with each other.*
3. *Post-negotiations, i.e. those which are necessary to sort out details of an agreement already entered into.*
4. *Pre-negotiations. Those negotiations which precede major negotiations and are usually to do with climate factors such as venue, agenda and timing.*
5. *Where one or both parties have a high perception of their own power.*
6. *Where one or both parties have a perception that they have no options.*
7. *Where one or both parties have a preconceived and single solution to the problem.*
8. *Where traditionally the parties negotiate in a competitive way.*
9. *Emergency situations where there is no time to come to some mutual understanding.*
10. *Where there is an absolute determination to win* (Rose, 1987).
    The *collaborative style* is characterised by:
1. *Where the on-going relationship is more important than the particular deal in hand.*
2. *Where it is perceived that the power of the parties is equal.*
3. *Where the relationship is on-going and in market place. (That is 'If I don't deal with that person I can easily take my business to some other person'.)*
4. *Where the parties are free and willing to explore alternatives.*
5. *Where there is a traditional stance of collaborative negotiations.*
6. *Where there is a wish to come to mutual satisfaction.*
7. *Where the perceived power by each of the other is so high that both parties fear war.*

These, of course, are only indicators. They do not necessarily apply in every instance. However, they are general tendencies and it is well worth working out whether the relationship that you have, or intend to have, is on-going. If so, is it in market place or monopoly? You can then make reasonable assumptions about the style that the negotiation is going to take. Forewarned is forearmed (Rose, 1987).

5.3.3 COMMUNICATION SKILLS

Good communication skills are important in negotiation as they are in most work practices. It is particularly important that the message sent by one party is in fact the message received by the other party and that nothing is misunderstood or interpreted out of context or in a way not intended.

Part of two-way communication is *listening*. Listening skills are highly regarded in a negotiator. Good negotiators only talk when they have something to say.

Use your two ears and one mouth in that proportion.

A whale only gets harpooned when it comes up to spout.

Observation of a Philosopher:
Nothing is opened more often by mistake than the mouth.

Henderson's Homily:
The less you say, the less you have to take back.

Henry Ford's Rumination:
Under pressure, the mouth speaks when the brain is disengaged, and, sometimes unwittingly, the gearshift is in reverse when it should be in neutral.

5.3.4 ATTITUDES

Every person will bring their own attitudes to a negotiation. By attitude is meant, *a person's predisposition* to act or think in a certain way.

The following is a suggested breakdown of what makes a good negotiator.

In politics and diplomacy negotiators are at work every day. Some appear to have a gift — the ability to ensure that their will prevails. These professional negotiators are successful because they can recognise a 'negotiation' before they become immersed in it, and are able to apply certain skills and talents.

- The essential skills are the ability to prepare and to plan. *Effective negotiators plan extensively. They assess their own position and that of their opponent. The [goals] of the negotiation are set at this stage, but are constantly reassessed. Effective negotiators begin any necessary research and, while building up a mass of detail, consider possible strategies to test their own assumptions. Appropriate concessions are devised and their opponent's goals are considered.*

- Good negotiators think clearly under stress. *They are aware of what is happening at any point in the negotiation. They have the ability to take an overview of the situation. Intuition is also important, as a good negotiator can sense when an individual is comfortable with the proceedings and whether it is productive to continue. Good negotiators are able to read their opponent's signals, both verbal and non-verbal.*

- Good negotiators need sound, practical intelligence. *They are flexible and can reject ineffective negotiating styles and switch to those which work. They can exercise options and alternatives, even changing some of their original goals when necessary. Good negotiators use different approaches and take the view that they are in the process of satisfying mutual needs, and are able to explore alternatives and use the resources of both parties to find solutions. Negotiators articulate issues as problems rather than as demands. They make broad statements which identify problems, rather than narrow statements which telegraph their own perceptions of the problem.*

- Communication is a central issue in negotiations. *Good negotiators can communicate clearly when clarity is necessary, and obscurely when obscurity is needed. They also have good listening skills which enable them to respond accurately to their opponent's argument. A major barrier to good listening is that we hear too fast. We think of a response to what we thought we heard, instead of responding to what was actually said.*

- Negotiators know their subject intimately. *They are perceived as experts, whether expert negotiators, experts in a particular field or experts on the issue being negotiated. Society places a great deal of faith in experts.*

- Successful negotiators manifest a sense of personal integrity. *They are personable, well liked and usually well respected. They are aware of their own needs and motivations, and they know what outcome is desired and what factors, if any, impinge on their abilities to negotiate effectively. Good negotiators put themselves in other people's shoes and use that knowledge effectively. They are patient, which means they can keep their frustration under control. Patience and tolerance are necessary in order to move methodically from one step to the next in the negotiating process.*

- Successful negotiators know how to exploit power. *They understand leverage and how to use it skilfully so that the other person is not put on the defensive. They have high aspirations and always expect success. Most notably, their concept of themselves is very sound. Confidence is contagious and success is associated with competent and self-possessed people. If you feel good about yourself and about the task in question, you will do well as a negotiator.*

Negotiating is the combination of all these skills. They require constant practice to keep them vital (Richards and Walsh, 1990).

5.3.5 BODY LANGUAGE

Effective negotiators are observers of the way others behave. Much can be learnt of others' feelings and thoughts by the way they behave. By observing *gestures, postures, facial expressions*, and so on a negotiator is able to 'read' people.

Body language is part of the communication process. Not all that is communicated is verbal. Effective negotiators observe body language in order to understand others, and use body language to convey their own message.

Hoffman's Rule:
Smile — it makes people wonder what you're thinking.

Body language, if 'read', can give the trained person a better understanding of a person's non-verbal communication. For example during a negotiation, the signs or body language of a person, who is interested in making or finalising a deal/bargain, may be open pupils and eyes and increased eye contact, the start of a smile, uncrossing arms and legs, and body position square on and leaning forward. Alternatively a person who is not interested in making a deal may have closed pupils, less eye contact and a stiff jaw, and crossed legs and arms, and maybe turning away and leaning backwards.

Different cultures convey body language differently, and so it may be difficult for a person of one culture to read the body language of a person from another culture.

5.4 TACTICS EMPLOYED

5.4.1 REASONS FOR TACTICS

Tactics are devices used to achieve a desired outcome. Some tactics are unscrupulous (and may be referred to by some writers as tricks); some are not. An *awareness of tactics* is necessary, even if it is elected not to use most of them, because the other side may choose to use them necessitating a counter action to either nip them in the bud or neutralise them.

Tactics are specific behavioural actions which are intended to provoke a specific result, preferably favourable. These are building blocks for the overall strategy. Tactics should be infinitely flexible.

Because their outcome is uncertain, high risk tactics in a negotiation are:
• Threat.
• Bluff.
• Personal blame.
• Telephone negotiations.
Tactics may be used singly or in groups. Some may be relevant to particular phases only. And to be effective they may require disguising.

There are many books written solely on tactics or tricks to use in order to improve a negotiation outcome for one of the parties. Some have been given quite fanciful names while some may be regarded as being unethical, immoral or dishonest. It is important, nevertheless, that negotiators have an awareness of possible tactics that can be used against them, even though they may care not to run their negotiations the same way.

The following is a list of possible tactic descriptions (without explanations) (original source unknown). The descriptions give some clue to the tactics used.

Deliberate Misinterpretation
Pretended Misunderstanding
Pro and Con Analysis
Agreement and Rebuttal
Ignorance and Obstinacy
Summaries
Make the Other Party Appear Unreasonable
Good Guy — Bad Guy
Appeal to the Emotions
Straw Issues
Recesses
Changing Members of the Negotiating Team
Attack and Place the Other Party on the Defensive
Blame a Third Party
Outright Lies
Bluffs
The Door Knob Price
The Tactical Use of Walkouts
Threats
Threaten to Keep Other Work From the Seller
Threaten to Blacklist the Seller

Examples
Deferring to higher authority

Case 1. On selling a pre-loved car to a car saleyard, agreement was reached on a price between the seller and a saleyard employee. The employee then stated that he had to get clearance from his supervisor, at which point the negotiation restarted, but this time between the seller and the supervisor. The advantage to the buyer was that the seller's initial starting price was now lower than when he first walked into the car saleyard, and of course the supervisor found additional reasons why the agreed price was not acceptable. The selling price was then negotiated down from this new starting price. This was a deliberate tactic used by the saleyard. The buyer was naive.

Case 2. Mediators always request, before a mediation starts, that the people present have authority to commit their organisations to any agreement reached. This is in order that the time spent in mediation is not wasted. However because the actual form the agreement takes is not known at the outset, some parties to mediation still seek approval for an agreement from a superior (generally not present at the mediation). This approval may or may not be forthcoming. It can mean the total mediation has been wasted; or it can be used as a tactic to restart the mediation with a new baseline, something which is to the advantage of the tactic user.

Case 3. A homeowner wished to have some window blinds made and installed. A 25% deposit was paid at the time of placing the order. The remaining 75% was to be paid on installation.

At the time of installation, some of the blinds turned out to have been made the wrong size and could not be installed. The tradesperson installer nevertheless asked for the remaining 75% of the payment. The homeowner objected — would the installer return to complete the installation if he had already been paid? The installer telephoned his head office which insisted on full payment straightaway — the higher authority had been invoked. This, together with being unable to remove the installer from the house empty-handed, intimidated the homeowner who reluctantly paid.

Example
One negotiator's view of tactics for a win–lose situation

['Tactics' here refers to any approach used in negotiation, rather than the narrower definition given in this section of the book.]

It is important that all negotiators are aware of the many tactics that could be used by an opponent, even if they choose not to use them because of a belief that they are unethical, immoral or dishonest. This is in order that the tactics used by an opponent can be countered or neutralised.

Tactics can be used to provide the leverage necessary to accomplish the negotiation goals, and if they are used correctly, can be a major source of power. Tactics are presented below in three groups. Tactics may not always produce the result intended.

It is important to recognise that this is not a complete list of tactics and that all tactics tend to be camouflaged and combined to increase their effectiveness.

Preparatory tactics
- The selection of the negotiator is a tactic. Important considerations include the individual's reputation, prior negotiation experience, hierarchical position in the company, confidence, self-esteem, pertinence and communication skills.
- Selection of the climate and setting.
- The careful selection of the right time can be a powerful influencing factor on the outcome of the negotiation.
- The use of an agenda has the ability to shape the negotiation.

Opening tactics
The use of opening tactics can convey information about each party's attitudes, aspirations, intentions, perceptions; they may explore the opponent's overall posture, establish each party's range or limits and establish style (cooperative style implies trust; competitive style implies suspicion). Some opening tactics are listed below.

- Demand a precondition, because agreement of this by the opponent generally gives an advantage.
- Have the opponent make the first offer because it avoids you making a miscalculation, allows you to discover the opponent's position and allows you to declare shock and demand more.
- Always make your first offer high because research indicates that a more favourable outcome is achieved by those who make extreme opening demands as long as they are backed up with some logical reason.
- Make your major demands at the beginning of the negotiation and the minor demands later, because major concessions are often made more freely early in the negotiation; it is no use discussing the minor issues if the major ones cannot be solved. The minor issues are best kept as trade-offs later, and minor demands tend to settle better after the major issues have been solved.
- Make a 'first and final offer' if your position is based on extensive research/ investigation and you have the means to prove that you are not bluffing.

General tactics
- Request participation by seeking your opponent's advice as to how you can comply with the opponent's demands. This changes the negotiation into a cooperative style.
- Nibble by making demands a slice at a time.
- Use hypothetical questions to make a high offer to test the opponent's reaction.
- Invoke competition by playing off the opponent with a real or imaginary opponent.
- Promote your achievements by citing past achievements.
- Change the time, place, negotiator etc.
- Argue a special case which deserves a favourable response.
- Split the difference can be used to close the gap when you are near the final settlement.
- Give a biased sample.
- 'Low/high balling' by making a unrealistic offer to lure your opponent and then find reasons to change your offer.
- Use an agent, with limited authority, who then needs to obtain the final approval, and this gives time to consider the offer.
- Signal surrender by pleading for pity when in a weak position.
- Make opponent appear unreasonable by claiming your opponent is lacking goodwill.
- Use anger to bluff your opponent.
- Make promise of future rewards if opponent concedes now.
- Be persistent by pushing your demands.
- Summarise position before moving to the final offer.

(Source unknown)

5.4.2 SOME GOOD PRACTICES

Some of the following suggested good practices are at odds with some of the tactics mentioned elsewhere. Obviously there are philosophical differences here. The suggested good practices relate to an ideal situation where both parties strive to maximise the return to both parties collectively. The tactics mentioned elsewhere relate to a single party striving to maximise its individual return independently of the other party.

- The key to successful negotiation is to shift the situation to a win–win situation even if it looks like a win–lose situation. Create options for mutual gain.
- Separate people from the problem; separate personalities from the issue involved. Never become personal in any negotiation.
- Focus on interests, not positions.
- Communicate clearly; use active listening skills.
- Respect the other person; show consideration and acceptance of each other's perspective, values, beliefs and goals; consider the other party's situation.
- Recognise and define the problem.
- Seek a variety of solutions.
- Collaborate to reach a mutual solution, a solution everyone can accept.
- Be reliable.
- Preserve the relationship; recognise the value of a relationship and have a mutual desire to continue it.
- Participate actively in the process.
- Be mentally alert to:
  what is being said.
  what does it really mean.
  why has the other negotiator said it.
  what is the right response.

Creating the appropriate (psychological) climate

There are situations when it may be useful to create different (psychological) climates:

The friendly climate
- The parties are trying to solve a joint problem.
- One party is in a weak bargaining position.
- The parties are attempting to build a long-term relationship.

The formal climate
- One or both parties are unsure of what to do.
- The negotiation is between groups.
- The issues are complex.
- The parties are hostile towards each other.

The adversarial climate
- Such a climate is part of the normal game.
- A very prominent person or perceived tough guy is involved.
- One party is in a powerful position.

5.5 THE AGENDA AND PHYSICAL ENVIRONMENT

Successful negotiators pay close attention to the agenda and the physical surroundings.

5.5.1 THE AGENDA

The agenda may be not much more than a list of issues for discussion or it can be more formally structured much like an agenda for a meeting. Each party may contribute to the agenda, and the agenda itself may be the subject of negotiation.

As well as issues for discussion there may be mention made of:
* *Conventions* to be observed.
* *Venue, participants, observers* etc.
* *Dates, times.*

The listed *issues* can give an indication of what the other party regards as critical. Amongst the items there may be, however, a number of 'straw' (false, inconsequential) issues.

The advantages (from the viewpoint of the agenda's author) of an agenda are that it can:
* Focus on the important issues.
* Introduce imaginary issues (to use when trading).
* Coordinate the agenda with proposed tactics.
* Establish ranges.
* Divide the issues to suit the situation.
* Place the issues in an order that best suits.

The disadvantages (from the viewpoint of the agenda's author) are that it:
* Reveals a position and assumptions before the opponent's position and assumptions are known.
* Allows the other party time to prepare arguments.

5.5.2 PHYSICAL SURROUNDINGS

The physical climate in which the negotiation is conducted can have a marked influence on the outcome of the negotiation, and the skilled negotiator pays careful attention to the environment before, during and after the negotiation.

Several venues are possible, namely home ground, the other side's premises, or neutral territory. The room may require communication, recording and recess facilities. Seating arrangements need to be considered with regard to the status, roles and numbers for each side.

The meeting room can be determined by the host, however the visitor may ask for a change.

Items to look for are:
* The room. Is it small or large, lavish or austere?
* The table. Is it rectangular which is more formal where the people on the end have more power, or round which leads to a feeling of closeness?
* The desk. The size gives an indication of power and if facing the door extends power, and hides body language.
* Chairs. Large chairs or an elevated position give power.

The ultimate power position is the person at the head of the table, back to the window, facing the door in an elevated chair.

5.5.3 TIMING

Skilled negotiators use all the elements of time to their advantage. Time elements in a negotiation are:
- Having sufficient preparation time.
- Pacing the negotiation by taking an adjournment or setting time limits or deadlines.
- Time expiration, because approaching the deadline builds up pressure for a solution. Therefore it is important to avoid disclosing any party deadline.
- Establish the right time for:
  - Commencing the negotiation.
  - Raising issues.
  - Using tactics.
  - Introducing counters.
  - Making concessions.
  - Breaking deadlocks.
  - Making a final offer.
  - Implementing an agreement.

Example
Consider a negotiation where one of the participants must settle by an early deadline and the other intentionally tries to extend the negotiation.

Eddie's First Law of Business:
Never conduct negotiations before 10 a.m. or after 4 p.m. Before 10 you appear too anxious, and after 4 they think you're desperate.

5.6 EXERCISES

Exercise 1
Most professional bodies have a code of ethics. Many of the tactics used in negotiating would be considered unethical by many people. How can the two be reconciled?

Exercise 2
Self-assessment questionnaire (Source unknown)
For the following statements indicate (by circling the appropriate choice) whether or not you: Often (O), Sometimes (St), or Seldom (S) do particular things in regard to negotiations.

a)	I negotiate at work / home	O	St	S
b)	I try to do as much research as possible about the other party prior to negotiation.	O	St	S
c)	I always plan ahead before going into negotiations regarding location, time and agenda.	O	St	S
d)	I concede into a win–lose position just to break the deadlock.	O	St	S
e)	The other party's initial offer shapes my negotiating strategy.	O	St	S
f)	I try to open negotiations with a positive action such as offering a small concession.	O	St	S
g)	I agree early on, rather than argue about the point.	O	St	S
h)	I tend to meet people halfway.	O	St	S
i)	I build an image of success on winning as much as possible in every bargaining situation.	O	St	S
j)	I admit I am half wrong rather than explore our differences.	O	St	S
k)	I make sure the person I'm negotiating with has the authority to finalise any agreement that comes out of the negotiation.	O	St	S
l)	I study all agreements drafted at the close of negotiations to ensure there are no loop holes.	O	St	S
m)	I observe and respond to others' body language.	O	St	S
n)	I use tactics.	O	St	S

[Score:
Answers to a, b, c, h, j, k, l, m, n score O = 3, St = 2, S = 1.
Answers to d, e, f, g, i, score O = 1, St = 2, S = 3.
The higher the score, the better it is suggested that you are at negotiating.]

Exercise 3
Some useful exercises that can be done to improve certain aspects of negotiations are:
• Next time you order some food, for example a pizza, negotiate with the seller on the type of pizza, size and where to get it from.
• Negotiate with your lecturer for a better mark in your assignment.
• Observe others at work on how they negotiate and note their style and the types of skills used.

Exercise 4
Consider a negotiation with which you have been involved. Were there identifiable phases that the negotiation went through, or was the whole process indivisible? What descriptive names would you give the phases that you observed?

Exercise 5
What form would you expect a negotiation familiarisation phase to take with people from America, Japan and the Middle East?

Exercise 6
List the advantages you see in using range charts in negotiations. List the deficiencies in using range charts.

Exercise 7
Consider the statement: 'If I drop the claim for liquidated damages then will that be OK?' What is wrong with such a proposal? Why is it bad negotiation practice to make such a proposal?

Exercise 8
How linear is the negotiating process and how much feedback occurs between phases?

Draw a block diagram of the negotiating process with each block representing a negotiation phase. Now draw arrows showing the connections including feedback between the phases.

Exercise 9
Identify a process by which you could appraise your performance in a negotiation. Given that your goals prior to the negotiation may have needed changing in the light of the negotiation, by what measure do you gauge your success?

Exercise 10
Under what circumstances would you use a collaborative or competitive form of negotiation?

List the critical points of collaborative (win-win) negotiation.

Exercise 11
In terms of skills and talents, how do you rate as a good negotiator? On what basis are you making your judgment?

Exercise 12
In your workplace, observe a range of gestures, postures, facial expressions and sounds of others. Associate a meaning to each of these forms of body language. Interview the persons concerned and try to establish whether your interpretations were correct.

Ask a colleague to repeat the procedure with you as the one being observed.

Exercise 13
One commentator lists the following traits of a good negotiator:
• Good at planning and preparation.
• Able to think intelligently under stress.
• Practical approach.
• Excellent communicator.
• High level of knowledge in the subject.
• Respected, has a reputation.
• Knows the value of power.
What is your view? Are all traits necessary? Are further traits necessary?

Exercise 14
Consider a negotiation of which you were part. List the tactics used by both sides:
• To encourage agreement.
• To break a deadlock.
• When a side's case is weak.

- When a side's case is strong.
- To gain some initiative.
- To wrap up the negotiation.

Identify the role questions played in the negotiation process.

   Categorise the different types of questions used by both sides.

Exercise 15

The best way to negotiate is presented in personal opinion form (as opposed to any rigorously justifiable form) by many writers. Much is based on anecdotal evidence and personal experience. Why is there no real objective treatment of negotiation available?

Exercise 16

Why do you think it is important to focus on the problem/conflict/dispute that the negotiation is trying to solve rather than the personalities involved?

Exercise 17

For a negotiation involving three people on each side, with each side having a leader, draw seating arrangements around a single table that you believe would:

- Promote confrontation.
- Promote cooperation.
- Do neither or both.

Exercise 18

(Source unknown) Complete the following table.

Statement	True	False
1 Negotiating skills require consistent practice.		
2 Good negotiators are willing to research and analyse issues.		
3 You must win at any cost in negotiation.		
4 Conflict is an important part of any negotiation.		
5 Compromise is a sign of weakness in negotiation.		
6 Your goals for every negotiation should be well thought out.		
7 Negotiators should be well versed in the techniques of conflict/ dispute resolution.		
8 It is possible for both parties to win in a negotiation because everyone has different needs and values.		
9 You must give, to get, in negotiation.		
10 What you achieve in a negotiation is related to what you expect.		

Exercise 19

(Source unknown) Select the most appropriate choice for each of the following statements:

(i)   In negotiating, it is beneficial to:
    (a) take a little time to get to know the other party
    (b) get down to serious business immediately
    (c) meet your needs regardless of the other party's needs

(ii)   Compromising in a negotiation:
    (a)   is a sign of weakness
    (b)   may be necessary to get what you need
    (c)   should only seldom be practised

(iii)   When conflict occurs in a negotiation you should:
    (a)   work toward its constructive solution
    (b)   stand firm in your principles or opinions
    (c)   ignore this issue and discuss issues that are in agreement

(iv)   It is a good idea to:
    (a)   learn the authority of the person you are dealing with in advance
    (b)   always show your authority in a negotiation
    (c)   assume the other person's level of authority is the same as your authority

(v)   The courage and confidence necessary to start a negotiation:
    (a)   are inborn
    (b)   come with a willingness to learn skills and prepare
    (c)   are only necessary in some negotiations

CHAPTER 6

Negotiation — Case Studies I

6.1 OUTLINE

A number of case studies illustrating the phases negotiations go through, together with various negotiating practices are presented from a range of industries and situations. A common culture was present in each case study.

6.2 CASE STUDIES

6.2.1 CASE STUDY — TUNNELLING

Background

This case study examines a negotiation which took place at a meeting between representatives of a contractor (Raepoid) and one of its subcontractors (Perch).

Raepoid, the head contractor for a road construction project involving tunnelling, engaged Perch to excavate half of the tunnel. Having completed all the subcontract work, Perch demobilised. Following that, certain outstanding issues were addressed in correspondence, however some of these were not resolved. As a result, Perch issued Raepoid with formal notices of dispute. In accordance with the terms of the subcontract, in this event, the parties were required to meet within a week to seek agreement. This case study looks at this dispute resolution meeting, which also considered Raepoid's outstanding claims against Perch. The subcontract required that, if agreement could not be reached at this meeting, the parties would proceed to a mutually agreeable alternative dispute resolution process.

The total amount in dispute was approximately one tenth of the value of the subcontract.

The negotiation

The meeting was arranged at Raepoid's site office and was attended by managerial staff who were empowered to make decisions on behalf of their companies.

Lasting about four hours, the meeting proceeded through the following stages:

Familiarisation. All the attendees knew each other, and so this phase only lasted ten minutes. Rapport was built by sharing information about recent activities (unrelated to this project).

Opening phase. The meeting chair briefly summarised the events that led to the meeting, culminating in Perch's notices of dispute. He expressed optimism that the parties could quickly agree on a settlement without having to escalate the disputes, because this would involve unnecessary expense and delay.

Perch stated its desired outcome by claiming that it had only pursued issues that it considered genuine. It believed that all its claims were fair and reasonable. Perch expected a settlement that would pay its claims substantially unchanged.

In reply, Raepoid assured the meeting that it had carefully analysed each of Perch's claims and considered that most of them had no contractual basis and were technically flawed. Moreover, Raepoid announced that it would be prepared to withdraw its outstanding claims if Perch withdrew its claims. Raepoid's preferred outcome was therefore an agreement to settle with no payments to either party.

Perch immediately objected to Raepoid's claims complaining that they were still unsubstantiated. Raepoid responded by stating that it had quantified these claims and that it was prepared to pursue them but it was hoping that this would not be necessary.

Proposal and bargaining phases. Each of Perch's claims was discussed in turn, with the proposal and bargaining phases becoming intertwined. Raepoid chose to deal firstly with the claims for which a minor concession could be offered. Accordingly, an agreement in principle was quite quickly reached for the minor claims.

Ownership of equipment, a major dispute item, was discussed next. The subcontract required that this equipment be left in place to be used by Raepoid, but was silent about what was to happen after the project. Intending to return the equipment to Perch but anticipating arguments about its condition, Raepoid stated that its interpretation of the subcontract was that the equipment now belonged to Raepoid. Perch objected and claimed the residual value of the equipment from Raepoid. Raepoid proposed to return the equipment at no cost to Perch on condition that Raepoid not be responsible for any repair costs. After pointing out that the subcontract made no reference to how long Raepoid may use the equipment, Perch reluctantly accepted this proposal.

With regard to a dispute item regarding work method, Raepoid pointed out that Perch had proposed the alternative work design. Furthermore, Perch knew the site conditions. Raepoid therefore could accept no responsibility for problems with the work method used. This took Perch by surprise and it requested more time to consider this matter, and so a follow-up meeting was scheduled in two weeks to take up this and any other unresolved claims.

In the light of contrary evidence put forward by Raepoid, Perch agreed to withdraw its remaining technical claims.

Perch's final claim was for loss of overheads and profit due to Raepoid deleting work from the contract. Raepoid objected on the grounds that the subcontract was on a schedule of rates, and that the subcontract did not contemplate renegotiation of the rates if the quantities changed. Furthermore, very few quantities had been changed substantially and Perch's claim had been artificially inflated by using an unreasonably high overhead and profit rate. Perch refused to proceed with further debate of this claim until it had sought legal advice.

In discussing the Raepoid claims, Raepoid agreed to forward details of its claim for rectification costs incurred due to defective work by Perch. Raepoid proposed withdrawing its other claims if Perch agreed to withdraw all its remaining unresolved claims.

Agreement phase. The agreement phase involved reviewing each of the claims and confirming and documenting the terms of agreement. A further meeting was scheduled for two weeks time to further address the unresolved issues.

Exercise
1. Whilst all the issues were not resolved during this negotiation, substantial progress was made. The style of the meeting was collaborative rather than adversarial, and this resulted in both parties believing that they had won something. The outcome was reached in such a way that the relationship between the parties was preserved. On this basis, would you judge the negotiation to be a success?

2. *Preparation phase.* Each party had prepared thoroughly for the meeting. Much of this preparation was necessitated by the considerable correspondence which had been exchanged prior to the meeting. Both parties had a clear understanding of the wants and needs of the other party. Raepoid was aware that Perch's motivation was to offset the loss it made on the work, whilst Perch resented the fact that its loss had contributed to Raepoid's considerable profit on the project. In the presence of such inequality, how could a collaborative meeting style be maintained?

3. Raepoid had four participants compared to Perch's three. In practice, this did not prove to be an advantage to Raepoid because one person played a passive role. However, it would have been preferable for the parties to have equal representation since Perch would have perceived unequal numbers to be an unfair advantage. Both parties adopted a similar style which facilitated amicable debate and creative problem solving. Neither party resorted to aggressive or intimidating tactics. What is the influence on negotiations of unequal numbers in each party?

4. *Opening phase.* Each party effectively presented its desired outcome. The parties at this phase hinted at areas where they were prepared to compromise by intentionally referring to them using generalities.
   Proposal phase. By starting with the easier claims and offering some minor concessions, Raepoid elicited a cooperative approach from Perch. During this phase, both parties were careful to extract alternative proposals from the other party before suggesting any counter offers. What general principle of negotiation is illustrated by these actions?

5. *Bargaining phase.* Some of the claims proceeded to this phase and concessions were exchanged. However, some of the claims ended in a deadlock. In these cases, the negotiation was postponed, allowing each party to consider the other's position in more detail. What general principle of negotiation is illustrated by these actions?

6. *Agreement phase.* This phase saw the items of agreement documented and signed by both parties. This process proved helpful since some aspects of the agreement were found to require further clarification before both parties could agree. Does this 'further clarification' issue recycle the negotiation to earlier phases?

7. Given the strongly held positions of both parties and the large amounts of money at stake, a confrontational encounter leading to an impasse was a likely scenario. In practice, however, the parties were able to understand each other's position and adopt a more collaborative style. This enabled a partial resolution to the dispute and a willingness to pursue a full settlement at a further meeting. What principles of successful negotiation does this demonstrate?

8. Albeit, based on the above given information in the case study, how might the negotiation outcome have been influenced by the fact that Raepoid was the more 'streetwise' in terms of commercial negotiation?

6.2.2 CASE STUDY — SOFTWARE DEVELOPMENT

Outline

InTecCo, an information technology company, had won a multi-million dollar software development, data conversion and maintenance contract with GovOSO (the owner), a large government owned service organisation.

The development and conversion part of the project was expected to take eighteen months but started to run late after nine months into the work. Due to the perceived poor performance of InTecCo on the work, GovOSO forced InTecCo to take action in controlling the schedule blow-out by directly contacting the head of InTecCo with a complaint about its performance. GovOSO stated that it had lost all faith in InTecCo to deliver the project to the terms of the contract and wanted to know what would be done about rectifying the situation.

In addition to its problems with the owner, InTecCo also had difficulty with its key subcontractor (SoftDev). SoftDev was responsible for the prime software development and data conversion portions of the contract for about half of InTecCo's total contract value. SoftDev was not the original subcontractor but, by way of a takeover, it had inherited this contract with InTecCo. Due to the takeover and the potential contractual problems, SoftDev had appointed its own people and a new project manager, and a technical manager had been brought in.

InTecCo was forced into a position of having to react quickly to reinforce the resources available to the project team, deal with SoftDev and most importantly deal with GovOSO. InTecCo supplemented the project team with an additional senior management team of three people, and then set about the task of renegotiating a project completion schedule with GovOSO and with SoftDev.

The causes of the schedule overrun were seen as being primarily attributed to InTecCo (SoftDev) but GovOSO had contributed to some of the causes. In order to prepare a more realistic schedule, a meeting was arranged to discuss and analyse a draft schedule put forward by InTecCo.

The negotiation

Because the schedule was associated with milestone payments and deliverables under the contract, care needed to be taken to consider these in negotiating the schedule between InTecCo, GovOSO and SoftDev.

InTecCo progressively presented a series of draft schedules to GovOSO and discussed these at the technical management level of all parties. The problem with these draft schedules was that the logic of trying to complete and implement the project in and around GovOSO's Y2K embargo periods made it almost impossible to ensure a new target completion date of April which was desired by GovOSO.

These drafts tended to be dismissed by GovOSO, as being a further indication that InTecCo did not understand how to manage and schedule the project.

In order to move the process along, the project managers of InTecCo (SoftDev) and GovOSO decided to hold day-long workshops involving the key participants. These workshops resulted in two schedules being developed:

Scenario 1. Based on previous draft submissions, Scenario 1 showed a first production software run before the end of the year and complete deployment by the end of April, as had been requested by GovOSO. More was now known about the status of the software development and the problems that existed and hence the potential risks. This schedule therefore was considered by some as optimistic.

Scenario 2 (Revision A). InTecCo's preferred schedule was also based on previous draft submissions that included a pre-end of year software deployment on a trial basis and a target completion by August. This schedule provided the lowest risk to all parties and a higher level of confidence of completion, because it allowed more flexibility on the critical path items. Critical issues requiring GovOSO input had the potential to impact on the schedule being achievable.

Negotiation then continued with two additional workshops. The first workshop evaluated Scenario 1 resulting in its rejection as a viable schedule, albeit not enthusiastically by GovOSO. Emanating from this workshop was additional information that needed to be incorporated into Scenario 2. These inclusions were done via additional discussions at the technical level of all three groups and the end result was Revision D.

In between the discussions at the technical level, both InTecCo (SoftDev) and GovOSO senior management met to agree on key schedule constraints and inputs to be provided to the technical groups.

Revision D was then put through another workshop, with senior people from InTecCo, SoftDev and GovOSO joining in to provide management direction and approval. The outcome, after a long session, was an agreed schedule that all parties could commit to and minimise the impact on GovOSO's business and give some surety of inputs required by InTecCo (SoftDev) to enable successful completion of the work. The schedule was named Revision E and was adopted as the new baseline performance schedule.

Exercise

1. In this case study, the preparation, familiarisation and opening negotiation phases were blurred by the circumstances of the need to agree to a revised schedule. The negotiating parties oscillated between each of these phases as they tried to work their way through conflicting priorities and agree an acceptable outcome. The parties continued to be 'out-of-phase' with each other as they grappled with issues presented in the draft schedules. Being out of phase essentially brought about 'confusion' and inability to negotiate a solution. Assuming that you recognise that the parties are out-of-phase, how do you adjust your approach to the negotiation in order to reach an effective outcome?

2. InTecCo (SoftDev) and GovOSO recognised the difficulty for their teams, and jointly agreed and proposed a workshop environment where issues could be thrashed out with the necessary authority supplied by all stakeholders, rather than going around and around in circles with smaller groups. Is it recommended that negotiation be mixed with solving technical problems in such a workshop environment, or could better outcomes be achieved by separating the two functions?

3. During this workshop the schedule 'proposals' were thought through and the two positions of both parties were crystallised into the two schedule scenarios. This part of the negotiation's success depended heavily on the project managers of InTecCo and GovOSO, who guided and directed their respective teams to sort out the detail, whilst they negotiated between themselves on key issues. They jointly contributed to the success of the negotiated schedule by creating the appropriate climate, separating people from the problem when personalities threatened the outcome, and they enabled clear communication by reviewing and discussing each side's position openly and by trading concessions. The project managers encouraged agreement and broke deadlocks. How much does the successful outcome of a negotiation depend on personalities to drive the negotiation?

4. Both InTecCo and GovOSO needed to reach agreement for the sake of the project and get things moving forward again. They were both looking for a win–win situation. The success of the negotiation was driven by the fact that an agreed outcome was better than the alternatives available (project failure and expensive litigation) and all parties felt that they had negotiated a schedule rather than being compelled to accept one. Would it be true to say that undesirable outcomes, alternative to that achievable through negotiation, drive negotiations toward some agreement?

5. Should negotiations only take place at the lowest level of interaction? When should higher level authority be used? In this case, the subordinates discussed the detail and then the project managers ruled on points of conflict.

6. InTecCo (SoftDev) and GovOSO recognised a need to arrive at an outcome rather than litigate over non-performance to deliver to schedule. GovOSO wanted the work completed and it wanted it done by the earliest possible date. InTecCo (SoftDev), although financially poorer from the contract work, realised that its best financial option was to finish the work rather than be sued for breach of contract; negotiating as much time as possible into the schedule was in its best interests. In this case, both parties were realistic in their thinking, but how in general do you make parties see 'reality', rather than think emotionally?

6.2.3 CASE STUDY — SWIMMING POOL FAILURE

Outline

Just before the handing over of a swimming pool, the ceramic tiles started to dislodge. This caused the pool to be drained for investigations into the cause of the failure, finally resulting in the contractor having to reconstruct the tiles. The owner's representative, after considering a report by a pool expert engaged by the owner, and submissions by the contractor, ruled that the contractor had not done the work in accordance with the specification. After considerable delay and the issuing of show cause notices on the contractor and the owner threatening to terminate the contract, the contractor, reserving its rights to sue both the owner's representative and the owner, reluctantly reconstructed the tiles.

On completion of the work the contractor, arguing that the original work was to specification, maintained that the owner's representative erred in issuing the contractor with a notice ordering the demolition of the defective work and reconstruction to the original specification. At practical completion, the contractor claimed the reconstruction work as a variation. The owner's representative rejected the contractor's claim and subsequently the contractor issued a notice of dispute in accordance with the contract conditions, disputing the determination of the owner's representative.

This case study looks at the negotiation that took place in an attempt to resolve the dispute.

The following points provide a background to the dispute and the negotiation:
- The conditions of contract required the parties to meet to attempt to resolve the dispute before proceeding to either arbitration or litigation.
- As the contractor had issued the notice of dispute, it had to instigate proceedings against the owner to recover its claimed losses.
- The owner had engaged experts to report on the causes of failure and to provide expert opinion in arbitration or litigation if necessary.
- The contractor was not prepared to engage experts when the failures were first noticed, preferring to bluff its way through and trying to sheet the blame to other parties such as the owner's representative, owner's consultants, and materials suppliers.
- The contractor's technical arguments were unsound and not supported by expert opinion. The consultants, engaged by the contractor, specialised in materials testing and non-destructive testing, and provided objective data. However, they were not briefed to advise on the cause of failure, with the contractor providing all its own interpretation of tests and technical arguments.
- The owner had freely provided copies of all technical reports to the contractor, together with rebuttals of its technical arguments.
- The contractor's claim was highly exaggerated. With the assistance of quantity surveyors, the owner had determined that the contractor was actually out of pocket by approximately one third of the contractor's claim.
- The contractor was a quality assured company and as such should have had systems in place to give the necessary quality to the original construction.
- The owner engaged an experienced engineer/lawyer/arbitrator to review the technical and legal aspects of its case and to give opinion of the contractor's chances in successfully proceeding against the owner. The owner was advised that its own case was sound, that the contractor would have little chance of success and that the contractor's best option would be for a negotiated settlement.

The negotiation

The dispute was not resolved by negotiation, with the parties then heading to mediation. This case study examines this negotiation.

Two meetings between the parties took place. The contractor was represented by its managing director (sole shareholder), general manager and an architect. The owner was represented by its chief executive officer (CEO), a solicitor and the project manager.

Preparation

Prior to the first meeting, the owner's people met to discuss its position, second-guess the contractor's position and to decide on tactics.
- The owner was confident that the technical and legal aspects of its case were sound.
- The owner knew that the contractor's claim was exaggerated and that any court or arbitrator would reduce the claim.
- The owner was confident that the contractor would threaten to take it to court, but in fact would avoid arbitration/litigation, preferring to attempt a negotiated settlement. An acceptable solution to the owner may be to call the contractor's bluff and force it to mediation or arbitration.
- Based on the paper warfare that had eventuated since the tile failure, the owner was aware that the contractor wanted to cloud issues and introduce 'red herrings'. The owner had to ensure that the parties stuck to the issues. The owner decided to use phrases such as 'we have been through that before' and 'that is irrelevant' in order to focus proceedings.
- The owner understood that if the dispute was to proceed to litigation or arbitration then the owner would be exposed to significant legal costs.
- The owner's staff ensured that the CEO was fully briefed, and whilst each party had to be represented by someone with the authority to settle the dispute, the owner decided that it would advise the contractor that any decisions regarding settlement would be subject to ratification by others.
- The owner would suggest that the meetings be held at the contractor's office, so that the owner had the option of 'walking out'.
- Based on the owner's assessment of the contractor's real costs and the possible future legal costs, the owner determined an amount of money that it would be prepared to offer the contractor as full settlement — about one tenth of the contractor's claim.

Meetings

The first meeting was held in the contractor's office and after the initial pleasantries and positioning on opposite sides of a large circular table, discussion commenced on the issues.
- As predicted the contractor wanted to expand the issues. However the owner was able to successfully refocus discussions.
- The contractor tried to install doubts in the owner's CEO about the actions of the owner's consultants and staff and the actions of the owner's representative.
- The contractor's managing director advised that he, as sole shareholder, was the only one that was suffering a financial loss as it was coming out of his pocket. In reply, the owner's CEO reminded the contractor that he was dealing with shareholder money and was not prepared to pay for the contractor's mistakes.
- The contractor advised that it had a barrister's opinion, that the contractor had a strong case, and threatened to sue the owner, and would ensure that the timing of legal proceedings would be chosen to embarrass the owner. The contractor was bluffing and no such opinion was forthcoming.
- The contractor advised that it was out of pocket by the claimed amount and it would expect a substantial fraction of that amount to make it go away. When asked what that amount would be, it was not forthcoming.

- After about an hour, and much table thumping, the meeting degenerated into a slang-ing match which could not reach any resolution. The parties agreed to another meeting in a couple of weeks time when things had cooled down.

The second meeting was also held at the contractor's office with the owner's tactics unchanged except for the fact that the owner was prepared to call the contractor's bluff by making him a 'take it or leave it' offer of one tenth of the claimed amount.

The meeting was short with the CEO coming straight to the point with the offer. The contractor made some unkind remarks about the quantum of the offer and advised that it would let the CEO know of its decision in a day or so. Two days later the contractor rejected the offer.

It was then expected that the contractor would proceed directly to either arbitration or litigation. However after six months it suggested mediation to which the owner agreed.

Exercise

1. The negotiation appears to have put the owner in a stronger position because it con-firmed that the contractor was most likely reluctant to proceed to arbitration or liti-gation. From the outset the owner believed it could accept an outcome that forced the contractor to either mediation or arbitration as the technical and legal arguments could then be tested. The owner believed that the outcome would be favourable to the owner and the costs of such an exercise would be comparable with the previous offer to the contractor which it had rejected. Would you agree with such an analysis? Would you agree that after the initial negotiating the owner's position was strengthened and the contractor's position weakened, or can nothing be said about such movements, or are such observations not relevant to the final out-come?
2. Would you agree that the most important aspects of this negotiation process revolved around preparation, tactics and bargaining? Why then were the other phases of famil-iarisation and proposal of lesser importance? Is it because everyone was aware of the personalities involved, having worked on the project for two years, or is it because the proposals were relatively clear cut, only relating to money?
3. In what ways does expert opinion help you to distinguish between matters of fact, legal or rule interpretation, and matters of personal or only partly qualified opinion?
4. How do you second guess the wants and needs of the other party? How do you predict the tactics to be used by the other party?
5. High risk tactics in a negotiation include threat, bluff, personal blame and telephone negotiations. The contractor engaged in all four. How have these tactics helped or not helped its case?
6. The owner was able to support its position with expert technical and legal opinion. The owner put the contractor on the defensive by topping any point the contractor made by stating one better. The owner set the agenda and was able to focus the meeting on the real issues by refusing to entertain discussion on the irrelevant. The owner was able to walk out of the first meeting. How have these tactics helped or not helped the owner's case?
7. Because there would never be the possibility of any future relationship between the owner and the contractor as far as the owner was concerned, the owner was only

interested in satisfying its needs of reducing any further costs to a minimum, and the owner had no sympathy for the contractor's losses, how would you classify the owner's style of negotiating? On the other hand, with the contractor indicating that it could cause political damage and could subject the owner to significant legal costs, how would you classify the contractor's style?

8. It could be argued, from the contractor's viewpoint, that mediation is not the way the contractor should be heading, after all it is only a continuation of negotiation. Why wouldn't the contractor engage an expert to give an opinion counter to that received by the owner, thereby nullifying an advantage that the owner had? Why wouldn't the contractor choose a binding form of resolution method such as expert determination?

6.2.4 CASE STUDY — CONSTRUCTION CLAIMS

Background

Work involved in a large construction contract had been completed. The contractor had carried out many variations to the contract, the majority of which had been accounted for, the quantum agreed and the costs paid.

The contract was very clear on the procedures to be followed for the claiming of extensions of time and variations. For the majority of these, the contractor had not complied with the contract procedures. However where work had been performed that was a variation and of additional benefit to the project, the owner had valued and agreed variation costs.

At practical completion, however, there were some claims from the contractor to the owner that were still outstanding. These were all claims with no basis under the contract and had been rejected by the owner. The contractor however was not willing to accept this situation and continued correspondence in an effort to receive compensation.

As well as these claims, there were others which the owner had accepted some cost for, determined the value and had made payment on. The contractor had not accepted the owner's determination and was now attempting to claim for the difference.

The negotiation process

The owner identified that the contractor was financially not in a position to withdraw its claims and so to resolve the situation a negotiated settlement would have to be reached or else it would be resolved at arbitration/litigation. The contractor's suggestion of a meeting to discuss the outstanding issues, before the instigation of a formal dispute mechanism, was therefore taken up and a meeting arranged.

The first reply to the contractor's claims positioned the owner. It was considered that the contractor had suffered a loss, such that it would not abandon its claim, and that the aim for the owner would be to reduce the claim's worth in the mind of the contractor.

Before the meeting, the owner's staff met and ran through various strategies and investigated the risks involved with each. It was decided that the meeting would be held at a senior level and therefore it was necessary to fully explain the claims and history to these people, who had until this time only a high level understanding.

The owner ensured the negotiation was held in the owner's offices, which were located many kilometres away from the work site. This was considered to be the best location for the owner, because attention would not be focused at the site.

The meeting began with the tabling of the claims that the contractor believed were outstanding, and it then began to work through them one by one. It became clear that it was the contractor's intention to try and draw the owner into discussing each and every claim in detail and to have the owner explain and justify the position that it had taken in response to the claims. This was contrary to the way the owner wished the discussion to go, because it had been decided to keep the negotiation at as high a level as it could, and so the owner mainly adopted a listening position at this stage. As was expected at this stage the contractor indicated that nothing short of full settlement in the amounts it was claiming was acceptable.

After the contractor had explained the claims, the owner reminded the contractor of its contractual obligations with respect to these matters. Further, before mobilisation to site, the parties had attended a teambuilding workshop wherein the processing and resolution of claims and issues in general had been discussed and agreed, in order for a smooth execution of the contract and specifically to avoid the situation that had now been reached. The owner further reminded the contractor of this undertaking and pointed out where the contractor had breached this agreement.

The owner then went on to remind the contractor that, should it now not accept the contract conditions it had agreed to and under which its claims were invalid, the only way forward was by invoking the dispute resolution mechanism, which ultimately would end in arbitration/litigation.

Knowing the financial status of the contractor, the owner further pressed the point of the costs of litigation and the inherent risk of, not only losing the case, but also being faced with the owner's costs.

Finally the owner made an offer of approximately one fifth of the value of the contractor's claims as full and final settlement, withdrew a previous ex-gratia offer for the other claims and gave the contractor a week to consider its options.

The contractor wrote back to the owner four days later and reduced its claim by half, neither accepting nor declining the owner's offer. The owner then wrote back rejecting the contractor's offer but raising its offer to one third of the claimed amount, which the contractor agreed to.

Exercise

Overall the negotiation followed identifiable phases. However the steps within each phase were not always so clear.

1. *Preparation.* Prior to meeting with the contractor, the owner went through a lengthy preparation phase gathering information and looking at the possible outcomes. The position of the contractor was also investigated in detail by the owner. During this time, the different positions the owner could adopt in response were established and approvals to proceed gained. What avenues to information about the contractor's position would the owner be likely to have?

2. *Familiarisation*. As both parties had been working on this project for the previous twelve months, most of the people knew each other already and therefore the personal familiarisation process was short. It began with general discussion on the progress of the project as a whole. Both the owner and the contractor expressed interest in the other's general business. Suggest why such topics might have been the basis for any discussion during the familiarisation process.

3. *Opening*. The owner stated that it would only discuss matters that the contractor had nominated before the meeting began.

   It was deliberately left to the contractor to outline its position first and for the owner to remain as general as possible in the discussion. Although the contractor wanted to go into the detail of each claim, the owner avoided this. What is your view on this way to proceed, because it set the scene for any further discussion and also set the limits of what was to be discussed? It also allowed the owner to understand what the contractor's aims really were. It was confirmed that the contractor was not really interested in individual claims but rather was looking for a monetary settlement to offset an overall project loss. Why might the contractor not have stated this interest at the beginning, rather than going through claim by claim?

4. *Proposal*. Having confirmed the general requirement of the contractor in the opening phase, the owner was then in a position to make a proposal. The proposal followed the 'if ... then ...' formula, and also set a week time limit for acceptance of the offer. What are the dangers of such an approach by the owner?

5. *Bargaining*. The bargaining phase of this negotiation was carried out by letter, with the contractor making a counter offer. This included concessions by it of a reduction in its claimed amount by half, with no further claims to come. This then reached a deadlock until the owner relented and reinstated the ex gratia offer, which the contractor accepted. Was the negotiation purely over the quantum of money at this stage? In what way did elevating the dispute to senior owner officials eliminate the possibility of any personality involvement?

6. *Agreement*. Agreement was reached and a deed of release signed by the contractor. What contractual force does such a deed have?

6.2.5 CASE STUDY — FIRE SUPPRESSION SYSTEMS

Background

The case study relates to a contract for the replacement of fire suppression systems within a factory. The contract was won by a reputable company.

The contract was divided into two separable portions as follows:

- A — Projects to be completed by the end of the year.
- B — Projects to be completed three months later (in March the following year).

The general conditions of contract were an industry standard and liquidated damages were to be imposed for late completion — there was a government regulation that required the replacement of fire suppression systems by the middle of the second year.

The contractor was late commencing work. However it was able to complete portion A within the required time frame.

From this point on, the contractor displayed a reluctance to get on with the project, reduced manning levels, and became tardy about problem solving. The deadline of March passed and still the project was far from complete.

Following an unsuccessful response by the contractor's site personnel to a direct approach by the owner's project manager, official written notification of incomplete work, unsatisfactory test results and erratic manning of the project was given by the owner's project manager.

The owner's project manager requested an urgent on-site meeting of both parties and suggested the attendance of the contractor's senior personnel.

The negotiation

The meeting was held on site and all grievances from both sides were placed on the table for debate, which was often intense, heated and emotional.

Following this 'blood letting', the owner's project manager then summarised what he believed was the current situation, and put to the contractor two main demands:
- It immediately address the issue of manning so that the project could proceed in an expeditious manner.
- The contractor prepare a realistic time frame, to complete the project, which could be monitored, commencing from the current point of progress.

The contractor was given two days to come up with this program. The owner's project manager explained that, if satisfactory progress was not achieved, he intended to invoke liquidated damages in accordance with the provisions of the contract.

Subsequent to this meeting, the contractor provided a program to completion as requested. However progress on the work continued to be sporadic and lacked positive direction.

The owner's project manager then advised the contractor in writing that liquidated damages had been invoked and were to be back-dated to the March deadline. This action brought immediate response from the contractor which requested a second on-site meeting.

At this meeting, the basis for the liquidated damages was explained, being the ongoing cost of contract supervision whilst the project continued to remain incomplete. An assessment of the liquidated damages to date was then given. The owner's project manager then proposed that the contractor complete the work by a date nominated.

The contractor made a counter offer:
- The contractor would complete the work by the new date nominated.
- A senior manager would personally manage the work.
- In lieu of liquidated damages, the contractor offered a monetary settlement of approximately one half of the calculated sum for liquidated damages, and an increased service warranty date.

The owner's project manager agreed.

Following this meeting, the contractor completed the project to the owner's satisfaction by the revised completion date.

Exercise

1. For the first meeting, the owner prepared a history of the contractor's time on site, the number of failed pressure tests, the promises not kept by the contractor's personnel, correspondence not replied to, and unresolved testing and installation problems. How might the contractor have countered such preparation on the part of the owner, or would it be considered unnecessary for a first meeting? Was the outcome of the meeting favourable to the owner because of good preparation and knowledge, or was something else at play here?

2. The owner requested that the contractor's senior management be present, as well as field staff, and the owner chose an important conference room venue. The owner's senior purchasing personnel also attended and this may have placed some pressure on the contractor's senior management because the contractor had other major maintenance contractual work on site at the time. In what way can the existence of other contracts affect negotiations associated with this contract?

3. The owner was prepared to listen to the contractor's argument, but had a bottom line of agreement to additional resources and a timetable for completion of the work by the contractor's senior management. Timing was arranged to suit the arrival of the contractor's senior management from head office. With a pre-established bottom line, does this imply that any negotiation was a sham?

4. No agenda was prepared. How important or not important was this?

5. The conference table was circular, with the owner's personnel occupying one half and the contractor the other, although it was not planned that way. Comment on the seating arrangements.

6. Having called the meeting, the owner gave an overview of the situation, stating clearly its main concerns. The contractor replied. Members of both parties interjected on specific points, and heated and lively exchanges proceeded. One contractor employee felt that he was being made responsible for the shortcomings of the work and gave an emotional address. All this was unsettling to the contractor. There was no attempt to stifle the exchanges as this was a time to clear the air. Would more formal control of this phase have been better? Would a better organised negotiation have drawn out more grievances on both sides? How better can you control emotional discussion in a meeting?

7. The contractor requested that liquidated damages not be imposed, but rather consider an extended warranty period in lieu. Liquidated damages was seen by the owner as a 'big stick' to wield if the revised program was not met. The contractor had grounds to counter the liquidated damages threat, and the owner was aware of this shortcoming. However the contractor was not contractually wise to this, and it further failed to capitalise on loopholes in the contract conditions. In negotiations involving contracts, what level of knowledge (commercial and legal) of contracts in general, and the specific contract in question, is necessary?

8. Detailed minutes of the meeting were prepared, checked and forwarded to both parties. The minutes clearly defined the proposal and its acceptance by the contractor. These minutes were prepared by the owner; what risk is there in this to the contractor?

9. By having the contractor's senior management present, decision making was done on the spot. What value is there in holding meetings if the people present do not have the necessary authority to commit their organisations to the meeting outcome?

10. The above account of the negotiation and situation is given from the owner's viewpoint, and in particular that of the owner's project manager. When you are personally involved in a negotiation, how realistic can you be in analysing the events objectively? Or do you colour the facts in your favour?

6.2.6 CASE STUDY — POWER STATION

During the upgrade of a power station, negotiation was undertaken to recover costs for changes to scope encountered throughout the project.

The design of all upgrading requirements was carried out by the owner. Procurement of all major components required for the project was also carried out by the owner.

Due to the poor control and coordination by the owner, of different design packages and purchasing of equipment, problems were encountered during the installation work. These problems led to extra work being carried out by the contractor and some design work being required to be carried out by the contractor.

Recovery of some costs by the contractor was achieved by the use of daywork. This was mainly for site work where quantities and hours could be easily identified and confirmed.

Some areas of additional costs could not be agreed, as to whether there was a basis to the claim for additional costs or whether they should have been allowed for in the original contract price. This led to lengthy negotiation over pricing and the validity of the claims presented to the owner.

Negotiation was mostly conducted on site, prior to practical completion and site demobilisation. A final claim was negotiated by more senior levels in both the owner's and contractor's organisations.

Because of the structure of the owner's organisation, it was often difficult to obtain answers or commitments from the owner's staff. Owner's staff, being present but not having the proper authority to make decisions or answer questions without referring to another person, added to the length of time involved in the negotiation.

The negotiation

Initial thorough preparation was carried out by the contractor and included the following:
• Gathering and reviewing all information with regard to the claim.
• Establishment of all costs involved with the claim.
• Establishment of a worst case fallback position.
• Discussion of all items and arguments that the owner's staff may raise.
• Some discussion of tactics.

This preparation period was carried out over a week, and involved all members of the contractor's site staff and the company general manager. This allowed the contractor to enter the negotiation well prepared, and gave the contractor a fairly strong position in the negotiation process.

The initial negotiating took place on site and was held in the office of the owner's project manager. The members of both negotiating groups were well known to each other, as at this stage work had been under way on site for some eight months, with regular monthly site meetings being held between the contractor's and the owner's staff.

The negotiation started with a review of the previous month's site meeting minutes and initially followed an agenda set by the owner's project manager. Very early in the meeting the contractor's general manager took control and endeavoured to conduct the meeting following his own agenda.

This allowed the contractor to prioritise its claims and to be in control of how the meeting proceeded. The contractor's general manager, being a seasoned campaigner and slightly overbearing in the negotiation process, felt it the best policy to take control of the meeting early and direct the meeting as per his agenda.

This tactic caused some friction in the meeting as the owner's project manager, conducting the meeting, tried to direct the meeting back to the agenda originally set. As soon as he thought that he had regained control of the meeting, the contractor's general manager would set off on his own agenda again.

The control of the meeting by the contractor's general manager, although causing this friction, did let the meeting proceed fairly quickly, allowed the contractor to identify the claims it was pursuing, and stopped the meeting digressing onto other matters not concerned with the negotiation.

The owner's project manager was known not to keep good control of the meetings held on site, as they often digressed into areas that were not really concerned with the matter at hand. This was something the contractor felt had to be avoided if it was going to be successful with its negotiation.

At the meeting, both sides were able to present their views and concerns on the matters involved in the negotiation and discuss alternative proposals. Some initial progress was made from this meeting, with some give-and-take being shown by both parties.

A number of the main issues raised could not be settled by the site personnel and, as such, finalisation of the claim was completed some months later after a number of other meetings held between more senior management in both organisations.

Exercise

1. The negotiation was considered to be reasonably successful from the contractor's viewpoint, and this can be largely put down to the experience in negotiating shown by the contractor's general manager. Does this imply that negotiation is more 'art' than 'science'? Comment.

2. The preparation carried out by the contractor was comprehensive. One area that was overlooked and created difficulties in obtaining acceptance of proposals, was the need to ensure that personnel with the appropriate authority to deal with matters raised at the meeting were involved in the negotiation. How common is this with large organisations, both in the private and public sectors? How can this be dealt with during preparation for the negotiation?

3. During the negotiation, it was difficult to bargain. This was due, as mentioned above, to not having owner's staff with the appropriate authority at the meeting. Minor elements of the negotiation were finalised at this meeting, but for the major items the bargaining continued, and agreement reached, at subsequent meetings. What is the point of conducting negotiations if the people present are not able to commit their organisations to any agreement?

4. The style of the contractor's negotiation was strongly influenced by the contractor's general manager, who was at times overbearing and aggressive. The members of

both groups involved in the negotiation were aware of this trait. Would this allow the negotiation to proceed without any negative effects, or would such a person always influence a negotiation outcome?

5. Under what circumstances would showing an aggressive approach to a negotiation be necessary? What difficulties can it lead to if one group feels threatened or feels that it is not getting an opportunity to be fully involved in the negotiation?

6. Tactics used in the negotiation involved the contractor setting its own agenda and using good guy–bad guy scenarios. Can you say, that whichever party controls the agenda in a negotiation goes a long way to controlling the negotiation?

6.2.7 CASE STUDY — AIRLINE REFIT

A firm Rehold was hired as project manager and designer for an aeroplane refit. Rehold worked for five months on the project with expected fees to be earned during the installation period. At the instruction of the airline, Rehold was asked to develop ideas with a mill supplier (the mill). This was expected as Rehold had several large ongoing projects with the same mill supplier. One month prior to commencement of the installation, the mill advised that it planned to supply directly to the airline, and as compensation would pay a royalty on supply. This was not to Rehold's satisfaction as it breached an unwritten agreement with the former mill managing director.

The mill proceeded and several royalty sums were received by Rehold during the next twelve months. Rehold continued to express opposition to the basic deal; the royalties received were only a fraction of a percent of the original transaction value

It then became apparent that the airline believed it owned the design ideas of Rehold, and as proof offered a letter sent by the mill transferring all rights. A month later, a manager of the mill on a routine visit to Rehold mentioned that the airline required it to transfer the Rehold ideas that were now registered designs. Rehold had to decide whether to challenge the airline for breach of copyright or negotiate the sale of the ideas to the mill for transfer to the airline.

The airline was a large ongoing client of Rehold and therefore it was decided to transfer the ideas, upon resolution of the conflict with the mill. For Rehold, this was a difficult time as the airline cancelled Rehold's sales visits. However Rehold believed the mill must settle. A team was sent to negotiate with the mill.

The negotiation

Meetings and telephone conversations were involved in the negotiation. Both group's lead negotiators were trained at the same university; this provided early humour to a serious meeting. In preparation, both firms had old files and diary notes at hand. A mill director contacted the Rehold marketing director to ask for a free transfer. The mill rang a second time and was angry about the unknown copyright. The Rehold legal team contacted the new mill managing director and advised the details of the original agreement. The new mill managing director advised that he would act promptly, and should have agreement prior to the next meeting in two weeks. The patent attorneys of both firms discussed the legal situation. The prior visit by the mill had established that Rehold would transfer the design, and the mill could purchase the rights or face possible legal action from the airline.

The meeting, with two managers and a marketing director on each team, opened by discussing recent holidays. The marketing directors led the meeting. The mill's initial offer was low with the royalty paid in full and in advance. The mill used the ethics and honour of an agreed deal, as the basis of its proposal. Rehold relied on the original deal, the loss of revenue, the ethics of the basic transaction, the legal right of copyright and lost marketing opportunity. The options to Rehold were established in a range chart involving 'must gets', 'should gets' and 'could gets'. Both parties used good guy - bad guy tactics and the approach was collaborative moving to competitive at times. The settlement involved double forward royalties, with Rehold naming rights to be retained and used by the airline. The negotiation was documented by Rehold that day, and it received a cheque with the design transfer within a week. Both groups got something from the negotiation.

Exercise
1. Both firms were fully prepared with project briefs, legal analysis, alternative costings, other's goals, and range analyses. Time was on neither group's side because two deadlines had been missed. How do time deadlines influence the course of negotiations?
2. The style was the same as in past meetings. What influences would this have on the course of the present negotiation?
3. The Rehold team were general about specifics. The deliberate understatement of marketing rights reflected the real goal of Rehold. What role would this play in the direction that the negotiation took?
4. The 'If I, then will you?', and the 'What would you do?' open questioning approach were used frequently. How is the bargaining process influenced by such indefiniteness?
5. Options were narrowed because litigation was not considered an option. The cost of a win–lose situation was unacceptable. The final agreement was slowly and carefully achieved without either party conceding anything until the playing field was established. The groups agreed only after a full package was settled. In general, is this preferable to groups agreeing along the way on issues in contention?

6.2.8 CASE STUDY — SUBCONTRACT ISSUES

Outline

All in a day's work.

Project completion was just over a week away, and everything was looking good for the contractor. The owner was happy, the work was on program and the margin was healthy.

The next activity was the delivery, and start of installation of the lawn pavers. Paving could have been complete, but the subcontractor had limitless reasons for leaving it to the last minute. And so the final stage of the external work was on a tight, yet achievable, program.

It was also time to start project finalisation. Half the project's subcontract work packages were complete and final values needed to be agreed. The finalisation of the mechanical subcontract was one example. This included the settlement of some variation claims that the subcontractor felt were justified. These variations were excluded by the strict letter of the contract.

Paving subcontract

The day that paving was meant to start, no lawn pavers had arrived. The subcontractor checked the drawings. Then the subcontractor was on the telephone, telling the contractor's representative that it had just found out that the lawn pavers, as noted on the drawings, were inadequate for the proposed traffic load. The correct paver for the application was called a BG lawn slab. It was twice the cost and if the contractor expected the subcontractor to pay the additional costs, the subcontractor would leave the project, leaving the contractor to find someone else to complete the external work. Additionally, insufficient stock of the BG lawn slab existed with the supplier, and fourteen days lead time was required to produce sufficient stocks. This delivery time exceeded the remaining project time.

The subcontractor had opened negotiation, while the contractor's representative had not started the preparation stage. The bombastic nature of the subcontractor and the way the information was delivered didn't help, but the contractor's representative, with effort, remained unemotional.

The contractor's representative expressed to the subcontractor that he felt let down by this late notice, and then directed the subcontractor to arrange samples of each paver for review by the owner later that morning. By this course of action the contractor's representative took ownership of his feelings, had not made a judgment on the subcontractor or the situation, and had defused the initial situation. The contractor's representative had gained some time to start the preparation for the negotiation with the owner and the subcontractor.

This negotiation preparation found the following:
- The drawings noted the use of lawn pavers.
- The specification called for BG lawn slabs.
- The subcontract was not clear on the distinction between the grades of pavers.
- The subcontractor had full and correct information at time of tender.
- BG lawn slabs cost approximately half as much again as lawn pavers.
- The BG lawn slabs would not be available until the week after practical completion.
- Half of the paving was covered by an owner variation, based upon the correct rate.
- The subcontractor's original claim for the variation work had been based upon the incorrect rate, that is funds were currently in hand.
- The subcontract price suggested that it had been based upon the cheaper paver.
- The subcontractor's quotation had been used within the contractor's quotation.
- The total purchase price of the correct slabs was about what was still to be paid under the subcontract.
- There were liquidated damages for external work.
- The owner did not require the area before two weeks after practical completion.
- Some work could be deducted from the subcontract.
- The paver/slab supplier would offer a small saving for cash on delivery.

The final stage of preparation for the contractor's representative was to consider tactics and stance. The relationships with both the owner and the subcontractor were long standing and of mutual benefit. To jeopardise either relationship unnecessarily would be inappropriate.

Negotiation with the owner

Dealing first with the owner, it was important that the contractor got an extension of time. This was a 'must get'. Additional costs would have been good, but contractually this was not supported and the contractor's representative knew that the owner was already fully committed. 'Should get' and 'could get' items included a formal extension of time and costs. The likelihood of jeopardising its 'must get' by chasing 'should gets' and 'could gets' seemed high. The relationship of the contractor's representative with the owner was strong and based upon trust. The owner's word would be sufficient. Upcoming projects added further incentive to finalise this project on the best possible terms. On a personal level, finalising the project with the owner happy was important as it would affect the career prospects of the contractor's representative.

Confident that a soft-sell approach would work, the contractor's representative approached the owner. Firstly the contractor's representative reassured the owner by saying that a nil-cost solution to the problem existed. Then the contractor's representative showed the owner the problem and gave him a brief history of the matter, noting the conflicting documents. This attempted to establish some sense of ownership of the problem with the owner. Then the contractor's representative made an offer: if the owner would give the contractor an extension of time, the contractor would take care of the additional cost. The owner replied that if the contractor would undertake to be complete by two weeks after the original practical completion, the required extension of time was available.

Exercise
1. The opening, bargaining and agreement phases of this negotiation lasted less than two minutes, and from the time of problem identification, the whole process lasted less than two hours. This negotiation was conciliatory in nature. Would other styles have worked?
2. The expectancy of a future relationship and the strength of previously established relationships enabled the contractor's representative to propose a solution acceptable to the owner. The preparation stage had informed the contractor's representative of the true position and allowed the contractor's representative to obtain a good solution without the possibility of confrontation. Would a confrontational approach have worked anyway, in such a situation?
3. Time was important for the contractor's representative. Money was important for the owner, while it was less important to the contractor's representative. How close is this to the win–win scenario in the classic textbook negotiation over an orange — one person wanting the rind, the other person wanting the juice?

Negotiation with the external work subcontractor

The next negotiation to face was with the external work subcontractor. This one was not to be so quick, or so straightforward. In the preparation stage it was identified that the subcontractor was currently overpaid. The supply price of the correct slab exceeded the outstanding balance of the subcontract value. And other significant costs were still to be incurred under the subcontract.

At this point the contractor's representative had several options:

- Terminate the subcontract and engage a new subcontractor.
- Postpone the disagreement until the work was completed, and then adopt a 'stone-wall' approach.
- Reduce liability by interrogating work performed under the subcontract, making deductions for work not performed.
- Accept the additional supply costs only.
- Accept the additional supply costs and overhead charges.
- Take the supply cost saving from the supplier, by paying it directly, cash on delivery (COD).
- Arrange for the subcontractor to pay the supplier COD and claim the available saving.

The first two options, both confrontational in nature, were ruled out on the grounds of cost — the first option because of greater expected costs from both the replacement subcontractor and time overruns; the second option, while potentially making a short-term saving, exposed the contractor to potential long-term negotiation, mediation or litigation costs including settlement costs. On a personal level the contractor's representative wanted to finalise the entire project without a dispute, enabling the contractor's representative to present his company with a project outcome that would be seen as successful on all accounts.

The last two options were also ruled out. The small saving available was not sufficient to offset the potential default by the subcontractor. The subcontractor would never be in a better cash position than he was in at that point. Use of either of these options would make this extremely clear to the subcontractor.

In the development of the 'could get', 'should get' and 'must get' options of the contractor's representative, the contractor's representative identified the following. The contractor:

- 'Must get' the external work complete by two weeks after the original practical completion.
- 'Should get' the work complete, paying only the additional supply costs and obtaining the savings for work not carried out by the subcontractor.
- 'Could get' the work complete on time, with no additional costs and obtain the savings for work not carried out.

At this point the contractor's representative rang the subcontractor. The contractor's representative explained that he had reviewed the documents and felt some exposure, the subcontractor was required to proceed with the work using the correct slabs, and to arrange a meeting with the contractor's representative. In the proposed meeting it would be necessary to agree the final split of costs and the final contract value.

In spite of several attempts, the contractor's representative was not able to organise the proposed meeting with the external work subcontractor before project completion. The subcontract work was completed a day ahead of the owner's requirements. This included using the correct lawn slabs and an additional paving variation.

Exercise

1. At this point in the negotiation, it would appear that the combination of the past relationship and the preparation have combined to identify a solution sufficiently within the subcontractor's range of needs. How could the contractor's representative be sure of the subcontractor's needs?

2. Money was important to the subcontractor. The achievement of completion within the owner's requirement, money and completion without drama were important to the contractor's representative. The preparation stage identified that in the best case scenario, additional funds would need to be expended. The early acceptance of this saved time, avoided upsetting the owner, satisfied the subcontractor's needs, and avoided potentially greater expenses. How does spending of extra funds affect the successful project view of the contractor's representative, or has the contractor's representative rationalised the situation in order to get a resolution?

Mechanical subcontract

In the mechanical subcontract, the issues related to stage one of the project. Due to time and supervision pressures, the subcontractor and the contractor's representative had overlooked making a variation claim for some modifications to the system. Under the head contract the contractor was now time barred from making any such claim, as was the subcontractor. The subcontractor based its claim upon the fact that the alterations were made on site, without its knowledge. The contractor's representative on the other hand had dealt with the subcontractor's representative on site, the ductwork installer (sub-subcontractor) engaged by the subcontractor. Any claim arising from that source should have been made at the time, and within the time bars.

While the subcontractor had no strong contractual claim against the contractor, the work had been carried out professionally and cooperatively. Design and sequence changes initiated by the subcontractor had impacted favourably upon the program, and the subcontractor had assisted with items outside its scope of work.

From the contractor's side, the contractor had assisted the subcontractor in many ways, including the refinement and presentation of design and coordination proposals to the owner. The subcontractor had made savings due to these changes and had included some extra work at no cost. At the time the changes were made, they appeared minor in nature, and larger variations were being prepared for the owner. It had appeared to the contractor's representative that the subcontractor had either included the work in its variation claims or it did not intend to claim for the work. The contractor did not have funds set aside for miscellaneous claims.

The problem faced by the subcontractor was that its ductwork sub-subcontractor was claiming the costs, and without some concessions from the contractor, the subcontractor would be embarrassed.

Discussions had opened some time earlier. The contractor's representative took an initial contractual position, with an attached proposal. If the work was completed without further dispute and in accordance with quality and time requirements, the contractor's representative would be prepared to discuss the claim.

To initiate the final discussions, the contractor's representative rang the mechanical subcontractor's project manager and asked for a full breakdown of the outstanding claims. The contractor's representative pointed out to the project manager that while he was prepared to talk, the project manager needed to realise the regrettable position that the contractor was now in. If the claims had been made at the appropriate time, the full costs could have been recovered from the owner. The rates it had previously used related to fully supervised personnel. Any claim now could not include supervision or overhead charges. Contractually, the subcontractor had no claim against the contractor. The contractor was not responsible for the subcontractor not having had a responsible person on site. It had chosen to take the saving in supervision cost. The savings made had not been shared with the contractor and generally the contractor had no funds available for such claims. However, due to the positive contributions the subcontractor had made to the project, the contractor's representative would see what could be done to help. If the subcontractor was prepared to discuss a three-way split, then the contractor may be able to help it.

In this opening position, the contractor identified a willingness to negotiate. In valuing this issue, relationship and future prospects were placed equally important as direct financial cost. This valuation enabled the contractor to take a conciliatory stance, while keeping a strong negotiating position.

The subcontractor's reply included an acceptance of a three-way split, and an additional claim, relating to an issue the contractor's representative felt had been previously settled. The contractor's representative now viewed the balance of fairness as having shifted; principle was now more the issue.

Telephoning the subcontractor's project manager, the contractor's representative said that he did not accept the additional item, and made an offer — if the project manager would guarantee that no additional claims would be made, and agree a final contract value, the contractor's representative would offer a total figure up to the limit of his authority, and within a small percentage of the subcontractor's desired final contract value.

In reply, the subcontractor's project manager suggested that he would settle if the difference was split. The contractor's representative, hiding behind the limit of authority excuse, undertook to see what could be done and to telephone the project manager the next day. The next day, an agreement was finalised just under the proposed split.

Exercise
1. Familiarisation was not contained within a specific period, with the participants dealing with one another regularly. Opening positions were conveyed over an extended period. Does this imply that early in any project, participants should be thinking of the possibility that negotiations may take place later and prepare accordingly?
2. The initial proposal of the contractor's representative was in the form: 'if you complete the work without further problems then I will consider your claim'. How could you make the term 'further problems' more definite?
3. The important issues in this negotiation were relationship, performance, face and money. How do you trade off such noncommensurate issues?
4. This negotiation was neither large nor critical in nature. Overlapping ranges of expectations allowed conciliatory settlements to be reached. Negotiation proceeded in a controlled manner. What controlled the progress of the negotiation?

6.2.9 CASE STUDY — ELECTRICAL VARIATION

Background

A major component of an on-site upgrade project was the letting of an electrical contract for all of the installation and commissioning work associated with the project.

Events

A broad scope of work, and delivery times of major pieces of equipment to be installed, was issued to several electrical contractors for tender. A fixed price, a timeline, approximate work hours and a list of supervisors were requested. The contract was awarded based on price and suitability of the contractor's skills base to carry out the work.

The work commenced on the nominated date. In the early stages, supervisory issues arose with regard to the designated supervisor not being on site. The work crew was being left to carry out the work unsupervised. A meeting was called to remind the contractor of the supervisor's task and the reasons for being on site full time. These reasons were: safety concerns, arrangement of clearances and the use of site systems, quality control and liaison with the construction project engineer. The contractual obligations with respect to having a supervisor on site were emphasised.

The major equipment arrived on schedule and work progressed well. A third of the way through the project, the project manager was informed of two major pieces of equipment that would be delayed by a period of three weeks.

The project manager attempted to relay this information to the site supervisor as soon as possible. However, the site supervisor was not on site (on the Friday) and the work crew was not being supervised. The site supervisor was not contactable. Messages and faxes were relayed to the contractor's office. The messages detailed the equipment delay and informed the contractor that the crew would not be required over the coming weekend or until equipment arrived on site.

A review of the contract documents revealed that there were no clauses relating to late delivery of major equipment, or for a documented procedure for informing the contractor of the information.

On the Monday, the site supervisor was contacted and informed of the equipment delay. The site supervisor informed the project manager that the crew would be put onto another job on site. That is, there would be no demobilisation costs. This was not conveyed in writing.

The equipment eventually arrived on site and the crew was back at work. The work was completed three weeks late.

Post-installation work

After work completion, the contractor's manager stormed into the owner's office and stated that the owner had no right to re-deploy the crew onto another job because major equipment was late. The contractor's manager stated that the contractor was out-of-pocket due to the re-deployment of staff and demobilisation of the crew half way through the work. The contractor's manager said that a variation was being prepared and would be submitted before the end of the week. The contractor's manager stormed out of the owner's office. The owner's project team was speechless.

The owner's project team informed the contractor's manager that a meeting to discuss the problems was to be held at the end of the week. The contractor was asked to prepare the variation, as well as a detailed log of events according to the contractor. The site supervisor was asked to be there.

The owner's project team prepared a detailed log of events and discussed contract issues relating to the project with the on-site contract management team. The on-site estimators were used to prepare a budgeted forecast of out-of-pocket costs that could be expected of the contractor due to the delay in equipment arrival. The project team reviewed the work done by the contractor and inspected all quality issues relating to the work. Prior to the meeting a detailed and precise document was prepared that covered:
- The sequence of events according to the project team.
- An estimate of out-of-pocket expenses of the contractor.
- Work quality deficiencies.
- Costing of quality deficiencies to be rectified.
- Contract issues relating to site supervision and late delivery of equipment.
- Proposals for a resolution.

A brief agenda for the meeting was forwarded to the contractor:
- Contractor presentation on information with regard to the variation.
- Project team information.
- Discussion.

The meeting

The contractor was represented by its manager, business manager and site supervisor. The contractor adopted an aggressive nature from the start. The owner was represented by its project team.

The contractor presented the facts from its point of view, the key points being:
- It was four days before the contractor was informed of the late delivery of equipment.
- The three weeks late on work completion resulted in a lost opportunity to work elsewhere.
- The demobilisation of the crew to another job.
- The re-deployment of the crew by the owner's project team.

No indication of emotion was shown by the owner's project team. The project team presented its view, the key points being:
- Supervision on site issues.
- The reason for the four day delay was that no contact could be made with the on-site supervisor.
- The re-deployment of the crew was to another on-site job and was initiated by the site supervisor.
- Quality issues with respect to the work carried out, and the time to rectify the work.

The contractor's manager obviously had some issues to resolve with the site supervisor, but the project team asked that these be resolved in their own time.

Because the contract documents did not say anything with respect to late delivery, but did state a machine delivery time, the project team accepted that some form of compensation was due.

The lack of supervision did breach the contract.

The work quality issues were minor but represented at least two week's work by the contractor to rectify the problems. After some convincing, the contractor's manager agreed that the quality issues were real and that they had to be rectified.

The issue of demobilisation was resolved. The project team accepted that a demobilisation cost should be incurred by the project. However, due to the contractor moving its crew to another on-site location, the demobilisation cost should reflect this.

The contractor was in a poor negotiating position because it didn't have all of the information prior to the meeting.

The final agreement was for the contractor to rectify the quality issues in its own time, and the costs of demobilisation was to be paid by the owner to recognise that some compensation should be paid for the late equipment arrival.

Exercise

1. The meeting presentations of both sides were open and up front about all information. The contractor's manager was given sufficient time to express his point of view. The project team consciously made an effort not to interrupt or debate the facts being presented by the contractor's manager. All information was taken on-board and notes taken. The project team's presentation of information was met with some apprehension, but all information supplied was verified and could not be questioned. What advantage or disadvantage is there in presenting your case first or second?

2. The proposal from the contractor's manager was based on the information available to him at the time. He made on-the-run changes to his proposal during the presentation of the facts from the project team. In what way is a proposal weakened or strengthened by adapting it as new information comes to hand?

3. The negotiation was heavily influenced by the fact that the contractor's manager did not have all of the information prior to the meeting. This led to a lack of credibility, and placed him in a very poor negotiating position. The proposal put forward by the project team was not a heavy-handed attempt at squeezing money out of the contractor and he recognised this. Thus the negotiation was relatively short because the proposal put forward was not met by much opposition from the contractor. What does this say about the task of preparation for a negotiation?

4. The final agreement was drawn up during the meeting and signed by both parties on the spot. This was done to minimise the administration time spent on the issue and to ensure that all rectification work took place as soon as possible. What other purposes does documenting an agreement at the meeting serve?

5. After the information was presented by both sides, the conversation flowed naturally towards an equitable outcome for both parties. The important factor here was that the baseline information was not questioned. Therefore, the conversation revolved around obtaining a suitable solution for both parties. With non-written records of some project events, there was the potential to contest information presented; why do you think this didn't happen?

6. A key factor was that the contractor had a relatively long relationship with the owner and this relationship was seen by the contractor as a key to its long-term survival. The owner's project team also believed that this relationship was in the best long-

term interests of the owner. Under such circumstances, is a satisfactory resolution inevitable?

6.2.10 CASE STUDY — CONCRETE SUPPLY

Introduction

This case study discusses two claims submitted by a concrete supply company to the owner resulting from variations made to the owner's existing scope of work. Both claims involved considerable negotiation to reach an agreement, with the majority of the negotiation being carried out by facsimile.

The owner was constructing an industrial plant. The project involved a large quantity of structural concrete, poured over a twenty-four month period. Accordingly, major concrete companies were invited to tender for the design, supply, installation and operation of an on-site computer controlled concrete batching facility which would provide the pre-mixed concrete at the project site.

In its tender, the contractor made allowance for the design and supply of four project-specific concrete mixes, each with an individual specification. For each specific concrete mix, the contractor was to supply a price which would hold for the duration of the project.

The required concrete batching plant capacity was stated, and it had to be capable of providing, over a two year period, concrete to the strengths and for the approximate quantities estimated by the owner.

To meet market demand the project was fast-tracked, with detailed design being done in two countries. As design progressed, it became necessary to change the number and types of concrete mixes, due to better definition of design requirements.

Claims

The contractor was relatively pro-active and entered into the spirit of developing a good relationship with the owner. However, a number of claims were submitted by the contractor, and these required negotiation. The claims were submitted because of:
- Revisions made to the concrete specifications.
- Modifications required to the existing batching plant.

Specification — revisions

Following the trial testing of each concrete mix, a number of revisions were made to the project concrete specification, which resulted in changes to the mix designs, and defined more clearly how the delivery and quality assurance testing would be carried out by the contractor.

Subsequently, the contractor submitted a claim in accordance with the conditions of contract, and advised that it would seek additional costs as a result of the changes to the specification.

Before a response was sent to the contractor, information from existing project files was checked by the owner and a strategy established for the response. The initial response was developed by the project engineer, and then vetted by the contracts engineer. Following agreement between these two engineers on the response that should be adopted, the strategy was then discussed with the project manager, from whom a direction was sought.

Items of importance to the owner were:

- Likely costs associated with the claim and methods of minimising them.
- Establishing the validity of the claims and hence whether they should be paid.
- Obtaining information from previous revisions of the concrete specification from which to base a response.
- The negotiation style to be adopted; this was agreed to be based upon written communication, with all initial negotiating being carried out by facsimile.
- The initial strategy to be adopted for the response to the claims would be to reject the items that were not considered valid.

Because the project had been running for eight months, both the owner and the contractor had already developed a good rapport and working relationship.

The owner's response was sent by facsimile to the contractor stating the owner's position with regard to the claim. Basically, the strategy of the owner was to reject the majority of the claim, referencing the relevant sections of the concrete specification as justification. However, there were a number of items which the owner considered were legitimate and for which it was willing to settle.

Following receipt of the owner's response, the contractor telephoned the owner's project engineer to discuss the proposal.

This discussion highlighted that many of the changes in the specification were made as a result of recommendations made by the contractor's technical staff, and as such the owner did not believe any payment for these to be appropriate.

The contractor agreed to speak to its technical staff and discuss the matter in more detail and then revisit its original claims. To achieve this, the contractor required more time to respond than what was indicated in the owner's response letter. Additional time to formulate a response was granted by the owner.

The response received from the contractor stated that its technical staff had not advised its operations staff of the changes made to the technical specification. The contractor's inspection and test plans (ITPs), which controlled its quality assurance on the project, had been revised and forwarded to its operations staff, with the comments that the changes had resulted from a revised specification received from the owner; no mention had been made that the specification had been changed based on recommendations made by the contractor's technical staff.

The contractor agreed to withdraw all claims conditional upon it revising some of the ITPs to produce a less rigorous testing regime.

This was agreed by the owner and subsequently the contractor faxed confirmation of this to the owner.

Modifications to the batching plant

For deep concrete foundations, a low heat concrete mix was specified so as to limit the heat of hydration during curing, thus preventing cracking. The exact volume of this mix to be used on the project was unknown.

Following award of the contract and ongoing design work, the required volume of this particular mix was established. This fact was conveyed to the contractor and a request made as to what, if any, cost implications would result in supplying this volume of concrete. The contractor responded that the existing plant would have to be modified to cater for the increase required, that is, the addition of another silo. This modification would attract a cost, for which a claim was subsequently submitted.

The contractor submitted a claim, in accordance with the conditions of contract, detailing the additional costs incurred to install the silo. The claim detailed the purchase, installation and commissioning cost of the silo as well as ongoing running costs incurred until the silo could be erected.

The owner requested a meeting to be held at its office to discuss the claim. Both the contractor's regional manager and state manager were requested to attend the meeting, at which the owner's project manager and senior engineering staff were also present.

The owner agreed in principle with the contractor's claim. However, it was not willing to pay for the full purchase price of the silo, because it considered that, at the end of the project, the contractor would be left with an asset which could be used elsewhere. As such, the owner would not pay for the purchase of the silo.

The contractor claimed that it could not possibly write off the cost of the silo on this project, because the project overheads were insufficient to absorb such an unforeseen impost.

The owner requested information as to how the contractor would typically cost and amortise plant of this nature. The contractor stated its normal practice was to write-off plant over three years.

The owner and the contractor agreed that the additional silo was most likely required on the project for only one year, after which time the silo could be demobilised. On this basis the owner offered to pay for one third of the cost of the silo.

The contractor agreed to this proposal, and stated that the claim would be amended to reflect this agreement. It subsequently revised its original claim and submitted it to the owner.

Minutes of the meeting were produced by the owner which confirmed the agreement.

Exercise

1. Comment on the owner's strategy of firstly choosing facsimile as the medium for the specification revision negotiation, and secondly an off-site face-to-face meeting for the batching plant modifications negotiation. Why do you think different approaches were adopted? If good rapport existed between the owner and the contractor, why couldn't both negotiations have been conducted face-to-face on site?

2. Why might the contractor not have attempted to bargain in the batching plant modifications negotiation, but rather accept the owner's initial offer? Why wouldn't the contractor have at least asked for mobilisation and demobilisation costs of the extra silo, in addition to part of its capital cost?

3. Where might the negotiations have headed if the contractor had insisted on recovery of the silo's full costs?

6.2.11 CASE STUDY — GRAIN HANDLING FACILITY

Introduction

This case study details the delivery and claims negotiation involving civil, structural, cells and mechanical packages in the construction of a new grain handling facility. The cells package involved the acceptance by the owner of an all-steel construction alternative design submitted by the contractor (instead of a reinforced concrete and steel composite structure).

Planning and delays

The contracts for civil, structural and cells packages were let at the same time. The three separate work packages were treated as one project by the contractor, and were planned accordingly due to the close interfacing requirements between each scope of work, and for resourcing purposes. The owner, however, treated the packages as three separate contracts for contract administration purposes and when granting extensions of time.

Short-term schedule updates were submitted to the owner at the weekly coordination meetings. The main construction program was updated on a fortnightly basis.

The contract package for the mechanical work was not let until six months later. Due to the major interfaces with the other three packages, the main construction activities had to be re-sequenced, and corresponding extensions of time claimed for each package from the owner's representative; these were subsequently granted.

The owner was late in issuing 'Approved for Construction' drawings for the mechanical work, and this resulted in the contractor having to expedite the workshop fabrication work. This involved offering an alternative design to the owner for construction of the conveyor galleries, from bolted to all-welded construction; this resulted in a saving of weight and fabrication time. An extension of time claim was submitted for this delay.

The conveyor galleries were all manufactured off site and transported to site by road, in a major exercise.

The owner also requested that the contractor accelerate the on-site erection of the conveyor galleries, so that the grain handling facility would be commissioned ready for the commencement of the grain season. This request involved the re-programming of the main construction on-site activities, the employment of additional people, plant and equipment, and the working of a night shift.

Delays were incurred on site with high winds preventing erection of the cell roof sections and gallery sections. Major delays were also incurred with the delivery of owner-supplied conveyor stringers and chutework, due to design and fabrication errors. Extension of time and disruption cost claims were notified to the owner in writing and at the weekly co-ordination meetings, at the time of the delays. Final quantification of the extension of time claims for each contract package, and the prolongation and disruption costs claim, took many months to prepare after demobilisation from site. The service of an independent consultant planner was used by the contractor to determine the cause and effect of the delays on the contractor's original program, and justify its disruption/loss of productivity claims.

Variation claims

On the first three contract packages, variation claims were of minor value at a few percent of the original contract sums. At project completion, there were variation claims still outstanding and the contractor intended to resolve these by negotiation with the owner.

The mechanical contract and last package had major variation claims raised for a number of reasons including: late issue of 'Approved for Construction' drawings; late delivery of owner-supplied equipment; daywork claims on modifications required to owner-fabricated materials; major disruption and loss of productivity on the original contract scope of work; and acceleration costs to complete the work ahead of schedule.

Contractual arrangements

The four contract packages were lump sum contracts, with schedules of rates included for pricing variations. The general conditions of contract were an industry standard. In the tender documents it stated that the general conditions of contract would be the latest revision, but the contract was signed using annexures from an earlier edition. Therefore, the conditions of contract could not be used, because some clause numbers in the annexures had changed.

The contractor prepared its subcontract documents to include the subcontract conditions which were a companion to the head contract standard conditions. The contractor engaged a number of subcontractors to carry out the work.

The owner managed the whole project and, in addition to the main contractor, had its own contractors on site for the installation of the dust collection system, and electrical, instrumentation and road work. It was the first time that the owner had managed and administered such a large project using its own employees. This was a cause of problems on site, because the owner's personnel were inexperienced with administering contracts, determining extensions of time and variation claims etc. The owner used the main contractor's construction program to control the activities of its other contractors on site, and in many instances expected the main contractor's on-site project team to carry out the project manager role for the total project, especially with regard to planning and coordination of the site work.

Contract administration

The first three contract packages went well both in time and cost.

The last and mechanical package incurred a large loss due to the owner delays and the contractor having to accelerate the work to complete the project within the owner's time frame. In hindsight, the contractor should have employed another contracts administrator and planner on site to enable a comprehensive detailed history of the project to be maintained, and to commence preparation of its contractual claims. In preparing the claims, the contractor found minor gaps in its records and found it very difficult to quantify and prove the amount of disruption to the contract scope of work as a result of the delays.

Personalities played a major factor in the success of the project and the resolution of variation claims with the owner. When the pressure was applied by the owner to accelerate the work, and the contractor was concurrently incurring major delays through daywork modifications to owner-fabricated materials, tempers and relationships were strained both within the contractor's project team and with the owner's staff (primarily through their lack of experience with administrating such a large project, rejection of straightforward variation claims, and reluctance to acknowledge variations when directing changes to the work).

The extension of time and variation claim rejections were motivated by budget considerations and not entitlements under the contract terms and conditions. The owner's representative also wore the hat of being the owner's chief engineer. When carrying out certifying functions under the contracts, he had a clear bias towards the owner when determining variations, and relied totally on his three offsiders for administering the contracts. When the contractor discussed the outstanding variation claims with the owner's representative, it was clear that he had been misinformed by his offsiders and that he did not have a clear understanding of the issues and matters in dispute.

Prior to the completion of the work, it was clear to the contractor when meeting with the owner's representative, that he was stalling on making any decisions on the contractor's extension of time and variation claims until he had received the contractor's main prolongation and disruption cost claim. The owner's representative was also reserving his right to apply liquidated damages to all four contract packages. The contractor could see that the owner was going to attempt to offset the contractor's claims with liquidated damages.

Claims resolution strategy

The contractor's first and preferred option from day one, was to negotiate as many as possible, if not all, of the outstanding claims with the owner's representative.

Its second option was to obtain an independent expert appraisal on the remaining outstanding claims.

Its third option was to recommend the other alternative dispute resolution techniques of mediation, conciliation or mini-trial.

Its last option was to proceed to arbitration, in accordance with the contract.

Following completion of the project, the order of priority of the contractor's claims submissions were:
- Settle all outstanding variation/daywork claims.
- Resubmit rejected variation claims, that had a strong contractual basis.
- Submit extension of time claims for all four contract packages. Also verify its claim using an independent consultant planner. This was to negate the owner's threat of applying liquidated damages.
- Prepare and submit its main prolongation and disruption claim on the mechanical package. (This was later submitted.)

During the claims preparation, submission and negotiation process the owner's representative went on three months long service leave. There were two different people acting as representative during this three month period, and the contractor found them reluctant to make any decisions, which had any financial implications to the owner.

The negotiation

The contractor's preparations involved various activities:
- A goals list (developed in conjunction with project, construction and contract managers).
- A range chart (worst, most likely and best cases).
- Make reasonable assumptions about the other party's goals (for example, the total project cost exceeded the owner's original budget, therefore the owner would be reluctant to concede any variations that were not straightforward, had a sound contractual basis and were priced at the contract schedule of rates).
- Consider appropriate style.
- Consider tactics.
- Fact finding — seek as much detailed information as possible about the reasons for, background and basis of the variations, the valuation, the owner's negotiation team and the likely responses.
- Consider options available — including what to do if the negotiation failed completely.
- Consider what options the owner had.
- Consider timing — deadlines, financial implications to the contractor and the owner.
- Prepare a suitable agenda and agree the final agenda with the owner (as well as the list of participants in the negotiation).

Although it was a long drawn out process (over six months), the contractor managed to negotiate all of its outstanding claims with the owner. The turning point with the negotiation was when the owner obtained an independent expert appraisal on the claims. A commercial settlement was agreed by the general managers of both parties prior to the claims in dispute being submitted for resolution by arbitration.

Exercise
1. How might role playing the negotiation by the contractor's team have helped the planning and anticipation of the owner's responses? How might playing the devil's advocate role assisted the contractor's preparation?
2. How true is it that negotiation preparations may contain 80% assumptions and 20% facts? Relate your answer to this case.
3. An independent person was instrumental in getting the owner to see reality in this case. What other functions can an independent person perform in negotiations?
4. The threat of arbitration, should the negotiation not be successful, focused the mind of the owner. Do all negotiations need such a 'stick' in order for them to progress?
5. The owner's staff changed during the project. How might personnel changes affect the progress of negotiations? How might personnel changes within your team and within the other party's team, be used to your advantage?
6. What advantages and disadvantages to the owner or contractor are there in bundling claims rather than dealing with each as it arrives?

6.2.12 CASE STUDY — MINING PLANT

This case study describes a situation between a construction contractor and a mining company (owner).

The contractor carried out design-and-construct work to upgrade the owner's plant. The contractor is said to have lost money on the project and sought to recover this from the owner.

The contractor and owner were already familiar with each other's organisations and businesses and were intimately familiar with the plant and its performance.

The contractor's negotiation preparation identified all real and not-so-real causes of delays and extra work, which were to be charged to the owner because of:

- Supply of faulty data.
- Supplying no data.
- Failure of the owner's existing plant.
- Failure of the owner to rectify problems in the existing plant and so extending the commissioning phase.
- Changes forced into the design by the owner's requirements post-agreement.

The review involved the collecting of information from all possible sources such as:

- Interviews with supervisors and consultants.
- All site diaries.
- All requests for information made to the owner and the responses in regard to timing and content.
- All details of the contract, to find any possible loopholes or areas of dispute.
- All submissions and subsequent approvals, to locate delays or changes.
- All drawings supplied by the owner.
- All drawings prepared by the contractor in areas affected by changes or incorrect data from the owner.

This enormous amount of information was put before the owner to substantiate a claim which was highly inflated. The presentation of the claim and the supporting data was the contractor's first move in what it knew would be a long and expensive process.

The first meeting resulted in the owner totally rejecting the claim and denying all responsibility. However the owner was now forced into its own defence preparation. It assembled a team of contract and technical engineers to refute all the claims, or to at least throw doubt on their validity.

The owner's engineers also began a detailed analysis of the plant with regard to compliance with the contract, and defects in design and workmanship. Every item from the trivial to the major was documented as a defect and presented to the contractor for rectification. This had a number of goals:

- To make the contractor look as incompetent as possible in the eyes of outsiders and to embarrass higher managers and directors now involved in the situation.
- To apply financial pressure on the contractor which now saw the possibility of losing yet more money.
- To get one of the goals of the work satisfied, namely, that the plant be a reliable and efficient unit producing to specification into the future.
- To prevent any finalisation of the contract and to maintain control over the contractor's bank guarantees.

Also, all further payments and the certificate of practical completion were withheld, and the owner threatened to carry out certain work and to deduct the cost of this from payments due to the contractor.

The contractor acceded to the pressure with regard to defects and mobilised a team of people back to site to carry out rectification work.

This was something that both parties realised was essential. The owner wanted what it asked for in the plant, and the contractor could not stand the stain on its reputation. As well, the owner had the power under the contract to enforce such rectification work, or to contract other parties to carry out the work and to backcharge the contractor.

Progress on negotiation over the main claim had stalled. To force the issue, the contractor hired a legal team to prepare a legally substantiated claim, which could be registered in the courts, and so force the owner to take the matter seriously. This also helped the contractor to see that some of the grounds for its claim were shaky indeed as the lawyers raised pertinent questions and demanded real facts. The 'he said, we said, you said we did, they did' issues began to fall away and the contractor got a better view of what the reality of its situation might be.

The owner began the same process with the same results.

Meetings continued in an attempt to produce a common position between the two parties. The plant continued to be rectified, and so the owner was having its main need satisfied. The contractor had lost a lot of money, but it had a way to go to justify what part of that was truly chargeable to the owner. Both teams put various proposals to each other, but to no avail. Meanwhile both parties were calculating the cost of running the dispute (the costs of engineering, administrative and legal input). These costs slowly eroded the difference in dollars over which they were arguing.

Exercise

1. What value would there be in calculating the number of hours of running a dispute at which the cost of running the dispute exceeds the original cost of the dispute? How could such a number be used to drive or stall negotiations?

2. Introducing legal argument into negotiations almost certainly escalates the dispute. How could such a tactic be used to drive or stall negotiations?

3. Disputes are commonly over extra payment wanted by the contractor versus defective work against the contractor. Why does such a common scenario continually arise in projects?

4. Presenting a claim to the owner with substantial documentation might have what effect on the owner?

5. The negotiation proceeded without a black and white analysis of the contractor's situation. How might this affect the contractor's outcome from the negotiation? Can you conduct negotiations without knowing the real facts?

6. An initial rejection of a contractor's claim is a common tactic. Where does this lead a negotiation?

CHAPTER 7

Negotiation — Case Studies II

7.1 OUTLINE

A number of case studies illustrating the phases negotiations go through, together with various negotiating practices are presented from a range of industries and situations. More than one culture was present in each case study.

7.2 CASE STUDIES

7.2.1 CASE STUDY — MICROCONTROLLERS

The situation

Stevens Moonymt had introduced a new product in its microcontroller range called the UY. This product was extensively marketed throughout the world and was aimed towards markets that already used Stevens microcontrollers as well as competitors' microcontrollers. Norbets took a particular interest in the UY, because it relied on high quality top-end components for its military based equipment. Eventually through many presentations and tutorials by Stevens Components (overseas) based on the UY, Norbets decided to design the UY into its equipment. The UY was the 'heart' of the equipment, and the remainder of the components were designed around the UY. Stevens gave samples to Norbets, and the prototypes were accepted by the local defence forces, resulting in a contractual agreement signed by the defence forces and Norbets.

Norbets placed orders on Stevens Components for this part and these orders were accepted by the factory. As it so happened, Stevens Moonymt had a directive to stop production of these UYs which were extended temperature components. This obviously shocked both Stevens Components and Norbets, and negotiation took place with regards to what would happen, considering the multi-million dollar contract with the defence forces. Several meetings involved accountants, the purchasing staff and engineering staff of Norbets. Stevens Moonymt was firm with its directive of not manufacturing the industrial temperature components, but were confident that the standard temperature parts would function properly in industrial temperature environments. Negotiation took place between Norbets and Stevens Components to discuss how the issue could be resolved. All Norbets wanted was to have the correct parts shipped to them in the quoted lead time. Stevens Components wanted to make Norbets happy, and to continue a strong relationship with the company. Stevens Moonymt did not want to produce the industrial temperature parts, as that would mean that it would have to invest in some expensive testing equipment which it did not have. This testing equipment was worth much more than the total order placed by Norbets for the UY components. Stevens Moonymt also wanted to resolve the issue and make Stevens Components happy with a result which would make Norbets happy.

How the negotiation proceeded

Considering that this negotiation consisted of three parties, where Stevens Components was not only involved but acted as a mediator between Stevens Moonymt and Norbets, a great deal of background information needed to be accumulated. Stevens Components held several informal meetings with Norbets to determine what Norbets' goals were, and what would be the minimum acceptable result of the negotiation. It was understood that there was no other option but to have the UY component delivered to Norbets according to specification.

Stevens Components had decided that it would be best to meet Norbets in its offices. Present at the negotiation from Stevens Components were a number of senior managers. Norbets had its key decision makers present. The meeting was held in Norbets' offices and proceeded by Norbets' director taking charge and getting straight to the point. The problem of not manufacturing the UY was discussed in terms of financial losses to Norbets as well as loss of business with Stevens in the future, and breaking a contractual agreement with the defence forces. This alluded to the fact that Norbets may have to use legal action if Stevens could not supply the goods. Stevens Components quickly offered a solution stating that 'If Stevens Moonymt are so confident that its standard temperature UYs will work in industrial temperatures, then Stevens Moonymt should provide proof of this in writing.' This was a sufficient solution for Norbets, and the meeting was adjourned so that Stevens could actually get the documentation stating the fact of industrial temperature specification compliance. This letter was finally acquired by Stevens Components and presented to Norbets which were duly appreciative with the solution.

Analysis

Preparation phase. Stevens Components gathered information on what it would cost Norbets if the UYs were not supplied. Costs also involved possible consequences in terms of breaking a contractual agreement with the defence forces. Stevens Components understood that no other part, from Stevens, had the exact same function as the UY. That is, there were no electronic component options available to Norbets. There were not many options for either party. It was decided that the negotiation was to be held at Norbets. This was done to give Norbets a feeling of control and security. No agenda was prepared, as the issues involved were quite succinct and both parties were keen to gain a resolution. An adlib attitude was preferred by Stevens.

The *familiarisation phase* involved introductions, and as there was an already established relationship, the director of Norbets promptly began the negotiation.

Opening phase. The director of Norbets introduced the topic of discussion by a brief review on why it was so important to Norbets that the UY be delivered and she discussed costs and legal aspects, that had been correctly anticipated by Stevens. The meeting proceeded with much open-ended questioning on Stevens' part, reinforcing much of the preparation phase information gathered.

Proposal phase. Due to the simplicity of the negotiation, a proposal was promptly suggested by Stevens to Norbets, and was duly accepted. If the proposal was not accepted by Norbets, then it was envisaged that a serious problem would occur because there were no other possible proposals or solutions for the issue. Stevens in this case was only prepared for one path for the discussion to flow. This did place Stevens in an awkward position, but similarly, Norbets had no choice but to accept the proposal.

A *bargaining phase* was not evident, although a condition for the proposal was stated by Norbets that proof must be provided to it of the UYs meeting specifications. Stevens agreed to the condition and wrapped up the negotiation by reaffirming the stance made by Stevens and Norbets and that relationships between the two companies would not deteriorate. This was a tactic ensuring trust.

A major role was played in *post-negotiation* by following up on the promises made and ensuring that the goods were delivered to Norbets, and also ensuring that Stevens Moonymt kept its word of testing the UYs at industrial temperature.

Exercise
1. Both companies involved in this negotiation were multinationals with employees from many countries. Besides the terseness of the negotiation, what else reflects the business cultural backgrounds of the participants?
2. What moves can you identify from each party signalling to the other continuing relations were to be an outcome of the negotiation?

7.2.2 CASE STUDY — LOCAL PARTNER

Introduction

This case study relates to a negotiation between two business partners (a software developer and a supplier), the outcome of which was not successful, and resulted in a parting of the two organisations.

Background

The software developer had its ASEAN regional headquarters in Singapore. As work was throughout most of the ASEAN countries, it was too difficult to have an office in each. Typically local partnerships were developed to assist in sales/marketing efforts and project implementations.

On a project in a neighbouring country, the company was selected by a power corporation as the prime contractor to provide computer hardware, network services and systems integration (SI) services in addition to normal software and business services. The contract also required the company to purchase a number of vehicles, office furniture, photocopiers, handphones, facsimile machines etc. To successfully tender for, and win this work, the company engaged a local supplier (the partner) for the hardware, networking and system integration aspects of the project.

The company had worked with the SI partner on previous projects in the country and had a good relationship with it, in particular with a number of its senior staff. For this project, in addition to the SI services it would supply, the partner would also be responsible for the coordination and purchase of all the vehicles and office furniture due to its local knowledge and contacts. Local law made it impossible for the foreign company to purchase such items itself because it did not have a local company presence.

The lead up to negotiation

At the time of being awarded the contract, the company involved the partner to ensure that it was agreeable to the terms and conditions which would apply to the contract being entered into. It was agreeable to the contract conditions because the conditions were standard conditions applied by the power corporation, and it had worked with the corporation on prior occasions.

The commencement date for the project was to have been January. Due to some internal re-organisations within the corporation, approval for the commencement of the project wasn't received until July. During the delay, several of the project implementation locations were changed. This required some variations to the contract. In preparing the cost variations, the company also had the opportunity to go back to its partner and discuss with it any need it may have to alter its original prices due to the six month delay. The partner agreed to hold to its initial price proposal.

One of the agreed terms with the SI partner was that it would purchase all the vehicles and office furniture necessary for the project, and would be reimbursed as payment was received from the corporation.

During the six month delay in commencing the project, the chief executive officer (CEO) of the SI partner was replaced. When the time came to commence work on the project, the SI partner said that it was no longer able to work on the project at the agreed hourly rates, and would not be able to purchase any of the project equipment, such as the vehicles and furniture, prior to the company paying it the funds.

How this negotiation proceeded

The initial reaction to the partner's request was one of disbelief. The company had been waiting for six months to commence the project and now with a start imminent, the partner was wanting to renegotiate the conditions under which it would work.

The key people involved in the project from the company were brought together for a meeting in Singapore. They looked at the options, at the partner's requirements and motives behind the request:

*   The partner was asking for an additional 10% in the hourly rate — was it justified?
*   It could well be, as it had been working with the company at the lower rate for the past two years.
*   Could the company afford to pay the increased rate?
*   If the company had known earlier it could possibly have factored the increase into the contract variation. If the increase was paid now the company would have to shoulder the extra cost, which would reduce the profit margin and the contingency margin to an unacceptable level.

- Could the company purchase all the project equipment without the partner?
- The company couldn't purchase the equipment directly itself, however it could do it through another partner. The company would have to wear the initial purchase costs plus interest until payment was received from the corporation.
- Why was the partner doing this now?
- It knew that the project was due to commence and the company had limited options if its request was declined.

The people, within the partner's organisation, with whom the company had worked before and had a good relationship, were contacted. What was happening? Why these changes at the last minute?

The changes were part of a policy set by the new CEO. All partnerships were now subject to new conditions and all projects that had not yet commenced were expected to meet these conditions. In order to cover a number of overruns in internal projects, the hourly rates for external projects were being increased. It had also decided that it no longer wanted to carry the overheads and risks involved in purchasing the vehicles and furniture.

The company agreed internally that it could bear some increase in the hourly rate, up to 5%, but it could not go past that. The company had no way of purchasing the equipment itself, and so the company must either work with this partner or find another partner.

A meeting was arranged later that week when the company's staff returned to the country. A request was made to meet with the new CEO and discuss requirements. At the meeting the employees, who the company had worked with in the past, were present and the company was told that the new CEO was not able to attend. They discussed the new policies and explained the directive from the CEO of the conditions that must be met for new projects. All talked about past projects together and a desire to maintain a successful relationship. It was agreed that its hourly rate had been very competitive and maintained at that level for the past two years. It was suggested that in future projects its new rates would certainly be built in, but as this project had minimal profit margin it could not accommodate an increase as large as the proposed 10%, and it could not carry the costs of the financing for the vehicles and furniture. The employees were sympathetic to the company's position but again stated the requirements of the new CEO. The employees agreed to discuss the offer with the CEO when she was available, and would possibly arrange a follow-up meeting.

Feedback from the meeting was that the CEO would accept no less than the 10%. As it had been agreed internally that that sort of increase could not be absorbed on the project, alternatives were discussed. The company had worked once before on a project with another consulting group that had established an office in the country, and had employed about twenty or so college graduates and experienced SI people. It was called and the situation explained; the people sounded interested because one of its projects had been delayed. The issues were discussed with the other group and it agreed that it could provide the services at a lower rate than was originally proposed by the SI partner. These lower rates would allow the company to take out a financing arrangement to enable this other group to purchase the vehicles and equipment. It was agreed that the company would go back to its original partner and if satisfactory conditions could not be agreed then the company would work with the other group.

A meeting was eventually arranged with the original SI partner and the new CEO. She was adamant that any new agreements that the partner entered into from now on would be following the new assessment criteria. During the discussions she explained how some additional work had been won and how very busy her people would be for at least the next year. The offer, past successful projects and future opportunities were again discussed, but she remained firm. The offer was then made to increase the hourly rate by the 5% as had been discussed, if the partner would agree to buy all the vehicles and office furniture. She again declined the offer. Knowing now that an alternative solution to the problem existed, the company was able to stick with its offer which surprised her. The company left the meeting explaining that it was disappointed that the partner was not able to perform the work as had been initially agreed and that the company doubted if it would work with it again.

Exercise
1. How much of the course of this negotiation could you put down to cultural influences?
2. There was, in effect, a subcontract agreement between the company and the partner. Would there be any point in pursuing the legal side of the breach or using this argument in the negotiation?
3. The new CEO was a local woman. Might this circumstance have influenced the way the negotiation proceeded?
4. What cultural sensitivities should the business partners have been aware of in the other?

7.2.3 CASE STUDY — DESIGN OVERLAP

In his role as a design engineer Edward was often involved in discussions related to obtaining additional recompense from owners in order to cover the cost of required additional work. This necessarily involved negotiation in order to come to a resolution that could be accepted by both parties.

This case study is from a project recently completed for a foreign-based company.

During the course of the project a misunderstanding developed about the design work being carried out on a blast-resistant building. Edward's company (EC) was employed by a foreign-based company (AC) to carry out the design of a building within its main project. AC was designing other parts. EC was awaiting blast loadings from AC so that it could complete its work. After some considerable delay, EC received from AC a package which contained designs for the building as well as the blast loadings, even though this work was clearly within EC's scope of work. EC had already carried out a certain amount of work on the building design believing that the information that EC received from AC would merely serve to confirm EC's own work. AC were, however, insisting that EC must adopt the structural arrangements as sent.

EC had been employed on a lump sum basis and had already consumed a considerable portion of its budget on producing drawings. EC therefore did not wish to change its design and documentation as a result of the work carried out by AC. Therefore, negotiation was required to resolve the conflict.

The first step taken was to collate all the background information involved in the issue at hand. This meant going back through all correspondence that had passed between EC and AC to ascertain that EC had acted in accordance with the agreement. It was important that EC based any arguments on accurate and as-complete-as-possible information. Having completed the compilation of this information, EC had a series of meetings internally to discuss the issues and to determine its position.

EC was then invited to the head office of AC to discuss the dispute. Edward travelled to the head office of AC with the project manager. The trip was also intended to be for discussion of other aspects of the project in order to avoid similar occurrences in the future. Edward spent time before the meeting with EC's project manager discussing the approach that EC would take. Whilst the project manager was primarily responsible for any negotiation over financial matters, Edward was involved in the preparation of the strategy. EC's basic aim was to take an assertive and matter-of-fact approach to the meeting. EC listed the additional work that would be required to redo the design and drawings that had already been completed, and estimated the additional costs involved. These estimates would be presented to AC as the basis of a claim for additional money required.

This approach put the AC personnel on the 'back foot' and in hindsight made EC's negotiating position very strong because AC was very keen to avoid additional costs. As EC did not wish to be forced to carry out additional work for no recompense, EC had two desirable outcomes. EC would require additional money to complete the work or AC should complete the work that it had already carried out. It was important that EC had more than one desirable outcome because this gave EC flexibility to respond during the impending discussions. Initially discussions centred around the extra money that EC would require to redo the design. This met with strong resistance and lengthy discussion over the items that EC believed had to be accounted for. The second option was then presented after the negotiation had come to something of an impasse. This was seen as a solution that both sides felt comfortable with. As AC was very keen to avoid any additional costs, it accepted this proposal. Cultural factors may also have influenced the decision, as AC personnel were very keen to avoid losing face in all of their dealings.

After agreement had been reached about the issues at hand, a great deal of time was spent documenting that which had been agreed to. This part of the proceedings possibly took longer to conclude than the actual negotiation. Edward wrote a large number of minutes from the meeting and all parties to the discussions signed. This document then became the basis of an agreement and future work.

There were different stages to the negotiation and also differing 'states of play' throughout. Initially EC was not aware of the likely attitude that EC would encounter. Equally AC did not know how EC would approach the negotiation. Therefore, when it became apparent that EC had the advantage, the 'state of play' changed from one of equal footing to one of control on EC's part. This allowed EC to put forward its views favourably. This negotiation could have been described as 'competitive' in nature, although it changed to collaborative discussions once common ground was found. This allowed a win–win situation in which both parties felt that they had achieved desirable outcomes.

Negotiation phases

The negotiation went through a number of phases.

In the *preparation phase*, EC spent considerable time and effort collating information and formulating strategies. EC was not able to control some aspects of the preparation, namely location and conditions under which the negotiation was to take place. However, this did not present a large problem and may have been a potential advantage as it would have served to put AC at greater ease, being in its own familiar environment.

During the *familiarisation phase*, the usual pleasantries were exchanged and people who had not previously met were introduced. This served to put participants at ease with each other as some considerable tensions did arise during previous discussions.

In the *opening phase*, EC's first approach was to take the lead and attempt to set the tone of the discussions. During this phase it was apparent that AC was desperate to avoid incurring significant additional costs. AC did not attempt to argue that EC had done anything wrong in the matter. AC appeared to accept that it bore the responsibility for the misunderstanding. This was a key indicator for EC that it had the upper hand from the outset.

In the *bargaining phase*, AC was quite keen on the alternative proposal when it was presented, and quickly agreed to it. EC's project manager had exercised considerable skill in gauging the tone of the negotiation and recognised the right time to offer the second proposal. By realising that AC would not budge on the original proposal, the project manager shifted ground to a compromise solution, thereby allowing AC to feel that it was getting a good deal.

Discussions then proceeded to an *agreement phase* and all the terms were documented. After the final agreement was documented, both parties to the discussion re-read the entire agreement to satisfy themselves that they were happy with it.

Progression of the negotiation

The negotiation passed from one stage to another through the need of both sides to feel that they were progressing towards a desirable outcome, and their preparedness to address the issues. When the discussions came to an impasse there was an incentive to find a compromise that would allow discussions to continue.

The discussions needed to head in a compromise direction because a point of impasse had been reached. Therefore, the proposal of a new option, which EC still considered to be a desirable outcome, had to be presented. Otherwise, the negotiation may have broken down altogether.

The recognition of when was the right time to progress to the next stage of the negotiation was also a very important part of the process which was closely tied to tactics and sensitivity to the discussions in such situations.

Exercise
1. What cultural traits (all cultures) were present in this negotiation?
2. Do you see this negotiation progressing in an alternative direction had different cultures been involved?
3. What stance could EC and AC have taken to end up with outcomes more favourable to them?

7.2.4 CASE STUDY — DELAYED PAYMENTS

This case involves the civil construction work for a power station.

The initial planned period of construction was nine months. The owner was an electricity department.

During this period the electricity department was facing a cash flow problem, and this resulted in non-payment of periodic certificates for work completed. The flow-on effect was a cash flow deficit for the main contractor, resulting in delays in material procurement, payment of subcontractors and payment of day labour. This culminated in delays to the project and consequent periodic claims by the contractor for production losses due to ineffective utilisation of labour and deviations in the construction program to use available material till finances were made available to the contractor for work completed. The claims were for approximately half the initial contract price and the expected project overrun was twelve months. Delays in payment of approved certificates exceeded six months, upsetting the main contractor's cash flow.

Other constraints included:
- Limited number of skilled local subcontractors available at that time.
- Labour had to be imported by the contractor, on a two-year individual labour contract. A delicate balance was required between labour requirements on the power station project and other projects.
- The possibility of shedding labour was small, because visas were limited and the contractor could not import labour again for the same project once the labour was repatriated.

The contractor's negotiating team consisted of members with commercial, technical, legal and financial backgrounds. To use cultural influence the contractor matched its team with members from the same cultural background that the owner was expected to have.

Background

The work commenced in April (year 1). All was well until August when the first delay in the periodic payments started and when the owner began to face cash flow problems. This resulted in cash flow problems for the main contractor. The result was an inefficient use of day labour, lack of materials, changes to the original construction program to utilise available resources and material so as not to stop the work, and additional bank charges on bank overdrafts for the contractor.

The contractor felt that there would be serious delays to the project. There were additional costs of housing and paying the imported day labour force for an extended period, plus all the other costs connected with a project faced with time overruns. The contractor acted fast and began presenting a claim for extra expenses caused by the owner not meeting its contractual commitment of periodic payments. The claim was logged with each monthly progress report. Thus documentation for the claim was progressively built up. The owner made no comments regarding the claims, and conveniently ignored them.

The situation worsened when the project was about 90% complete. The contractor had only received 60% of the due payment and the project overrun was six months. The first meeting to discuss and negotiate the claims was called in August (of year 2). The contractor had prepared summary documentation of its claims including:
- Planned and actual progress curves; cash expectancy and actual cash receipt curves.
- Projected budget analysis up to December (of year 2).
- Projected labour payment statistics up to December (of year 2).
- Projected status of payment certificates up to December (of year 2).
- Projected payment status of money received up to December (of year 2).

These documents were revised periodically before each negotiation meeting to reflect the current state of affairs.

Common local practice was to conduct a lot of business on trust, with disagreements settled through negotiation and arbitration. There were very rare cases of litigation. Many negotiations were settled part officially and part in-kind 'off the record' in terms of a favour, some consideration or a word in the right place at the right time.

Phases of the negotiation

Preparation phase. The contractor had anticipated negotiation and had prepared well when each periodic claim was submitted by inflating the claim values to give a margin for concessions during negotiation.

The contractor had also planned for an eventual loss and began to use the day labour force on other projects.

The main goal for the contractor was to get the claim and periodic payment expedited to enable it to complete the project. The alternative was to negotiate and trade-in part of the claims for extra work from the owner. The worst alternative was to stop work and walk off, thus cutting the losses, and to settle in an international court.

The owner's primary goals were to get the contractor to complete the work with the prevailing conditions of payment delays (due to the owner's cash flow problems) and not pay any claims. The alternative was to negotiate the claims and drag out the periodic payments to help its cash flow.

The order of priority of the owner was to get the work completed as soon as possible, because other projects and stages of development were being delayed, and to drag out the periodic payments. The next on the priority list was to settle the contractor's claims at least cost and retain international goodwill.

The contractor's order of priority was to get as much money in claims, get accelerated periodic payments to enable it to complete the work, and enhance its goodwill in the country.

The owner endeavoured to check the validity and value of the claims of the contractor.

How far could the owner push the contractor before it would walk off? The contractor knew that the owner's financial position was bad and there was little the owner could do to improve it immediately. The contractor had to evaluate the risks involved in the contract being terminated and the contractor being blacklisted. The contractor would one day receive the payment, but at what cost? How far could the contractor push its claims?

The owner could threaten the contractor by terminating the contract if the project was not completed immediately and blacklist the contractor. The tactics would be bluff, threats, tactical use of walkouts and make the contractor look unreasonable.

The contractor, on the other hand, would start from a collaborative style and now and then use a competitive style, threatening that its limit had been reached. The tactics employed would be good guy–bad guy, changing members of the negotiating team, and inflation of the claims by double to enable the contractor to make concessions in lieu of additional work. Another tactic was to utilise the labour on other projects while claiming on the power station project. This would provide a buffer for negotiation.

Options to the owner were to negotiate the claim, pay swiftly the money due or have the contractor walk off or terminate the contract.

Options to the contractor were to make concessions, complete the work, maintain goodwill and, as a last resort, walk off the site.

The deadline for the owner was June (of year 3) when the power station extension was to go live. If the civil work was delayed, this deadline could not be met even though the mechanical and electrical work was being run concurrently. The deadline for the contractor was March (of year 3) when another large project was to start and the labour was required to be diverted to that project. Then there was the further extension of the power station — the next stage of development; quotes for these would be closed in May (of year 3). The contractor had to complete the project to be eligible to quote for this extension work.

Each team had members from various cultural backgrounds. Each nationality had its own style and body language.

The location for the negotiation was to be the electricity department's meeting room. The first meeting was to be held in August (of year 2), and thereafter on a monthly basis to coincide with the monthly progress meeting held on the first Wednesday of each month.

The claims meeting agenda was to follow the progress meeting agenda, and only the negotiating team members from each side would participate in the claims meeting.

The members in each negotiating team were senior management personnel with the necessary authority.

Familiarisation phase. Team members from both sides were known to each other culturally, as the contractor had matched its team members with the cultural background of the owner's members.

Opening phase. The owner's prioritised wants were to firstly get the project completed, secondly to extend the period to disburse the payments in order to manage the poor cash flow, and thirdly to minimise the claims. The contractor's priority was to get payment first and be able to complete the project as soon as possible, to maximise the payment of the claims, and lastly to negotiate the claims to a minimum of about two thirds the original claim, and thus maintain goodwill.

The contractor swamped the owner with documentation, properly tailored to justify the claims.

Bad international publicity put pressure on the owner to expedite the settlement of the claims, plus the delays on the project would affect further expansion of the station.

Being a matter of claims, the contractor had to state its wants first. A written proposal was submitted including all the individual periodic claims.

Proposal phase. The contractor presented one claim at a time, and provided documented justification for each claim separate from the extra work variations that were approved, discussed each individual claim with the owner's team, compromised to prevent a deadlock, and obtained an agreement from the owner in principle.

For some of the claims, the owner asked the contractor to re-present them, and decisions on these claims were left to the next meeting.

It was clear to the contractor that the owner had a serious cash flow problem. The owner would ultimately make payment. The owner's primary concern was to get the project completed. The owner was bluffing to apply the penalty clause, terminate the contract and blacklist the contractor. Though unstated, the owner realised and appreciated the contractor's position. The owner would consider negotiation of the claims. The owner (to save face) was required not to accept all the claims. The owner would ('off the record') negotiate part of the claims for other additional work. The contractor would have to wear some claim deduction.

Bargaining phase. The contractor asked for additional due payment to enable it to complete the project, failing which the contractor would be forced to stop the work as it could not get any more overdraft from the banks. With regard to the claims, the contractor was willing to settle some of the value in kind through additional work.

For the contractor, the proposal and bargaining phases overlapped, as each claim was individually presented and followed with bargaining.

The contractor felt that the first concession would be to accept part payment for the money due and to agree to complete the project swiftly.

The contractor negotiated each claim independently indicating when it had reached its limit of bargaining. As each claim was individually agreed to in principle, the contractor knew that the owner would finally bargain again on the total value of the claims.

The contractor new that the owner would become more desperate, as time passed, to get the project completed and would agree to the claims of the contractor. Delay tactics were used by the contractor after the November meeting to force the owner to agree to the claims.

The contractor had inflated its claims to provide a margin for negotiation. During the negotiation, the contractor convincingly stated that it was making a big loss and pleaded for the owner's sympathy.

The contractor used the technique of a stage-managed walkout, during the December (of year 2) meeting when the claims were being discussed at a summary level. By this time the claims were down to approximately two thirds the original claim. The contractor displayed its disagreement, staging a walkout, and asked the owner to think it over. The real concerns of the owner became evident when the owner asked for an additional meeting within a week. The contractor agreed.

The owner of the contracting firm took the place of the team leader. He was locally born, well-travelled, educated in Europe and spoke several languages. He was well known as a fair negotiator and, being the owner, his presence carried clout. Above all he was a good dramatist.

The owner agreed in principle to pay half the original claim on a summary basis, although on an individual basis it had agreed in principle to about two thirds the original claim. The owner stated that it could not approve such a large percentage and did not want to set a precedent. (However it would make it up to the contractor in-kind in another situation — this was an 'off the record' deal with the owner in private.) The contractor then agreed to split the difference between these two figures.

The contractor agreed to complete the project swiftly on receiving at least half of the due payment and did not mind to wait for the claim money. This made the owner very happy.

Agreement phase. The meeting in the second week of December (of year 2) was fruitful, as the owner was forced, due to time constraints, to agree in principle to the contractor's offer after negotiation. Final agreement was reached during the meeting of January (of year 3).

The owner agreed to release within two weeks half of the money overdue to the contractor to enable the contractor to complete the project, and pay the agreed amount in claims as soon as the owner's cash flow position improved and at the latest within a few months.

A verbal 'off the record' agreement was made with the owner to give the contractor extra work, and due consideration to the contractor's cooperative behaviour would be taken into account in future development work.

Post-negotiation. The owner, as well as the contractor, was satisfied with the result — a win–win situation. The contractor broke even on the project. The contractor got compensated with extra work through 'off the record' considerations.

Exercise

1. What aspects of this negotiation are peculiar to local practices as opposed to being common elsewhere?
2. The negotiation teams were made up of people from many countries. Could there be such a thing as international practice in negotiation?
3. The intrusion of the owner of the contracting firm seemed to make a significant contribution to the negotiation settling. Why might this be?
4. How important do you assess the 'in kind'/'off the record' part of the settlement?
5. Is it possible to have the above situation and for both parties to still part amicably with a retention of good business relationships?

7.2.5 CASE STUDY — MECHANICAL SERVICES

Background

This case study outlines aspects of a negotiation on a hospital construction project between a mechanical services subcontractor and the owner's representative (consulting engineer) and the main contractor (builder).

After the mechanical services contract had been awarded, clarification by the architect and the consulting engineer of the specification and several addenda were required. The original building management control system (BMCS) scope of work and contract price had changed throughout the tender negotiation.

The consulting engineer advised the builder to allow a provisional sum for load shedding of future generators, as the future stages had not yet been finalised. The builder included such a provisional sum in the original tender price. As the price was altered through the tender negotiation, the provisional sum somehow became omitted. The work was to be done under subcontract to the builder.

The negotiation

After the subcontractor's estimating division had accepted the subcontract with the builder, the project was handed over to the subcontractor's construction division. At this stage, clarification of the scope of work and the associated cost was required.

It was found that the load shedding for future generators, as described in a contract addendum, had not been included in the final BMCS quote. The consulting engineer was informed that the load shedding had not been included in the final quote.

The consulting engineer instructed the subcontractor that load shedding was required and to look at areas where cost savings could offset the cost of the load shedding variation.

While the contractor was working out the variation for the load shedding, the consultant submitted cost reductions from the reduced scope of work negotiated between the consulting engineer and subcontractor.

At this stage of the negotiation, the consulting engineer was replaced, due to the expiration of his contract. A conflict arose when the new consulting engineer did not approve the submitted variation as previously agreed by the first consulting engineer, and asked for the BMCS scope to be reduced to allow for the variation.

Two possible approaches were available to the subcontractor at this time:
- Enforce the agreement that was made with the first consulting engineer.
- Redefine the existing control system with possible cost savings.

After discussions with the contractor, it was believed that there were areas of the control system which were unclear and had the potential for possible cost savings.

At this stage the subcontractor believed the second option would be the better. The second option gave the contractor and the subcontractor the opportunity to redesign the control system creating a more appropriate and economical control solution, whilst satisfying the second consultant's requirements of no extra cost to the owner and producing a more reliable control design.

Exercise
1. The subcontractor had a good understanding and working relationship with the original consulting engineer; familiarisation was already established. However, little effort was put into becoming familiar with the new consulting engineer to gain a better understanding of the new engineer. How might this have affected the change in the course of the negotiation?
2. The same proposal offered to each engineer resulted in two different responses:
   - The first engineer was open to negotiation and the possible reduction of mechanical or BMCS scope of work to offset the load shedding variation.
   - The second engineer responded with a blunt 'No' to the additional cost and specifically wanted the BMCS scope of work reduced.

   How might the contrasting responses have been the result of different negotiation skills or perceived pressure from the owner?
3. An element of good bargaining is to trade that which is cheap (relatively) for you to give and valuable to the other party to receive, in return for that which is cheap for the other person to give and valuable for you to receive. The trade-off for the load shedding control was relatively cheap from the subcontractor's perspective, because the BMCS contractor was already committed to the BMCS system. From the consultant's perspective, the modification to the control system was a cheap trade for the valuable load shedding. Is it always the case that each party knows which is the best trade for the other party? Under what circumstances would it not be the case?

7.2.6 CASE STUDY — CONSULTANT TO A CONTRACTOR

The case study involves a negotiation between a contractor and its consultant.

The contractor and the consultant were part of a team working on a design-and-build project, with the contractor in the role of a client for the consultant. The consultant's organisation included units responsible for the design of different parts of the work such as traffic management, structures, and electrical and mechanical work.

The contract between the consultant and the contractor, and the contract between the owner and the contractor, were lump sum. The scope of work for the design was defined without an exact indication of the number of drawings. Other required submittals were also described without clearly indicating their number. Therefore a certain experience was needed in the planning phase to assess the quantity and the duration for each design package. A construction program (based on the tender program) was prepared by the contractor. It was used to define the latest finish date constraints for the design program. The idea behind these constraints was that all drawings needed for certain work on site had to be approved by the owner before the start of any related work in the construction program. Although no exact approval time was defined in the contracts, an average value for the approval time was assumed based on previous experience.

The consultant's work (and program) was broken down into a number of packages according to the location, for which the drawings were needed, and the structure/type of work they related to. For example, the design for the diaphragm walls was divided into a number of packages; one package was a complete set of documents for the diaphragm walls in one location. This design program had been submitted for checking by the contractor. If approved by the contractor it would then be submitted to the owner.

The design work started and in a short while it became clear that the consultant was behind its own schedule. After updating the design schedule with the current rate of progress, it was found that without acceleration or some changes in the sequence of the design work, achieving construction start dates could be a problem. A meeting was held where the contractor's programming staff explained the above concerns to the consultant and proposed some possible changes in order to ensure that all target construction start dates would be met. The consultant promised to look into the matter and give the contractor a quick response. The contractor was not aware that this meeting was just the start of a negotiation that was to end up in an unexpected way.

After approximately one week, the consultant came back to the contractor with some proposals, which were explained to the contractor's scheduling team during the next meeting. The consultant proposed to split several design packages into a number of smaller ones, assuming that a more detailed breakdown would help in optimising the resources (mostly draftsmen) available, because the consultant's schedule was basically resource-driven, with constraints derived from the construction program.

When the contractor checked the newly proposed schedule, the contractor found that there were several smaller design packages in it where the construction start dates were not achieved. In addition, some of the new packages were, in the contractor's opinion, not necessary, because they and their purpose were not clearly defined, and this could lead to the confusion with the owner and make the overall approval time longer.

Among the items which delayed the start of construction work were activities related to traffic management design. Accordingly, the next meeting with the consultant was attended by the consultant's person responsible for traffic management design.

The contractor prepared thoroughly for this meeting, analysing the consultant's program. The contractor also investigated some alternative sequences of design work, trying to understand the reasons for the delays mentioned above. The contractor found several possibilities to resolve the conflict between design completion and start of work on site, without changing activity durations and relationships between activities. The idea was to change some sequences and to start some of the design activities earlier.

At the beginning of the meeting it was mentioned by the contractor that the present design submittals ended up behind the existing schedule and this trend seemed to grow. The consultant's reaction to this statement was unexpectedly very negative. It felt that it was an insult, and it took the contractor some time to convince the consultant that the contractor was just pointing out the problem, and that the contractor was very keen to find a solution together with the consultant. Later the contractor reported on drawbacks it had found in the design program and showed how it could be modified to achieve all construction deadlines. The contractor's approach was met negatively. The consultant said that the contractor was trying to intrude on the consultant's work and that according to the consultant, with its available resources, the proposals appeared to be realistic. However, the consultant promised to consider them and give the contractor an answer at the next meeting.

For that meeting, the consultant prepared a proposal which included earlier start dates, some sequence changes and even the reduction of some durations. The contractor was surprised by and initially suspicious of such an about-turn, but after a short check the contractor found that all of the construction deadlines were achieved and the time for the owner's approval of the drawings had not been changed. Both the contractor and the consultant were satisfied with the present changes in the program and the whole disagreement disappeared.

Several days after the last meeting on the design program, the contractor's scheduler talked informally to a person working for the consultant. The scheduler wondered what was the reason that made the consultant change its mind so abruptly. The answer was that the consultant, at first, didn't understand the contractor's ideas. It thought that the contractor just wanted to compress its design time and 'push' the consultant without really understanding the consultant's problems.

Exercise
1. The contractor and the consultant knew each other well and this negotiation was not the first one between them. Proposals and opinions were expressed openly and directly, often without considering the emotional part of the negotiation and its influence on the participants. Why did emotion play such a negative role in the negotiation?
2. The contractor stressed the technical side in its preparation, often neglecting the human aspects of negotiations. For example, the attendance of a new person (responsible for the traffic management design from the consultant) was almost unnoticed by the contractor, although he was responsible for the issue of concern in this negotiation. The impression that the contractor's team was just trying to apply pressure was possibly the first one this person had of the contractor. Why did the human aspects play such a low role in the negotiation?
3. The contractor's team was not interested in the consultant's opinion and areas of common ground were not looked for. The reasons for some of the consultant's solutions were not clear until the end. How might communication practices have been improved in this negotiation?
4. The contractor's tactics during the negotiation were straightforward and its style was aggressive throughout the negotiation. At what point (the second meeting?) should

the contractor have understood that something was going wrong? How could realising that a conflict was slowly arising in the negotiation have helped in developing further actions? Should the contractor have seen that the consultant was confused by the number of contractor proposals without explanation? The contractor's behaviour put the consultant on the defensive and made its behaviour unpredictable, and this did not give any advantage to the contractor.

5. Was the contractor's team aiming, without realising it, at a win–lose solution? Why wouldn't all negotiations between a contractor and a consultant in a design-build team be totally collaborative? How would the following have assisted collaboration:
   - Clear explanation to the consultant of the problems connected with the present design program.
   - An explanation of the importance of finding a solution (a delay with the start of work could lead to an overall project delay).
   - An indication of possible solutions and their impact on changes to the design program.
   - Asking the consultant about its present and future resources.
   - Requesting the consultant to make some solution proposals.
6. Would defusing the consultant's defensiveness have facilitated a quicker solution?

7.2.7 CASE STUDY — FITOUT PROJECT

Documentation / players

The architectural drawings were prepared by a foreign consultant.

The mechanical and electrical design, as well as the specification and tender documents, were produced by a foreign property management company based locally. The company was also contracted to project manage the site work; this role was later reduced to that of assisting and advising on local contracts management practices and costing of variations.

The owner had other major work, property purchases and developments proposed.

The local building and fitout contractor, newly formed and engaged to do the work, was keen to establish a relationship with all parties, with the view for future work.

The project supervisor/manager/owner's representative was from the owner's organisation.

Project details

The project involved an office fitout on several levels of a multi-storey office block occupied by a number of groups/businesses. Temporary relocation of each group was needed while sections were refurbished/upgraded.

The dispute background

A dispute occurred over the number of power outlets/supply.

The owner's brief to the mechanical and electrical designer included upgrading the number of outlets in each office and at each workstation to four double power points.

The electrical drawings produced by the consultant were not specific. They didn't identify the locations of switchboards or show the increased number of power points. The owner's brief was included (cut and pasted) in the specification, as work to be carried out.

The contractor had only a very limited time to inspect the site and provide a tender. The contractor stated that it quoted on the electrical drawings.

The negotiation

The contractor submitted a claim for additional costs at the completion of the first of the stages for upgrading/increasing the number of power points. The contractor had installed outlets as shown on the drawings provided. This claim was rejected based on the grounds that the number of power points installed was not as specified. The design consultant agreed on this rejection. (Its motivation was possibly due to the fact that if the contractor didn't wear the cost then the consultant's firm may be in dispute with the owner for not following the brief.)

The contractor realised the implications of the rejection and the additional costs associated with the electrical work, and how this would affect its profit margin. On close examination of the contractor's submission, it became evident that no allowance was made for extensive electrical work. The contractor cried unfair decision, became very emotional, and talked of not being able to complete the work. It was also found that the location of switchboards would not suit partition layouts.

The contractor was requested to carry out a detailed survey of the electrical services and advise of its implications — how the different project stages would be affected by the locations of the switchboards, capacity of boards to cater for the number of increased outlets, and what boards were 110 V or 220 V. This detailed survey should have been carried out by the consultant prior to tender.

The contractor was led to believe that all items were up for negotiation and discussion, and work must proceed.

Halfway through the project, the contractor forced the issue to the surface by threatening to leave the site if a clearer picture was not provided. The owner agreed to pay for switchboard relocation due to partition relocation. The contractor agreed to continue to wire to international standards at no cost. Both parties agreed to review the costs associated with the increased number of power points as the project neared completion.

The cost provided for switchboards to be relocated was relatively high. A breakdown of costs revealed an excessive number of hours to perform the task. Generally the contractor performed very well throughout the complicated project. Being a new company, with this project used as a stepping stone, it became obvious that little profit was achieved. To walk away losing money would not be easy. A final meeting to resolve variations was held after completion. At this meeting many issues were resolved including power, but the owner would not pay the costs to increase the number of power outlets.

At this meeting it became obvious that all parties were keen to ensure that future relations were good. This helped tilt financial decisions towards the owner.

The following lists the steps in the negotiating:

Preparation phase. Meetings were held by the owner with the consultant prior to meeting with the contractor, to analyse the documentation and to confirm entitlements.

The consultant was not advised how flexible the funding was, or of the extent of funds that were available to resolve the dispute. The full extent of the relationship between the consultant and the contractor was never known. The timing of the meeting ensured little to no disruption to the owner's workface. The owner's representative advised that all issues would be discussed openly and fairly.

Opening phase. To save confusion, and to overcome a language barrier, the owner started the meeting by providing an agenda outlining the contractor's claims with the relevant specification clauses rebutting the contractor's claims. This silenced the room, although many of the disputed items had been well documented. Previous meetings became emotional with the contractor's representative crying on several occasions. (Was this her tactic?)

Proposal phase. Individual proposals were offered to the contractor and these tied groups of items together, including the power issue. Each group was treated separately on its merit. All discussions were directed to the company's managing director for a response.

Bargaining phase. The contractor, softened by earlier rejections, accepted fairly quickly what was considered by the consultant as reasonable.

Agreement phase. Minutes of the meetings were issued, as well as details of the different proposals, including what costs were associated with each item, and items rejected. All parties signed the minutes and the proposal. The contractor wrote a letter stating all claims had been resolved and no further claims would be made under the current contract.

Exercise
1. The negotiation was very long winded and extended for many months. The tactics used by the contractor included:
   • Ignorance.
   • Obstinacy.
   • Make the other party appear unreasonable.
   • Good guy — bad guy.
   • Appeal to emotions.
   • Blame the third party.
   • Bluff — threat.
   • Tactical use of walkout.
   During the negotiation, the contractor fell back to a 'make the other party appear unreasonable' position which was a surprisingly softer approach. Why would this be so?
2. The contractor went through a steep learning curve when negotiating and using each of the above tactics to reach an agreeable result. Would you expect this to be more common with international projects?

CHAPTER 8

Methods of Dispute Resolution

8.1 INTRODUCTION

> *Let us not be blind to our differences, but let us also direct attention to our common interests and the means by which those differences can be resolved.*
> John Fitzgerald Kennedy (June 1963)

> *There is a fatality about all good resolutions. They are invariably made too soon.*
> Oscar Wilde (1854–1900) Phrases and Philosophies for the Use of the Young, 1894

The challenge on any project is to resolve disputes that occur, generally as quickly as possible with minimal disruption to the project in terms of delays and costs, in a way acceptable to all parties involved, without relationships deteriorating. An equitable outcome is expected, while preference is to not release control of the dispute to an outside party.

The true cost of any dispute includes any time spent by staff and consultants in preparation and attendance at any resolution attempt, in addition to the direct cost to the project and indirect costs to the organisations involved. A major factor in the total end cost of any project will be how effectively disputes can been avoided, managed and resolved. As the cost of any project is one of the most important issues to any owner, it is paramount that all parties prescribe to meaningful and structured procedures to deal with disputes.

By far the most common and preferred way of resolving disputes on contractual matters is by *negotiation*. It is also the cheapest and quickest method available. However, negotiation may not always be successful, and something else has to be tried. *Arbitration* has always been a favoured alternative to a court hearing in many contractual disputes. Arbitration however commonly occurs after the project is complete and in some instances mirrors the formalities and costs of *litigation*. Courts can refer matters to *external referees* and mediators. Various *alternative dispute resolution (ADR)* methods have been contrived and are available. The term 'alternative' is in the sense of alternative to the confrontational, adjudicative and adversarial traditional methods of arbitration and litigation. Some people use the term in the sense of 'organised dispute resolution outside the courts', but this usage is not adopted here. The use of organised dispute resolution outside the courts has a tradition in many cultures.

> *Discourage litigation. Persuade your neighbours to compromise whenever you can. Point out to them how the nominal winner is often a real loser: in fees, expenses and waste of time.*
>
> Abraham Lincoln, July 1, 1850

Various figures are quoted by writers on the percentage usage of each of these various methods and the average duration each takes, but how reliable these figures are is unknown, because most disputes are resolved in private. A consensus seems to be however is that by far the majority of disputes are resolved by negotiation, and many of these are at the workplace level, in spite of a reported lack of negotiation skills in project people. The issue causing the dispute is only one of many considerations determining the selection of a resolution method (and also a solution). For negotiation and ADR to work, both parties have to have an interest or incentive in seeing the dispute resolved; mutual understanding and respect between the parties is needed. When one of the parties does not want to resolve the dispute, then the less understandable, and more complex, formal and expensive options of arbitration and litigation become necessary.

Negotiation

Negotiation is the only approach which doesn't involve a third party. As such, it may be described as a non-intrusive approach. Genuine discussion and negotiation should be promoted as the preferred way of resolving disputes, and to the mutual advantage of both parties.

ADR

There has been a push in recent years by certain groups to promote *ADR* methods. These generally use an independent third party, are non-binding (with some exceptions) and are intermediate in cost between negotiation and arbitration. For verbal contracts, they are particularly useful.

ADR includes methods under such names as:

Non-binding methods
- Mediation.
- Conciliation.
- Expert appraisal.
- Senior executive appraisal.
- Disputes review board.

Binding methods
- Expert determination.
- Disputes review board.

These methods have developed over time to cater for the array of dispute types and disputing parties' peculiarities. 'If one only has a hammer, one will treat all problems as a nail.'

The following outlines the more popular ADR methods. However *hybrids* and other approaches, without commonly accepted naming, are possible. In fact the various ADR approaches, when used in practice, tend to merge and any boundaries between the common types disappear. As well, the terminology is used loosely, the meaning varying from one person to another. Often definitions can be mutually contradictory. The labels given to the various approaches are therefore only useful to describe a type of approach. Combinations of ADR approaches (for example, mediation with expert appraisal) are possible. ADR (for example, mediation) may occur within the framework of arbitration. An ADR approach can be tailor-made or flexibly fashioned to suit the parties' needs and situation. Any ADR approach adopted should be kept flexible and designed to suit a particular application. The given approach selected for any situation is only limited by the parties' imagination and creativity.

A general recommendation is that negotiation and ADR should be tried at the earliest opportunity after a dispute arises.

Reference to ADR in the following excludes reference to the binding forms, for example expert determination and disputes review boards, unless specific note is otherwise made.

Arbitration and litigation

Arbitration and litigation are sometimes described as intrusive processes, in that a third party decision is imposed on the parties. (In this sense expert determination, and the binding form of disputes review boards could also be called intrusive.)

In principle, arbitration can be a quick and cheap way of resolving a dispute, using a neutral person with technical and legal expertise. In practice, though, it rarely seems to deliver such benefits.

A common view is that the legal and related systems are costly, time consuming and could possibly not lead to a result for a number of years after the project is completed. The financial liabilities of the parties cannot be predicted. Also, the interpretation in the judgment or award, that prevails, may have less to do with the contract's original intent, and more to do with the capabilities of the legal counsel. Sympathetic expert witnesses may be 'purchased' ('hired gun') (rather than being objective and helping the tribunal) by either side to reinforce its case. Arbitrators may be perceived as not being impartial, particularly where the contract provides for a particular nominating body (for example, an industry association). It is for this reason that many different, alternative forms of dispute resolution are tried.

However, in some cases, litigation may be the only thing left that can resolve a particular dispute; this may arise for example where:
- There are substantial legal issues.
- The proceedings are multi-party.
- The proceedings are likely to be difficult to control.
- There are allegations of dishonesty.
- One party refuses to acknowledge that a compromise may be necessary, and wants a court ruling totally in its favour.

Interestingly, the majority of cases that head in the litigation direction are reported to be resolved out of court by negotiation or other means. Some reasons given for this include:

- There is uncertainty about the outcome.
- The costs start to increase alarmingly.
- The publicity of the hearing is feared.
- A party discovers that its case is not as strong as it first thought.
- Litigation was being used as a tactic to force the dispute issue.

Generally, contractors and owners are not in a hurry to get into arbitration and litigation. From the contractor's viewpoint it is considered bad for business and reputation. In many cases, the owner may be stronger financially. This is an important factor, because litigation and arbitration are expensive pursuits, and the outcome is always uncertain.

Only methods other than arbitration and litigation are discussed here, because both arbitration and litigation practices and controls vary from one country to another.

The following article (extracted from The Arbitrator, August 1990, p. 83) is a humorous account of how arbitration was formerly practised.

... a form of arbitration which flourished in County Down during the last century. The parties agreed on an impartial chairman, who sat at the head of a long table with the parties on either hand. Down the middle of the table a line was drawn, and grains of oats were placed along it at intervals of a few inches. A foot or so from the head of the table the line stopped, and two grains of corn were placed a few inches from the middle, one in front of each party. Then with the chairman as umpire, a hen turkey was gently placed on the table at the far end. The turkey would then delicately peck her ladylike way all up the table until, when she reached the two grains of corn at the top, she delivered her award in favour of one party or the other by taking first the grain nearer to him.

It is, however, recorded that on one occasion the loser in such an arbitrament was a litigious character who refused to accept the decision as just, and brought a civil bill in the county court against the winner. On the facts being proved, the county court judge dismissed the action, whereupon the plaintiff exercised his right to appeal to the assize judge. This was an aged and learned equity lawyer, Lefroy C. J., who unlike counsel for the defendant knew little of local customs. During cross-examination of the plaintiff the following passage occurred:

Counsel: 'Tell me, wasn't the turkey for the defendant?' No answer.
Counsel: 'Tell my Lord the truth, now. Wasn't the turkey for the defendant?'
Chief Justice: 'What on earth has a turkey to do with this case?'
Counsel: 'It's a local form of arbitration, my Lord.'
Chief Justice: 'Do you mean to tell me that the plaintiff has brought this case in disregard of the award of an arbitrator?'
Counsel: 'That is so, my Lord.'
Chief Justice: 'Disgraceful! Appeal dismissed with costs here and below.'
Counsel (sotto voce): 'The Lord Chief Justice affirms the turkey'

It will, I am sure, be at once obvious that this form of arbitration, although perhaps unattractive to professional arbitrators, has in large measure most of the merits claimed for this form of dispute resolution. It is very inexpensive, the more so since the bird can be used again. It is private. It enables the parties to select an expert tribunal. It minimises, as the story shows, the opportunities for judicial intervention. And ... it promotes the expeditious determination of references.

A step/staged approach

Dispute resolution clauses in conditions of contract may provide for a step/staged approach, where the parties are walked through a deliberate series of resolution methods. For example, the steps/stages may be:

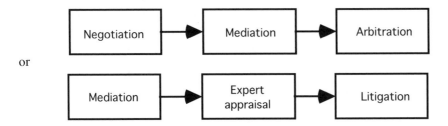

or

The choice of staging can be fashion driven, but the underlying idea is a good one and recommended by many writers. Of course, if any of the stages involves negotiation or non-binding ADR, there is little compulsion on either party to take these stages seriously, but commercial interests may convince the parties to take advantage of the chance to resolve their dispute.

The staging of resolution methods is adopted by those owners, amongst others, wishing to not jump straight into but possibly avoid arbitration or litigation wherever possible. Arbitration and litigation should be regarded as a last resort for resolving a dispute, when all other methods have failed. Arbitration may be included as an option, rather than being mandatory.

Negotiation is considered by most people as the most appropriate way to resolve any dispute. Should negotiation not be successful, ADR can be attempted. So as not to drag out the process, one recommended practice is to make submissions once off, followed by deliberation and decision making; the scope of the dispute is not allowed to expand. Another recommendation is to jointly identify those matters that are in agreement, and to only take forward matters that are not in agreement.

Disclosures in negotiation and ADR attempts should be 'without prejudice' in case arbitration or litigation of the issues does eventuate.

Senior management should be involved before progressing to a subsequent stage.

The final stage is one that provides an enforceable outcome. It elevates the dispute, if it has not been taken seriously, to where it will be taken seriously. It forces the parties to properly appraise their positions, and may rekindle negotiation, but this time with more focus. The fear of anticipated financial consequences of arbitration or litigation may force a settlement, rather than have a settlement by genuine desire.

The courts and private organisations

Some judges actively encourage the use of ADR (mediation, expert appraisal, referee, ...) within the court system, in an attempt to shorten the court waiting lists.

Courts provide an overseeing influence on the behaviour of most dispute settlements, ADR or otherwise, and shape and constrain the processes. The settlements and people's behaviour occur 'in the shadow of the law'.

A number of private organisations have established themselves, domestically and internationally, to assist in most forms of dispute resolution. Their services include the provision of advice, dispute resolution clauses for contracts, recommendation of a preferred ADR approach, procedural rules and guidelines under which any dispute resolution process is to be conducted, forums, lists of trained professional third parties, and the conduct of dispute resolution sessions. Where parties choose to follow rules developed by these organisations, there would generally be no need to establish any other agreement between the parties governing the conduct of the dispute resolution process.

No Dispute
"No Dispute" (NPWC/NBCC, 1990) gives:

[Goals]
[Goals] in dispute resolution are generally the same for both contracting parties. They are:
- *Timely resolution.*
- *Understandable and simple processes.*
- *Equitable outcome.*
- *No third party involvement.*
- *Low handling costs.*
Successful disputes resolution therefore requires:
- *Mutual understanding and respect between the parties.*
- *Genuine desire to resolve by the simplest means possible.*
- *A staged resolution process.*
- *A progression of [alternative] resolution methods.*
- *The involvement of senior management before adoption of the next stage.*
- *Incentive to resolve by agreement.*
- *Enforceable resolution when agreement fails.*

Summary of guidelines for disputes resolution
- *Encourage, facilitate and expedite genuine negotiation.*
- *Avoid legal representation.*
- *Avoid arbitration and litigation processes.*
- *Specify compulsory conferences of senior management of both parties before embarking on formal third party processes.*
- *Concentration on cost mitigation of the problem area, rather than procrastination about negotiation and resolving the dispute.*
- *Be cost-conscious, contemplating end financial implications of resolution processes once genuine negotiation has failed.*
- *Encourage use of alternative dispute resolution processes.*

Discussion
Expedient and successful dispute resolution requires competent personnel with clearly stated limits of authority to manage and determine the process.

Currently in most forms of contract, provisions exist for dispute resolution. Too often, however, resolution at the working level has been frustrated, resulting in the

> *use of more formal adversarial forums for judgment such as arbitration and litigation.*
>
> *Current experience and research indicates a satisfactory outcome in such forums is precluded by the excessive costs associated with those methods. The problem is exacerbated by the excessive number of commercial disputes set down for hearing, causing long waiting periods with attendant costs.*
>
> *An added by-product of these long lead times is the consequential cost arising from the attrition and mobility of staff ... who possess the 'hands on' knowledge of the particular dispute.*

8.2 CHARACTERISTICS OF ADR METHODS

Most of the ADR methods share some common characteristics. Important amongst these are: early settlement, low cost, and possible preservation of a working relationship between the parties. They are driven by commercial imperatives.

The characteristics of ADR apply equally in the national as well as the international pictures, though the cost and time saving benefits of ADR over, say, arbitration, could be expected to be greater in the international picture.

Getting to ADR

The dispute resolution clauses in contracts can be used to trigger attempts at ADR. As well, by other ad hoc agreement, ADR can be attempted if the parties so want. Providers of ADR services have standard wording for inclusion in contracts. Such clauses would make mention of:
- Procedure and time limits for the notification of a dispute.
- The type of ADR approach to be utilised.
- Time limits for dealing with the disputes.
- How the independent third party is to be appointed.

Whether such clauses are enforceable is unclear, but at any rate this is not that important an issue, because a party may only make a token effort towards satisfying the requirements stated in the contract, with no real intent of pursuing ADR. In fact, for ADR to be successful, both parties have to want ADR. Reluctant parties need to be counselled on the expense, delay and loss of goodwill consequences of going to arbitration or litigation. The court may grant a stay of any arbitration or litigation proceedings in favour of forcing a reluctant party to ADR, but this would achieve little, because the ADR process would almost certainly be unsuccessful without both parties wanting ADR.

Informality

The approaches are characterised by a low level of formality. They are not threatening to the parties involved.

Pragmatism, innovation

The approaches lead to pragmatic solutions. They are not restricted to remedies obtainable through the courts. The outcomes are flexible. The focus tends to be on commercial issues instead of legal ground rules.

An innovative solution can be fashioned by the parties to fit the needs of both parties. A solution is not forced on the parties. Any agreement reached may possibly be amicable or of the win–win type.

Independent third party

The independent third party brings specialist expertise to the settlement. The combined skills of more than one third party may be used. ADR stands or falls on the quality of the third party.

This person is expected to respect the privacy and confidentiality of information that comes to hand during the proceedings. The parties mutually select a person of respect and status, though each party may adopt some tactical thinking in this selection. The ideal third person is one who understands the technical aspects of the matters in dispute, while having a working knowledge of legal issues. This avoids the parties having to educate the third person in the issues of the dispute.

The third party is totally impartial, equally concerned with the interests of both parties. For international disputes, a neutral person, familiar with both cultures and languages, is necessary; so as to bridge both cultures, more than one independent third party may be useful.

The independent person assists the parties, but has no authority to impose a decision or settlement. Any outcome is that of the parties.

The third party attempts to dissociate him/herself from the emotive issues surrounding a dispute so as to ensure as much objectivity as possible. Where emotive issues are involved, the dispute resolution may become protracted and several attempts may be needed.

This third person acts as a 'reality tester', to break down each party's entrenched position. To help the parties see realism, the neutral third party can indicate what the likely outcome of a dispute might be if it went to court or arbitration; or the best and worst outcomes that could be expected, the so-called BATNA (the best alternative to a negotiated agreement) and WATNA (the worst alternative to a negotiated agreement) of Fisher and Ury (1983).

There may be personal *liability* issues for this person, in terms of losses suffered by either party because of advice or actions of this person. The consequences for this person may far exceed the remuneration received. Parties contractually agree to limit their rights of recourse against this person. Agreements engaging third parties would include exclusions of liability, indemnities from the disputing parties, and an undertaking not to include the third party in any subsequent proceedings. This protection enables a third party to act independently, without fear. However, there is the opposing view that, without accountability, unfair treatment may be given to one party.

With a dispute settled or put to one side, the parties may lose interest in paying the third party for his/her services. For this reason, independent third parties consider security for their fees.

No dispute-handling *qualifications* are needed of this third party, though various short courses are held by ADR organisations to familiarise mediators, expert appraisers etc. with accepted practices. Some dispute settlement organisations, which provide names of mediators, expert appraisers, ... for consideration by disputing parties, may insist on attendance at these courses, and passing some assessment before being listed as suitable candidates.

Timetable
The parties select a timetable, that best suits their purposes, for dealing with their dispute. Time lost in commencing ADR is small. Outcomes can be achieved speedily.

Venue
A neutral venue is recommended. For international disputes, a neutral country, equidistant between the home countries of the disputing parties, may be suggested, or a neutral venue in the country of the project may be suitable.

The venue needs suitable meeting, communication and eating facilities, with ready access to transport and accommodation.

Binding?
Nothing is binding, unless agreed to by both parties that it be so. Expert determination is an exception. Disputes review boards may give decisions that are binding for the duration of the project, but not thereafter.

With the non-binding forms, any outcomes can be ignored, allowing the dissatisfied party to then proceed to perhaps arbitration or litigation. However if an outcome is acceptable to both parties, this may be given contractual force through establishing a signed written agreement.

Attendance
Generally, attendance is not compulsory. Legal representation is discouraged.

A party may withdraw from the process at any time, should it feel that it is not going as wished, should the parties' representation be unbalanced, or other reason.

Confidentiality, privacy
Attempts may be made to prevent the use of information disclosed in an ADR approach. Anonymity and confidentiality may be important commercial reasons for choosing ADR over going to court.

Parties agree to go into an ADR approach on a 'without prejudice' basis. Information tabled, and statements made in relation to the particular dispute, could be expected not to be used as evidence by the other party in any subsequent court proceedings.

Cost
The cost of the third party and facilities (rooms, audio/visual, ...) is shared between the parties. The hourly or daily rate payable to the third party will vary depending on that person's standing in the industry as a dispute resolver.

The methods tend to be quick. Costs are minimal. Employees' time lost attending the settlement is small.

Ongoing relationships
Business relationships and goodwill can, in some cases, be preserved. Straight out negotiation is the most likely way of preserving a business relationship between the parties. The parties' ability to resolve one dispute, without going to arbitration or litigation, may provide a capability to resolve future disputes.

Outcome

Although the process is non-binding, a successful ADR outcome would be converted into something binding — a written, signed document, outlining any agreement reached. This ensures that the dispute does not resurface at a later date. The courts would be expected to support such an agreement reached by consensus.

Care would have to be exercised in drafting any agreement, in order that it extinguishes any avenue for further claim or dispute. Some legal input may be helpful in this regard. (Elsewhere, legal input should be restricted to advice only and giving a legal perspective, and not to act as advocates, else the process becomes adversarial.)

Example

... the [owner] may claim that certain work is defective. The contractor might offer, instead of repairing the defect, to give the [owner] a reduction in the contract price of $1000. If the [owner] agrees then there is accord. The contractor thereupon reduces the contractor's invoice by $1000. There is satisfaction. Thereafter the [owner] cannot claim damages for defective work.

Now imagine that the defect was an insecure hand rail on a balcony. Imagine that someone unaware of the defect leans on the handrail and it gives way causing injury to that person. The injured person might have a claim directly against the contractor. The injured person is not bound by the agreement between the contractor and the [owner]. In the agreement with the [owner], the contractor would have been prudent to include a provision that the [owner] will make the rail safe and indemnify the contractor against any claim by any third party.

Uher and Davenport, 1998

Example

A frequent occurrence is where the [owner] and the contractor agree upon a price for a variation and later the contractor claims 'prolongation' of delay costs resulting from the variation. It is a matter of legal interpretation to determine whether the agreement upon a price for a variation extinguished the right to make a claim for 'prolongation'.

Uher and Davenport, 1998

ADR approaches aim to reach an outcome by consensus and collaboration. Where consensus is achieved, it is expected that the parties would honour such an agreement.

Usage

Given the advantages of ADR over adjudicative methods, it would appear that ADR would be adopted quite widely by industry. However it is unclear why ADR hasn't been taken up with greater speed and enthusiasm by industry. Some suggested reasons for its slow uptake may be:

- People have a preference for the familiar.
- The result is generally a compromise ('splitting the baby').
- Some parties wish to have someone else (judge, arbitrator, expert determinator) to blame for their dispute outcome, rather than accepting responsibility and the possible associated stigma themselves.
- The legal profession has a conflict of interest in recommending speedy dispute resolution.
- The legal profession is educated in traditional adversarial practices.
- Disputants are informed that they are foregoing their legal rights with ADR.
- ADR is thought to represent 'quick and dirty' justice; 'real' justice only occurs on going to court; ADR is second class justice.
- There are few published detailed results, compared with publicly available court decisions; most published information is superficial leaving users little indication as to what their outcome might be.
- Arbitration and litigation demonstrate that 'no stone has been left unturned', compared to the more superficial approach of ADR; this may be important to someone answerable to, for example, shareholders, creditors or the public.
- Where the process is non-binding until a final agreement is reached, a party may walk away from the process at any time. A party may have no commitment to reaching settlement. The binding forms are used where a non-binding form might be considered to lead to a waste of time and money.
- A party may use ADR as a fishing expedition in order to find out how strong the other party's case is, before proceeding to some adjudicative method. This may make the parties reluctant to freely and openly communicate, for fear of any information disclosed ('showing their hand') being used in some subsequent tribunal.
- One of the initial strengths of ADR was its relative informality. However, some people take the view that the process must be guided and overseen by experienced operatives, and the hands of the disputing parties need to be held. Promoter-organisations consequently train formalism and legal consultants encourage formalism, and the processes become entangled in formal structures. People with disputes then seek less formal methods, which in turn attract the attention of promoter-organisations and legal consultants, and so the cycle repeats. The extreme example of this is the evolution of arbitration, which in principle should serve the needs of all disputes, but which many people won't even go remotely near today, and shudder at the mention of what has happened in the hijacking of arbitration over time.
- For international disputes, there are cultural barriers and language issues. Each party may not understand what is driving the other party, and a meeting of the minds may be difficult to achieve.
- For contractors which have tendered at cost or less (or have an unprofitable job), it is necessary for their claims, valid or otherwise, to be successful in order to make a profit. To achieve this, the threat of litigation and its attendant costs are necessary to force the owner to concede some of the contractor's claims. A reverse situation can also occur where the contractor is forced to concede some of its legitimate claim by the owner threatening litigation.
- Statute and common law commonly intrude on a dispute. Matters which are not of a

commercial nature, for example where fraud is involved, would not be suitable for ADR.
- The bureaucracy in large organisations can prevent experimentation in different dispute resolution methods.
- Large corporations are not deterred by the cost of litigation.
- ADR is unable to address any power imbalance between parties.
- There may be concern that the third party may turn out to be not impartial.
- A party may fear that by proposing ADR, it will signal some weakness.

Failure of ADR

A number of reasons can be put forward as to why ADR might fail in any particular case (in addition to some of the reasons advanced for the slow uptake of ADR) are:
- Either party may not see that a prompt solution is desirable.
- The situation may be not suited for a compromise. For example, there may be held deeply personal values, one party's economic survival might be threatened, or one or both parties may have fixed views.
- Either or both parties may not have done adequate preparation for the resolution attempt.
- There may be an unwillingness of either party to try to understand the view of the other party.
- Either party may not wish to resolve the dispute. ADR requires both parties to want a resolution.
- The chosen independent third party may not have a working knowledge of the industry or profession concerning the dispute. This person is unable to identify and question key issues relevant to the dispute.
- The representatives of both sides do not have the authority to settle the dispute. Authority is commonly checked up front by the third person, though even this doesn't provide total certainty, because the authority may only exist based on the assumption of a particular settlement; when the outcome turns out to be different, extended authority may not exist.
- With the non-binding forms, any outcomes can be ignored, allowing the dissatisfied party to then proceed to perhaps arbitration or litigation.
- The parties may have a different perception of the matters in dispute. This could be based on different information, or on a different assessment of the same information.

8.3 NEGOTIATION

Schematically, negotiation is illustrated in Figures 8.1a and 8.1b.

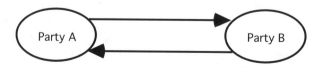

Figure 8.1a. Communication channels, direct negotiation.

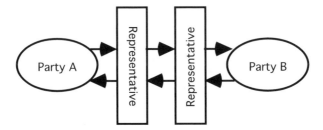

Figure 8.1b. Communication channels, assisted negotiation.

Negotiation is the most common way of resolving disputes, and presumably the oldest way. It is commonly employed as a first attempt in resolving a dispute. In general terms, negotiation consists of each party attempting to persuade the other party to its point of view.

Courts may prescribe that negotiation takes place as a pre-hearing. Contract dispute resolution clauses may prescribe negotiation as a first step towards resolution.

Advantages

- It only involves the two parties, and in that sense, out of all resolution methods, it gives the parties greatest control over the process.
- It is flexible and has no structure, though the parties can mutually agree on any structure they like.
- The process can commence immediately, with associated potential savings in costs and time.
- It is potentially the quickest and cheapest method available. Potential delays can be reduced.

Disadvantages

- Individuals may lack negotiation skills, and this may hamper a successful outcome.
- The parties may not be able to see any shared and compatible interests without an independent third party. There is a common assumption that because the other party has an opposing position, then its interests are also opposed. Some writers take this view further and contend that a close examination of underlying interests will reveal that parties have more interests that are shared or compatible than are opposed.
- It requires preparation and a knowledge of the legal and factual issues.

8.4 MEDIATION

Schematically, mediation is illustrated in Figures 8.2a, 8.2b and 8.2c.

Mediation is a form of third party assisted negotiation with structure. It is an extension of negotiation. Mediation, as practised, may involve varying degrees of involvement of the mediator, with the range limits being:

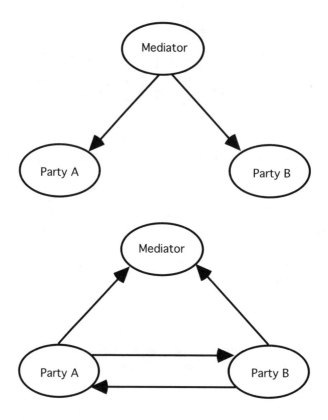

Figure 8.2a. Communication channels, opening phase of mediation.

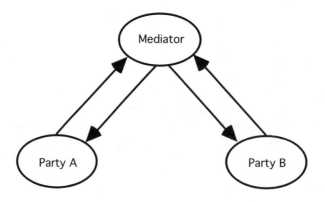

Figure 8.2b. Communication channels, caucus phase of mediation.

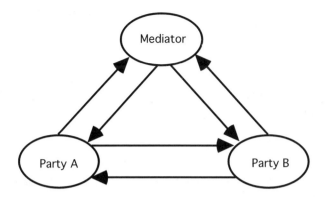

Figure 8.2c. Communication channels, joint phase of mediation.

- As a facilitator of a meeting between the disputants, but thereafter the disputants are left to work out their own resolution.
- As an intruder on the process, contributing his/her own opinions and assessment, and suggesting or proposing solutions (either by request of the parties or voluntarily).

In the second form, the mediator provides (usually in private) an assessment of the parties' respective rights, and a likely outcome were the matter to proceed to arbitration or litigation. This assessment is couched in the context that the mediator is not privy to all information, as well as the presence of uncertainties in any matter outcome. If this assessment is needed in writing, perhaps to justify a resolution to others in the parties' organisations, then some form of ADR other than mediation (for example, expert appraisal) is needed.

Mediation can take on a very broad meaning. In fact, mediation is used, by some people, in a generic sense to mean any third party assisted ADR approach. The terms *facilitation* and *neutral evaluation* may be used by some writers instead of mediation.

A version halfway between the above two dot points is described here.

Attendance at a mediation is voluntary. It aims, like negotiation, to achieve an outcome acceptable to both parties. The mediator does not impose a solution on the parties, but rather encourages the parties to reach their own solution.

Having an independent, impartial or neutral third person, enables the parties to meet with the mediator in private ('caucus'), where information, not wished to be known to the other party, may be disclosed in confidence to the mediator. Hidden issues, straw issues and emotions can also be discovered. This assists the mediator to understand each party's position.

Joint sessions involving the mediator and both parties are also usually held, though this may introduce some adversarial nature to a process that attempts to be non-adversarial.

Because the mediator is not an inquisitor, some issues relevant to the dispute may remain dormant, if the parties do not raise them.

Mediation may be useful for multi-party disputes. Also, more than one mediator may be employed.

As an extension of negotiation, mediation retains the flexibility and informality of negotiation in the process. There are no ground rules, though some dispute industry groups try to insist on structure to the proceedings and behaviour of the mediator. The parties have control over the process.

The mediator

- The mediator is a neutral expert who assists and encourages the parties in their negotiation, and could be expected to act as a catalyst or facilitator. The mediator provides an atmosphere conducive to settlement.
- The mediator tries to create doubts in the parties' minds as to the validity and weakness of their positions and the strength of the other party's position ('make them see reality'), and encourages them to adopt more realistic positions and understand the other party's position.
- The mediator may suggest alternative views, joint benefits and gains, and possible directions that the parties might consider moving in, or explain objectively the other side's position, or ask pertinent questions in order keep the negotiation progressing. Some mediators become actively involved, others take a more watching and listening role, only commenting when asked to. The mediator would generally be open to approaches from the parties.

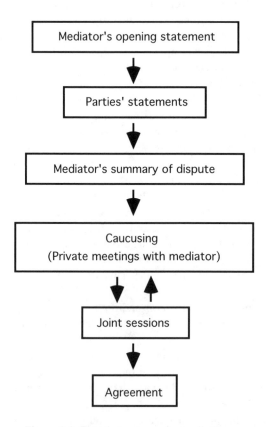

Figure 8.3. Broad structure of a mediation.

- Mediators are frequently asked to provide legal advice, technical advice, solutions, opinions, and recommendations or a decision. Some mediators oblige, but this is not the intent of having a mediator, and can compromise the independent status.
- Good mediator traits are that s/he must be impartial, be a good listener, keeps the parties talking, and establishes credibility and trust with the parties, and a positive attitude towards a settlement.

Mediation structure

Some dispute industry groups try to insist on structure to the proceedings and behaviour of the mediator. Such a structure to the proceedings may follow Figure 8.3.

Mediator's opening statement. The mediator builds credibility and trust, and educates the parties in the process. Behavioural guidelines are set.

Parties' statements. The parties can vent some emotions. Communication starts between the parties. The mediator gains a better understanding of the dispute. Issues and needs of the parties are defined.

Mediator's summary of the dispute. Issues are summarised; interests are identified; an agenda is set.

Caucusing and joint sessions. Interests and priorities are reviewed; issues are processed and explored; reality is tested; issues are resolved; agreement on issues and interests is obtained; options are generated; recommendations are made; impasses are broken; negotiation occurs.

Agreement. An agreement is finalised.

8.5 CONCILIATION

There are several uses of the term conciliation:
- In a general sense to mean any third party assisted ADR approach.
- As a facilitation of a meeting between the disputants, but thereafter the disputants are left to work out their own resolution.
- Instead of, or synonymously with, the term *mediation* as described above.
- As an approach like mediation but involving the third party contributing his/her own opinions and assessment, and suggesting or proposing solutions (either by request of the parties or voluntarily).

In practice, the two terms, mediation and conciliation, may be used interchangeably, and their distinction is blurred. The terms *facilitation* and *neutral evaluation* may also be used instead of conciliation.

The fourth dot point form, believed to be the most common, is adopted here. Clearly, different conciliators conduct themselves in different fashions, and hence the possible overlap among all four forms. Possibly some of the reason for the varying definitions is that the term is used in both the commercial disputes arena and the industrial disputes arena. Only commercial disputes are talked of here.

In mediation, the outcome derives more from the interaction of the parties as facilitated by the mediator. In conciliation, the outcome is more influenced by the conciliator (Figure 8.4), but ideally should be that constructed by the parties. But unless otherwise agreed or stipulated by a court, the conciliator's opinion or determination is not final or binding.

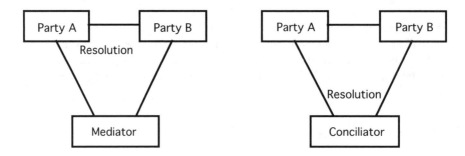

Figure 8.4. Mediation and conciliation compared.

Unlike a mediator, a conciliator may form a view which may assist or harm a party, and this view might be reflected in the conciliator's recommendations.

The advantages and disadvantages of conciliation are similar to mediation.

8.6 EXPERT APPRAISAL

Expert appraisal may also be termed appraisal, independent *appraisal, independent expert appraisal*, and *non-binding determination*. It is suited for disputes of a technical nature. This implies that the appraiser has a detailed technical understanding of the matters in dispute. The approach has been used for valuations of work, work measurement, fixing of hire/rental charges, quality issues, extension of time claims, specifications, demarcation of responsibilities, interpretation of a party's rights, as well as resolving complicated technical and legal disputes. Appraisal may be used as an adjunct to mediation.

Appraisal involves the neutral third party formulating an objective assessment or opinion on the matters in dispute, and providing this in writing to the parties (within the limits of the parties' request). To do this, the appraiser takes an active, inquisitorial role to identify and understand the dispute. The appraiser may be provided with documents only, or partly written and partly oral submissions may be made. There may be a hearing to enable the appraiser to ask questions and clarify information. The appraiser is not concerned with whether the assessment is agreeable to the parties or not. See Figure 8.5.

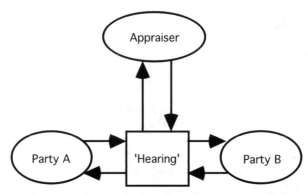

Figure 8.5. Communication channels, expert appraisal.

Two experts may be engaged if there are substantial issues of both law and technical expertise involved.

The assessment is non binding, but is used by the parties to assist a subsequent nego-tiated settlement. In many cases it satisfies the need for an independent assessment of liability, and for accountability purposes, it satisfies others in the parties' organisations as to why a particular settlement was reached. The oral nature of mediation may not satisfy these needs. If the expert is a respected person, the assessment can be persuasive in leading to a settlement.

Compared to mediation, expert appraisal could be expected to take longer and be more expensive, essentially because of the additional time for the appraiser to sift through the given information and present a report with rigorous arguments, and the parties' time involved in developing the written submissions.

The rules for proceeding to expert appraisal and the rules, under which the appraiser operates, are contractually agreed between the parties. Some typical rules relate to:
• Confidentiality.
• Venue.
• Expert appointment procedures.
• Timetable for submissions.
• Timetable for appraisal.
• Costs of the expert and the process (shared equally).
• Non-binding nature of the decision.
Whether any evidence or discussion is treated on a 'without prejudice' basis, that is unable to be used in any following legal proceedings, is by agreement between the par-ties.

Structure

A typical structure to an appraisal may take the form of Figure 8.6.

Written, oral submissions. Each party submits brief written submissions to the appraiser and the other party. This may be supported by oral statement. No witnesses are usually involved.

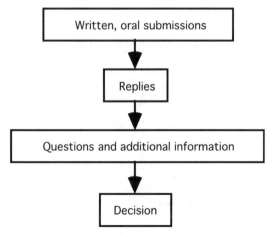

Figure 8.6. Structure to an appraisal.

Replies. Each party gives a brief written reply to the other party's submission.

Questions and additional information. The appraiser may have questions to ask of the submissions, and may ask for additional information.

Decision. The appraiser gives an objective assessment or opinion, together with reasons, based on the information presented. This assessment is not binding, but rather used in an advisory or persuasive manner.

8.7 EXPERT DETERMINATION

Expert determination may also be termed *independent expert determination*. Occasionally the terms *expert appraisal* and expert determination are used synonymously; this is not the usage here. Schematically, expert determination is illustrated in Figure 8.7. (A similar arrangement exists for arbitration and litigation.)

The types of applicable disputes are the same as in expert appraisal. The process structure mirrors that of expert appraisal, except that the determination is agreed by the parties to be binding. Reasons may or may not be required. Appeal on the decision, to the courts, may not be possible, unless the parties agree otherwise.

The choice of the determinator is important, because the parties have to live with the determinator's decision. Expert determination may not be attractive, therefore, where the quantum of the dispute is large.

Expert determination is attractive to those who don't want the costs and time delays associated with formal arbitration, but want a similar end result. The main differences between expert determination and arbitration are that no legislation applies to expert determination while it does to arbitration, and expert determination dispenses with the requirements of procedural fairness and rules of evidence.

Care has to be exercised in the way expert determination is carried out, such that it is not arbitration in disguise, and hence subject to any arbitration legislation. Where there are issues of law involved, the jurisdiction of the expert determinator is unclear, because of the courts' overriding jurisdiction; the courts cannot be excluded by any private agreement. It is unclear to what extent the courts would interfere with an expert determination, following a process agreed to by both parties.

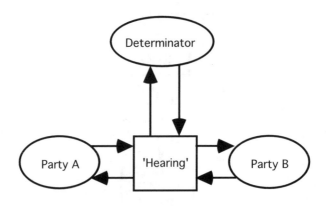

Figure 8.7. Communication channels, expert determination.

The language used in a contract dispute resolution clause, to describe the particular type of dispute resolution method, may be open to interpretation by the courts, depending on the intention of the parties, the nature of the dispute and the process intended to be carried out. Reference to expert determination may be interpreted as arbitration, and vice versa.

The determinator's costs can be contained by limiting the amount of time for the process. Dispensing with the requirement for procedural fairness is the biggest help to speeding up the dispute resolution.

Agreements engaging an expert determinator would include exclusions of liability (including liability to aggrieved fourth parties), and indemnities from the disputing parties.

The rules for proceeding to expert determination and the rules, under which the determinator operates, are contractually agreed between the parties. Some typical rules relate to:

- Confidentiality.
- Venue.
- Expert appointment procedures.
- Timetable for submissions.
- Timetable for determination.
- Costs of the expert and the process (shared equally).
- Binding nature of the decision.

8.8 SENIOR EXECUTIVE APPRAISAL

Senior executive appraisal may also be termed a *settlement conference, case presentation,* or a *structured information exchange.*

Typically each party would be represented by an authorised decision maker (an executive authorised to settle the dispute) and representatives familiar with the matters in dispute. As well, a neutral adviser or facilitator selected by the parties, would be present. See Figure 8.8.

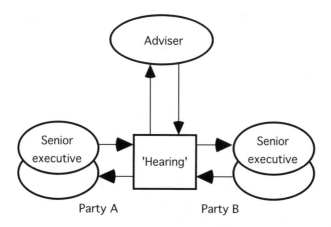

Figure 8.8. Communication channels, senior executive appraisal.

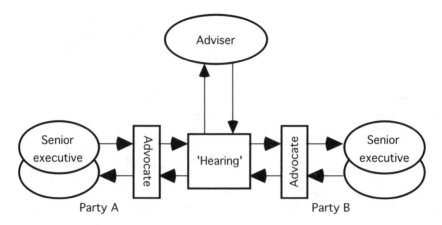

Figure 8.9. Communication channels, mini trial.

Each party may give a short presentation including argument, after which the senior executives, possibly hearing the facts and arguments for the first time, assess the strengths and weaknesses of both cases, and negotiate a resolution. The neutral adviser facilitates or moderates the resolution, may give an appraisal, and may also act as a mediator. How the resolution is achieved is not bound by anything other than what is acceptable to both parties, and it is expected that it would be one reached out of commercial wisdom.

An extension of the method, where additionally each party has an advocate (much like counsel in litigation), who presents a summary case, may be termed a *mini-trial* (even though it has no connection with the courts). See Figure 8.9. This could be expected to be more adversarial and less consensus-oriented than senior executive appraisal.

The parties agree on the procedure, form of presentation and timing. A restricted time frame would apply to each party's presentation. With a mini-trial, there would also be agreement on cross examination, and pre-trial procedures such as discovery.

Senior executive appraisal and mini-trial raise the dispute to senior management level, a level more concerned with commercial outcomes rather than the trivia and emotive aspects which would occupy the minds of the employee level at which the dispute occurred, and which get in the way of a speedy resolution.

8.9 DISPUTES REVIEW BOARD

A *disputes board of review, board of review* or *disputes review board (DRB)* would normally contain a representative from each contracting party (owner and contractor) and a mutually agreed impartial third party. All board members would have to be acceptable to both parties for the board to function effectively. An alternative composition is three neutral people acceptable to both parties. The board is constituted for the life of the project, is given regular reports on project progress, and meets regularly to keep abreast of the project performance; otherwise the board convenes when the parties are unable to resolve a dispute.

When a dispute occurs, the board is given appropriate documentation. The board may meet informally, or there may be presentations and arguments made. See Figure 8.10. Legal representation is discouraged; open and free discussions are encouraged. Solutions are given quickly, and the project is allowed to progress without any of the usual problems that occur when a dispute arises. The parties don't have time to become set in their positions, and ill-will, that could undermine the project's performance, is not allowed to develop.

Because of the cost with three people involved over the whole project, such boards are more commonly used on larger projects. The cost of the board, primarily the cost of the board members to appraise themselves of progress and any issues that arise, is shared by the contracting parties. However the cost should only be a very small part of the total project cost, and this is argued to be offset many times over through reduced tender prices, improved project efficiency and lower other disputation costs.

Where an impartial third party is used, this person can act as both an expert appraiser and a mediator and, to help the parties see realism, can indicate what the likely outcome of a dispute might be if it went to court or arbitration. It is unlikely that the parties would not accept and make use of the board's recommendations, because any other third party in any other forum would more than likely give similar conclusions. Information provided to the DRB would generally be able to be used in any subsequent arbitration or litigation. The board's recommendations may also be available to the arbitrator or judge.

Where the party representatives on the board are senior management, they can bring the full powers of commercial judgment, company authority, and decision making into the process. Commercial decisions, even though they may involve compromises, can be made in the interest of the project and the contracting parties. Issues of little consequence, that can cause immense negotiation problems and have emotional involvement at the workface level, can be quickly dismissed. People at the workface level may pursue an issue in the hope of finding a solution, to vindicate an error, out of emotion, or in ignorance; the DRB can bring a quick end to such issues.

Decisions of the board can be non-binding, or can be made binding for the duration of the project (*interim binding decisions*), after which time the dispute may be reopened, or can be final and binding, depending on the wishes of the parties. In the second case,

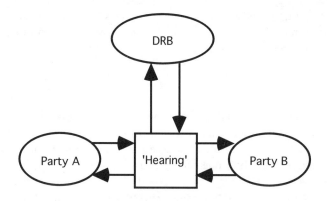

Figure 8.10. Communication channels, disputes review board.

it is very rare for a dispute to be reopened, because the decision would be similar to that which an arbitrator or court would make anyway. Making the decision binding for the duration of the project lets the project progress unimpeded by unproductive haggling over disputes. Where the decision is non-binding, the decision becomes part of the project record, and this contributes to its acceptance by the contracting parties.

On smaller projects, a *disputes resolution person (DRP)* can act in the same way as a DRB on larger projects.

8.10 CHOICE OF METHOD

Different cultures will be more relaxed with some methods of dispute resolution than with others. A party in one country may be reluctant to accept the law of another country; hence there may be a need to write some alternative dispute resolution method into international contracts between those parties.

Each method has a number of pluses and a number of minuses.

Decision trees

When selecting a dispute resolution method, a decision tree approach may be helpful. Figure 8.11 shows a possible general structure, but specific projects and disputes would benefit from a more specific tree tailored to the particular situation.

The squares represent points at which decisions are made, the circles represent chance nodes, and the triangles represent outcomes following particular decisions and chance events. Chance events will have estimated probabilities, typically based on informed opinion. For example in Figure 8.11, p1, p2 and p3 sum to 1; p1 is the probability that the matter will settle before reaching court, p2 is the probability that the matter will settle some time during the court hearing, and p3 is the probability that there will be a judgment (which could be further divided into a favourable judgment and an unfavourable judgment). Outcomes will be in terms of cost (both direct and indirect), but could also be in units other than money, for example in terms of work hours spent. Using expected cost or expected utility as the criterion, would indicate the most appropriate decision (choice of dispute resolution method) to make.

Adopting such an approach gives informed decision making as to the most appropriate dispute resolution method to follow. It removes much of the subjectivity commonly involved in making such a decision. Of course the estimated probabilities could be manipulated by someone wishing the tree to indicate a particular resolution method.

Adjudicative versus ADR methods

All methods of dispute resolution have their advantages and disadvantages. They have applicability for different circumstances. The methods may be broadly grouped as adjudicative (arbitration and litigation) and alternative dispute resolution (ADR) methods. Expert determination is the odd one out, because it is adjudicative yet more closely resembles ADR thinking. Disputes review boards are intermediate. The term 'alternative' is in the sense of alternative to traditional confrontational and adversarial methods of arbitration and litigation. Negotiation is more closely linked with ADR methods.

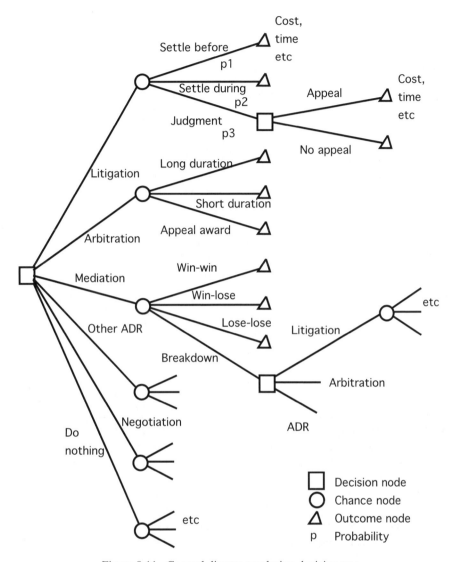

Figure 8.11. General dispute resolution decision tree.

When deciding which particular method to use, consideration might be given to a range of issues, including:
- The nature of the dispute, and its complexity. The range of issues.
- The desire of the parties to settle. The success expectancy. Susceptibility to tactics.
- The relationship between the parties. The ability of the parties to talk to one another. Is an ongoing relationship desirable? Commercial interests.
- The amount in dispute. The cost of running the settlement process.
- Time pressures. Delay potential.
- Winning legally versus losing financially.
- Confidentiality and privacy expected.

- The choice of the third party.
- The need or not for a binding decision. Enforceability of a decision.

Summary advantages and disadvantages

Non-legal (negotiation and ADR) methods

Advantages	Disadvantages
• Can be fast. • Can maintain a good relationship. • Relatively cheap. • Objective assessment possible.	• Generally not binding. • May be compromising. • May be difficulty in selecting an acceptable third party.

Arbitration

Advantages	Disadvantages
• Can be fast. • Can be a simple procedure. • Binding. • Has legal status. • Can be cheap.	• Could be abused with delay. • May not suit complex legal problems. • Can be expensive in long disputes. • More suitable for two parties.

Litigation/court trials

Advantages	Disadvantages
• May suit more complex disputes. • More thorough legal analysis; can address substantial legal questions. • Can accommodate multi-party disputes. • Useful for difficult to control proceedings. • Useful where there are substantial allegations of dishonesty.	• Can be significant waiting time. • Lengthy procedure. • Technical disputes may be unpopular with judiciary. • Could involve more than two parties • May need expert evidence. • Could be costly. • Deep-pocket syndrome — 'only who can afford pays'. • Strict construction 'precise meaning of language'.

Circumstances

Table 8.1 indicates the circumstances under which the various dispute resolution techniques might be used.

Cost and time

Generally the direct cost and time involved in resolving a dispute increases from top to bottom in Figure 8.12. Generally it could also be said that those methods at the top are

Circumstances relevant to the dispute	Appropriate dispute resolution method
A. Both parties wish to resolve the matter and the parties wish to make their own agreement.	
(i) The authorised decision makers of both parties know the details of the dispute and are available to attend meetings to resolve the dispute.	Conciliation Mediation
(ii) The party representatives who know the details of dispute are not authorised to resolve the matter. The authorised decision makers of one or both of the parties are not available to attend meetings to resolve the dispute.	Expert appraisal
(iii) The parties require a written opinion regarding the dispute to provide auditable evidence that a subsequently agreed resolution is properly based on the merits of the matter.	Expert appraisal
(iv) The party representatives who know the details of the dispute are not authorised to resolve the matter. The authorised decision makers of one or both of the parties do not know the details of the dispute, but the decision makers of both parties are available to attend meetings to resolve the dispute.	Senior executive appraisal
B. Both parties wish to resolve the matter but wish a third party to make a binding decision.	Expert determination, Fast-track arbitration, Disputes review board
C. One of the parties does not wish to resolve the matter.	Arbitration Litigation

Table 8.1. Choice of dispute resolution method (after Collins, 1989).

Figure 8.12. Comparison resolution methods.

more consensual (less adversarial), and more informal than those at the bottom; at the top, parties have control over the process, while at the bottom the control is with the third party.

For example, some writers suggest the direct cost of mediation is about 10% of the cost of litigation, but such figures tend to only be 'gut feel' figures.

If disputes are not settled in their infancy through negotiation, or later through an appropriate ADR method, then the cost of settling the dispute will rise and involved people's time will rise, in some cases unreasonably.

Employing legal advisers and preparing for court or arbitration adds significant costs to the settlement. The counter argument advanced by the legal profession is that the result may be a better one for the winner, because it will not involve a compromise. But winning is very uncertain, particularly when the outcome is based on the interpretation of case law and the manner in which a case is presented.

ADR offers the least potential for delay, though even here, the processes can be cleverly manipulated with delay tactics. But still the delays would be expected not to be great.

In ADR, typically the costs of the third party and any venue used are shared. In litigation and arbitration, the losing party could expect to pay the legal costs of the winning party.

Desire for settlement

Generally the ADR methods are only suitable if the parties are genuine in their desire to resolve the dispute. A certain amount of trust is involved in hoping that one party is not using the process to stall for time or to establish the strength of the other party's case ('prospecting'). The non-binding nature enables either party to walk away from the process whenever it wishes to.

ADR is susceptible to tactics, but not to the same extent as arbitration or litigation. In the adjudicative methods, for example, one party may deliberately delay the proceedings by manipulating the rules to give it more time, or to increase the losses to the other party; one party may be financially stronger than the other party and may cause the dispute-running costs to increase so much that the weaker party eventually gives in before reaching the tribunal. The lack of a requirement to adhere to fixed rules, frees ADR from such tactics, but is still susceptible to conventional negotiating tactics.

Compromise

All ADR approaches work best if parties are prepared to compromise. With compromises, however, no disputing party may be totally happy. The so-called win–win solution may not be possible unless solutions alternative to compromises are found.

The ADR approaches are regarded by many as being fairer. If the settlement is negotiated and acceptable to both parties, then a presumption might be that the result is fair. However this may not be so; as in court cases and arbitrations, the complete background and records to a dispute may not be made known to all, and so a decision is reached based on restricted knowledge.

Flexibility

All ADR approaches offer flexibility. The parties have control over the proceedings, and resolve the dispute in a manner which best suits the parties' interests. Discussions can include business motives as well as technical and legal argument (if necessary). As well, the ultimate solution might be able to be tailored to the circumstances at hand. There is the potential to be creative in the approach and the settlement, which may include reference to a whole range of concessions and services without involving money. The width of remedies tends to be more restrictive in litigation and arbitration — a win–loss outcome tends to be the usual.

Arbitration can be flexible, but as practised it tends to be very structured.

Technical issues

Appropriate technical experts can be used in ADR, although such experts could be made available to arbitrators and judges. The chance of misinterpretation of technical issues by using ADR is reduced.

Range of issues

Arbitration may be restrictive in the range of issues and the number of parties that can be dealt with.

Adversaries

Adversarial attitudes can exist on any project because of the differing interests of the parties. Such attitudes come to the surface, for example, when the contractor is dissatisfied with an inherently biased determination or direction given by the owner's representative or owner, and the attempted resolution of the following dispute. ADR methods attempt to diffuse this adversarial attitude, while the adjudicative methods further enhance the adversarial attitude.

Business relationships have a better chance of being preserved using ADR. However compromise solutions may still leave people with bad feelings.

Appeal

The right to appeal issue only applies if the method leads to a binding result, such as through expert determination, arbitration and litigation, and even here there may be only limited grounds for appeal. Where an agreement is reached by consensus, an appeal is unlikely. Agreements reached at the end of ADR can be converted to a contractual type status, and as with arbitrators' awards, enforced through the courts.

Privacy and confidentiality

Litigation, and court appeals of arbitrators' awards are the only tribunals which are open to public scrutiny. Nevertheless, industry grapevines seem to make public the nature

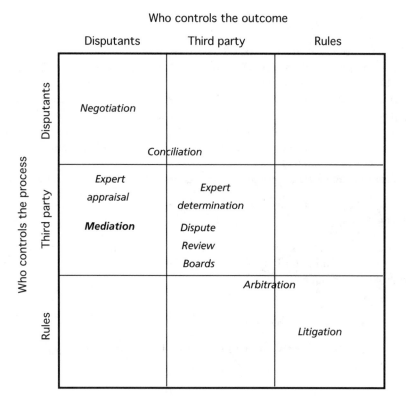

Figure 8.13. Process and outcome control (after Bryson).

of many disputes, particularly those involving large sums or extensive arbitrator sitting days. Negotiation, because no third party is employed, offers the most privacy.

Contractors and owners, concerned about reputations and future relations, may be reluctant to pursue their rights publicly.

Control by the parties

Figure 8.13 indicates approximately where the control of the process and the control of the outcome lie. In ADR and arbitration, the choice of the third party is by agreement. In litigation, there is no choice.

8.11 MECHANISM FOR SETTLING DISPUTES

If a dispute cannot be avoided, the next best solution is to manage it effectively so as to minimise its impact on the project and the parties involved. To this end, it is important to have effective dispute resolution procedures in place prior to letting the contract. This acknowledges, at the time of entering into a contract, that disputes are possible.

Three suggested mechanisms for settling disputes are:

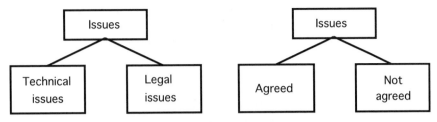

Figure 8.14a. Possible issue separation.

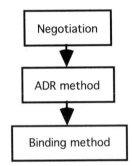

Figure 8.14b. Example step/staged approach.

- Separate technical issues from legal issues (Figure 8.14a).
- Separate issues in agreement from issues not in agreement (Figure 8.14a).
- Evolve the dispute in a step/staged approach (Figure 8.14b).

Step/staged approach

One suggested step/staged approach involves:
- An attempt is made to resolve the problem at the level at which it occurs, in a timely fashion ('nip it in the bud').
- If this fails, involve people at a higher level with decision making authority, and the potential to compromise in the interests of a commercial solution.
- If this fails, proceed to an ADR approach using an independent third party.
- If this fails, use arbitration or litigation.

See Figure 8.15.

Attempts at resolution involve a time commitment in amongst people's already busy working days. There is economy in committing this time early in the dispute. Letting the dispute resolution proceed to later stages involves an even bigger time commitment.

Often contractors and owners leave the resolution of disputes to the end of projects. Such a practice can lead to the worsening of a dispute, because the exact details of why the dispute occurred are often forgotten and may become distorted, deliberately or otherwise, with time. Disputes should be resolved preferably as promptly as possible. Time dulls the mind. Later conflicting recollections cause all sorts of troubles.

Generally, it could be said that most parties to contracts would prefer to avoid legal conflicts, because they are costly, time consuming and a result may possibly not be obtained for a number of years after the project is completed. For most people, disputes

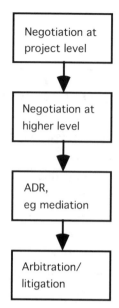

Figure 8.15. An example step/staged approach to dispute resolution.

are seen as undesirable occurrences that impact on the cost and schedule of the project. Owners are usually keen to see disputes settled quickly so that project work can resume or continue without incident. Contractors are generally keen to see disputes resolved and the contract work resume, so that they can make their profit in the shortest necessary time, and move onto the next project without incident.

All steps are attempted in as timely a fashion as possible. Alternatively, time limits are stated applying to each step. Procedural reasons and other causes of delays are to be avoided if possible. An early attempt at resolution also means that the people connected with the dispute should still be available to consult or be directly involved in the resolution. Personnel with first hand knowledge of the dispute will not have moved on. The second and third steps may be qualified, such that matters discussed may be classed as legally privileged in an attempt to make the parties more open, and not fear that the process is being used as a 'fishing expedition' by the other party.

Generally the parties are expected to continue performing their obligations under the contract while a resolution is being pursued. The contract should include 'safety nets' which will enable the project to continue irrespective of the presence of the dispute. For example, there could be inclusion of day rates, and retentions, and time limits for progressing the dispute.

Disputes should be resolved when they arise by the persons directly involved. They are the ones with the best knowledge, experience and intent of the situation. Initially the basis of the dispute should be discussed by the project people in an atmosphere of honesty with a view to finding a fair and reasonable solution. If there are matters that can't be resolved, then these items should be agreed in writing so that the scope of the dispute is clearly defined. However, sometimes personalities intrude, in which case the personalities have to be removed from the problem.

> *Learn ... how to fight without your ego participating.*
>
> *... When a wind of personal reaction comes, do not go along with it.*
>
> Maulana Jelaluddin Rumi, 1207-1273,
> ('The Essential Rumi', transl. Coleman Barks, Harper, San Francisco, 1995)

The next step would be for senior staff from each party to meet and again try to reach a resolution. At this level the commercial consequences of maintaining a hard line will be understood and the wisdom of compromise hopefully will prevail.

Unfortunately, the dispute resolution process is only as good as the commitment of the parties. There are many reasons why one party may not wish to see a dispute settled. For example:

- The owner may see some advantage in delaying payment to the contractor.
- A member of the project team, for example the owner's representative, may be unwilling to admit to a mistake.
- The contractor may be engaging in 'claimsmanship' with a 'straw' claim.

The particular ADR approach may or may not be specified in the contract. Not specifying the particular approach acknowledges the different suitability of different approaches for different circumstances, and allows for the possibility of choosing the most appropriate approach for a particular circumstance. Mediation (non-binding) and expert determination (binding) have some current popularity.

Such a step-approach aims at quick internal resolution of disputes and discourages external intervention. It is also cost efficient and not likely to damage the relationship between the parties. The early steps force each party to listen to the grievances of the other party.

The traditional use of the courts, or the threat of litigation to resolve disputes, puts people on the defensive, with them spending as much time on preparing for failure and the reduction of the chance of any future litigation as they do on the project itself. It becomes analogous to a chess game where each side is making moves to win whilst looking for ways to protect itself. Many standard conditions of contract used to go straight to arbitration as soon as a dispute arose, converting the parties to combatants instantly.

Some contract conditions have dispute resolution clauses that are antagonistic to the contractor. For example, on receiving notification of a dispute, the owner's representative makes a decision as to accepting or rejecting the issues at hand. If rejection occurs, then the matter goes straight to arbitration. Such an approach fails to allow for resolution without a facilitating third party intervening, and discourages the parties from resolving the issues between themselves.

Some dispute resolution clauses cause the delay of the resolution of disputes and entrench both sides in their respective positions on the issues at stake.

Many projects are completed without disputes having to be settled through the formal dispute resolution procedures contained within the contract. Often disputes or differences of opinion are settled by negotiation between the parties during the project. Different interpretations of the scope of work or specification can be settled with joint discussion of the documents and the conceding to the other party's perhaps more logical interpretation.

Negotiation is the least expensive resolution option, and the one most used. Legal advisers are generally not present in project-level negotiations. Both parties should put their cases forward in a logical and chronological format. Often a party will exaggerate its case knowing that the negotiation process will reduce the value of the case, or through offering some of the excess as a compromise. This can jeopardise a good working relationship between the parties.

A goal of a competent contract administrator for the owner is to ensure that procedures are in place so that a reasonable contractor has no reason to declare a dispute. This implies that contract administrators have a good knowledge of the rights and obligations of the parties and understand the difficulties that often face contractors in tendering and contracting activities. Genuine claims resulting from difficulties and increased costs caused by deficiencies in the contract documents should be settled sympathetically and quickly.

8.12. EXERCISES

Exercise 1
Some people include arbitration as part of ADR; others do not. What is your view?

Exercise 2
What skills would you expect an independent third party to have in ADR procedures?

Exercise 3
ADR has some popularity in some countries but does not appear to be as popular as many reports state the case to be in America. Why do you think there are differences between the American and other countries' experiences?

Exercise 4
What are your views of having a dedicated third party (or disputes review board or disputes resolution person) for each contract? The third party would maintain interest in the progress of the project and would make non binding decisions on claims as they arise. Work on the project continues meanwhile with the parties having the option of, say, later arbitration on any decision with which they disagreed.

Exercise 5
Do people wish to be masters of their own dispute resolution, or do they wish a third party to make a decision for them? Discuss.

Exercise 6
What is the situation if the contract contains no statement on how disputes are to be settled, should they occur?

Exercise 7
One way a contractor can submit a non-conforming or alternative tender is to modify the dispute resolution clause in the conditions of contract proposed by the owner. What are the dangers in such a practice?

CHAPTER 9

Disputes and Verbal Contracts

9.1 INTRODUCTION

Both written and verbal (oral) contracts are generally valid, though legislation and prac-tice may insist on written contracts in certain circumstances.

In some situations, people decide not to use a written contract, but rely on a verbal or oral agreement (and possibly implied terms). No written record of the agreement/resulting contract exists. Non-written contracts are widely used in everyday life, in spite of the limitations of human memory over time. Every day people make contracts without putting them in writing, or without putting all of the contract in writing — buying food in shops, travelling by bus or taxi, booking a hotel room by telephone, or calling a plumber out to do a repair. All these involve valid contracts. Multimillion-dollar deals, as well as numerous small deals have been done on a handshake.

Mrs Greenberg and Mrs Friedman, two grandmothers, sat poolside at Grosinger's and chatted about their children.
'My son is a lawyer!' bragged Mrs Greenberg.
'Yeah?' said Mrs Friedman. 'So how's he doin'?'
'My son is so brilliant he could look at a contract and immediately tell you whether it's oral or written.'

L. Wylde, The Official Lawyers Joke Book, Bantam, 1982

While business contracts are generally in a written form, whether it be a detailed document or a note of the overall bargain between the parties referring to the contrac-tual terms, some contracts may be partly verbal, partly written and partly implied. The implied terms might come from conduct, trade regulations, standards, common customs, awareness of normal terms of engagement etc.

With subcontracting practice, oral contracts are preferred in some circumstances. For example, the procedure for obtaining subcontractor quotations may be informal. Quota-tions may be provided at the last minute due to time constraints. They may be provided over the telephone and used as a basis for the contractor submitting its tender. There is nothing in writing.

Labour and equipment may be needed at short notice and for limited periods, and recruited often on a word-of-mouth basis. No written contract is made before the labour starts working.

Advantages

With verbal contracts, it might be argued that disputes are less likely to go to court, as the contract, although binding, may be difficult to evidence. Parties are more likely to agree on a resolution themselves rather than spend money trying to enforce an agreement in court.

Because the basis of the agreement is trust between the parties involved, it might be argued that verbal contracts are possibly less likely to end up in dispute. This saves time and money from both sides.

Where the contract is initiated face-to-face (conversation, meeting, ...), both parties hopefully have a clear understanding of their own expectations of the contract in the same place at the same time. Ambiguities, contradictions or omissions can be raised in the course of discussions under a collaborative approach. This facilitates easier negotiation and counter offers until unconditional acceptance is reached. There may be less likelihood of a requirement for negotiation or changes after the work is commenced.

There can be no claims resulting from discrepancies in the contract documents.

Disadvantages

The disadvantages of verbal contracts over written contracts generally can be observed when disputes, disagreements and problems arise.

It might be argued that there is a greater propensity for disputes because of the lack of written agreed terms of reference. No record of what was agreed upon can make it extremely difficult to resolve the dispute satisfactorily for both parties.

The difficulty lies in ascertaining the precise terms of the contract. Such elements as delivery or practical completion dates, scope or extent of work, price and practically all other conditions can, to a certain degree, be disputed. The terms, that would be set forth in black and white in a written contract, may be the subject of a disputed or contradictory testimony in court. Even with witnesses present at the acceptance of a verbal offer, their interpretations at a later date can be markedly different. For example, inflexions in the voice can change the emphasis of a point, jargon can be understood by some and not by others and people just plain forget exactly what was said. This leads to greater uncertainty as to what the judge or arbitrator will rule.

McKenzie's First Law:
If it's not on paper, it doesn't exist.

The results of any dispute can be very unpredictable and therefore enforcement of the original conditions of the verbal contract may not be possible. This reduces the certainty of verbal contracts and limits their use.

Any goodwill developed with a 'handshake' agreement can be quickly eroded should a dispute occur.

If the parties are unable to resolve their differences, there is generally no predetermined method for dispute resolution or a person nominated to act as an independent umpire.

Dispute resolution

The resolution of disputes may be difficult in any form of contract but verbal contracts create their own difficulties.

Disputes may arise in the quantities being ordered, the wrong description of the materials, or even the wrong delivery to another location. For example, disputes may arise because of extra delivery charges incurred; the supplier may claim for extra handling and delivery charges; the other party may refuse to pay the extra charges.

It may be very difficult to achieve a win–win situation in disputes with verbal contracts. The main reasons for this are that, with the dispute comes a loss of trust, and there is an inability to easily reproduce the details of the contract intertwined with personalities.

Trust is often the first thing to be damaged when a dispute is allowed to escalate. The fairness and reasonableness of one or both parties disappears and along with it the chance of a satisfactory resolution.

Where disputes arise, the situation can often be delicate and in many cases it is due to a misunderstanding between the parties from the outset. There is a need for both parties to satisfy themselves, at the outset, that they completely understand the requirements of the contract.

An approach is to proactively search for the underlying issue and understand each party's position on the matter. It is important that this process is undertaken by both parties together.

In the evaluation of the settlement for the dispute, some of the more important costs to be considered by the parties are the prospect of dissolving the relationship and the cost of seeking external dispute resolution.

In written contracts there are usually detailed procedures on how to resolve any dispute if one arises. But in verbal contracts there is no such reference to dispute resolution. In these cases, all the effort of both sides may be be focused on resolving the differences through negotiation.

Dr Conklin's Summations:
- *Gentlemen's agreements can get very ungentlemanly.*
- *Corollary: Verbal agreements lead to verbal disagreements.*
- *Corollary: In an argument, it is a mistake to allow your opponent to establish the definitions of the situations.*

9.2 DISPUTE RESOLUTION

In an effort to resolve a dispute that has arisen in a verbal contract the following issues are relevant.

Reason for dispute

To resolve a dispute it is necessary to understand the reason why the dispute occurred. This would identify if the dispute originated from the contract content or as a result of its administration. Starting at the beginning and working through to establish where things went wrong can work.

What was agreed

What was the offer? What was accepted? What was the consideration? Do the two parties agree on what was the offer, what was accepted and what was the consideration? During the work stage, were there any witnesses to any conversations that may lead to more information and possibly resolution? And so on.

What the parties agreed upon needs to be established and the issues that they discussed need clarification, for example payment, technical details, and completion time, even though there is no reproducible record. It can be difficult to prove or determine what was actually agreed to. It can be one person's word against another's.

After listening to the viewpoints of each party, there is a need to examine the contract intent and qualify it with the actions that were taken before or after the acceptance of the offer.

A third party asking questions to each party individually (whilst the other party is not present) and then asking more questions whilst both parties are present may help.

Records

A review of records kept by both parties is made.

Having good contracts administration procedures in place would help to support aspects of the contract that may be in dispute.

Expectations, understanding

The expectations and understanding of the roles of each party to the contract are quizzed to establish any commonality.

Personal relationships

Most people believe that they enter contracts with honesty and good intentions. If an issue in the contract is put in dispute, people may consider the dispute as an affront to their integrity or an accusation that they may be lying. With emotions present, resolution is difficult.

Actions

The actions or conduct of each party have to be examined to ensure that they are consistent with the original contract intentions. (In fact their actions may be used to determine their intentions.) This would include variations. It is possible that a variation will have modified the original contract intentions.

Parties to the contract will inherently do this if they negotiate a resolution of the dispute between themselves. If an independent third party is involved, then a more formalised examination of the above would need to be undertaken.

Extrinsic evidence

Extrinsic evidence may be used to help solve problems in verbal contracts. Extrinsic evidence includes: information from witnesses who attended the meeting when the verbal contract was made, records about acceptance of any offer, the conditions that were provided by the offerer, and receipt of any expenses that have been accrued in carrying out the work under the contract.

The surrounding circumstances, which may assist in the interpretation of the contractual terms, are examined.

Witness

Was a third person a witness to the discussions that took place? This person could corroborate the accounts of the discussions that took place.

The presence of any witnesses that can substantiate opinions made by one party can prove crucial in resolving a dispute.

Prior dealings

Whether the parties had dealt with each other before may have influenced the parties as to the way they interpreted any new agreement.

Intentions

Did the parties intend to be legally bound?

Industry practice

An investigation of what would normally be expected within that particular industry may reveal accepted industry practice. For example, a dispute regarding quality could be resolved through a comparison of the product/service with other similarly priced items. This will assist in determining the likely expectations of each party.

The nature of a typical contract of the same type may influence an interpretation of the present agreement. What would it say about the current circumstances?

Is there a relevant statute or code?

Other

What would be an equitable solution to the dispute? Have each of the parties suggested a solution and tried to reach a compromise?

Is an appropriate term to be implied, that is a clear, reasonable and equitable term, which is required in the contract to give it meaning and effectively 'goes without saying'?

Is a term of due care and skill to be implied, where professional services are involved?

Resolution process

Possibly the most suitable method to resolve any dispute would be through negotiation. If not, is there an appropriate neutral person that could better establish a suitable settlement for the dispute, or express an informed opinion on the dispute?

At the time of any contract formation, parties should agree on a dispute resolution process. However as verbal contracts are generally made in a spirit of trust, this issue may not be discussed at the time of formation. Resolving a dispute in a verbal contract has the added burden that each party may sincerely feel that they are correct, because their perception of the original agreement differs.

In the end, if the parties can't find a resolution and the value of the loss is not major, then it may be better for the parties to walk away and learn from the experience, or senior management of both sides can look at the issues and try to reach a common understanding or a compromise.

Ideally the decisions made should be documented in order that both parties have a clear understanding of the result.

When both parties will not accept any compromise, arbitration and litigation are available, but these present difficulties because of the uncertainties present. There are so many uncertainties, that nobody can be sure of being correct. There is also the cost of arbitration and litigation.

9.3 CASE STUDIES

9.3.1 CASE STUDY — REFITTING OF WAREHOUSES

This case study relates to the refitting of warehouses for a client. The developer (A) did not have the necessary expertise and was essentially relying on the designer to define the scope of work. The design was developed by the designer in consultation with the client (B), the ultimate owner of the site. The original contract was documented and, in general, the project was run in a best practice manner. The dispute arose due to some poor practice and misunderstandings.

Bitumen pavement was to be laid as part of the upgrading. Developer A had provided guidelines as to which areas could be paved as part of the work. The designer was responsible for the design and supervising the contractor. During the construction, client B felt that the bitumen should be laid to a further extent than that intended by the designer (using developer A's guidelines). The designer had discussions on site with client B about the extent of paving. At the completion of the project, the extra paving had not been provided. The exact nature of the discussions was conflicting and there was no written record. Client B threatened to sue developer A for the cost of providing additional paving.

The initial negotiation of the variation for the provision of the paving was done verbally. On site, the designer and client B discussed the feasibility of various options for the work.

Although both parties at the time came to an agreement, the dispute arose many months after the discussion (when the bitumen was actually laid). The discussion was not documented. In addition, client B misunderstood the contractual relationships and felt that a verbal contract had been entered into with the designer.

As the dispute arose some time after the discussion, both parties could only recall their perceptions and different understanding of the conversation, as opposed to the actual words used during the conversation. Because both parties had alternative views of the discussion and there was no documentation, detailed examination of the actions and intentions was required, and this was expensive and time consuming. Documenting the conversation and transmitting it to each party in the beginning would have highlighted the misunderstanding, and allowed its resolution before it became an issue. The dispute was resolved in a lengthy process of negotiation. Client B had no legal basis for the action, but developer A had an interest in keeping good relations with client B.

Exercise
1. What practices would you have adopted to avoid the dispute in the first place?
2. What practices would you have adopted to resolve the dispute?
3. Given the nature of the dispute, was it the correct decision to go with a verbal contract in this case? If not, why not?

9.3.2 CASE STUDY — RESIDENTIAL BUILDING

Company Xform had a good reputation in the execution of residential building projects. Pmac was a new construction company that was interested in entering the residential building market.

Pmac was interested in tendering for a residential building project. However the company did not have any experience for that kind of project, one of the requirements for tendering. To solve that problem Pmac asked to use Xform's name, in order that Pmac was eligible to participate in the tender. Xform agreed to Pmac's idea, with one requirement — if Pmac won the tender, Pmac had to pay Xform 10% of the contract price. That agreement was verbal. Pmac, which used Xform's name, won the tender. Because Pmac had used Xform's name, the person who signed the contract had to be one of Xform's directors. In this case, the contract was between Xform and the owner. However, Pmac would do the work and all payment would go to a project account number which Pmac had full access to.

Unfortunately the project did not run smoothly. Pmac had early financial problems because of its overheads, the owner of the project had not made any up front payment, and only a small amount of expenses could be claimed.

To solve the problem, Pmac entered into an agreement with a company Bodary. There were two requirements from Bodary. Bodary agreed to finance the project before obtaining payment, if Bodary had financial control and several of Bodary's staff were involved in the project. Pmac accepted those requirements. After reaching this agreement, Pmac and Bodary did not put it in writing.

The problems started when the payment from the owner was released. Because the owner made the payment to the project account, that only Pmac had full access to, Bodary did not get access to the payment. As a result Bodary could not control the project finances. This situation caused Bodary to stop funding the project until Pmac gave it access to the account. However Pmac did not give access to the account to Bodary.

The results of this dispute were:
- The project went behind schedule.
- Xform lost its reputation with the owner.
- There was uncertainty on the project site.
- There was labour payment trouble.
- The project went over budget.

The side effect for Pmac and Bodary was that they lost their good reputations because the media published their dispute.

A side effect caused by the dispute for the owner, was that the owner had to pay for accommodation for its staff who should have been housed in the new residential buildings.

The case went to court. The court decided that Xform was responsible, because Xform's director signed the contract. Bodary was successful in obtaining all its expenses for the project by showing evidence of all receipts, and having several witnesses who attended the meeting where the verbal agreement was made. Xform had to pay all of Bodary's expenses for the project. Pmac lost its business licence.

Exercise
1. What practices would you have adopted to avoid the dispute in the first place?
2. What practices would you have adopted to resolve the dispute?
3. Given the nature of the dispute, was it the correct decision to go with a verbal contract in this case? If not, why not?

9.3.3 CASE STUDY — SALE OF PROPERTY

Tan in Hong Kong was to advertise, market and sell, to people in Hong Kong, a group of new townhouses built in Australia by Choong (a developer of Hong Kong descent).

Choong had been recommended to Tan, and so the arrangement between Tan and Choong was friendly. They had a verbal contract, whereby Tan would receive a commission of 10% for each townhouse sold.

Tan advertised and marketed the properties, and managed to sell all the properties without the developer's help. When Tan travelled to Australia to collect the commission, a dispute occurred. The amount was less than agreed. Choong explained that the 10% included the selling costs (advertising, marketing and setting up a display in Hong Kong). Tan had assumed that the selling costs were normally excluded from the commission. Both parties had made different assumptions. There was a misunderstanding.

Exercise
1. What practices would you have adopted to avoid the dispute in the first place?
2. What practices would you have adopted to resolve the dispute?
3. Given the nature of the dispute, was it the correct decision to go with a verbal contract in this case? If not, why not?

9.3.4 CASE STUDY — NON-PAYMENT OF CONSULTANT'S FEES

This case study involves an experienced architectural designer. The designer enjoyed her work and took it as a philosophy that once a design brief was stated, she would work until the design was completed, in a professional manner.

The designer did not advertise her services. All requests for services came from clients who had been built up over many years. This was normally the case with builders, with whom the designer had developed ongoing relationships, and from word-of-mouth with past clients passing on a recommendation to prospective clients.

History

The designer had done many jobs on a personal basis and generally had no problems obtaining payment using a verbal agreement. Most clients realised that the designer was doing the work unencumbered by large overheads and was able to provide a cheaper service than an architectural company. It was common to have deals done on handshakes and word of mouth.

The designer had only once before encountered problems with a client, in a situation where the design ended up outside local government guidelines, but was encouraged by the client to go ahead despite the inevitable outcome. The outcome was that the local government rejected the scheme. Advice was given to the client on how to alter the proposal, and thereby enable it to be passed. But the client refused to pay for the work. Numerous unsuccessful telephone calls to collect fees were made. This situation was dismissed as a bad debt.

The client's residence

The client (a married couple) heard of the designer through word-of-mouth. The client couple telephoned and organised a meeting on site to discuss the extension and refurbishment of an existing house. At this meeting a design brief was established, even though there was a slight conflict in ideas between the married couple — one seemed bent towards a quality 'Vogue' style, while the other had a main concern of cost (maximum $100k cost). A selection of fee proposals was discussed:
a) Work done on an hourly basis and charged $500 progressively (to keep the client aware of further expenditure).
b) Fee based on a percentage of the cost of the building work (based on quotes from an assortment of builders).

The agreement was verbally made between the client and the designer that the fee would be based upon a percentage of the final cost of the building work. (However the total cost and the percentage figure were not agreed.) A cheque for $500, showing goodwill, was forwarded and the mood was set for a working relationship that would provide, as an end result, a successful design and a designer rewarded for her work.

The design was worked and manipulated in an ongoing interaction between the designer and the client. A further cheque for $500 was forwarded to the designer.

After months of consultation, the drawings and documentation were produced, approved by the client and subsequently submitted to local government for approval. The designer estimated the cost of the construction ($250k) and multiplied this by a percentage (6%). An invoice (Figure 9.1) was submitted to the client.

Invoice

Dear ....

I now advise indicative fees for architectural services for your consideration which would be subject to adjustment based on the final cost of the building work carried out. The approved design proposal is, as stated on the building application, an increase in floor area of 250 square metres. This increase means that, based on indicative guides for building costs, the proposed work will cost approximately $250000. This figure will need to be adjusted as further information is obtained from the Quantity Surveyor and Engineer and the quotes from prospective builders.

Fees, in accordance with the Professional Association recommended fee guidelines would be 6% of the cost of work outlined above (subject to adjustment on final cost of building work), that is:

6% of $250000 $15000

This fee does not of course include for any other consultant services, fees payable to the various Authorities or any other building levees.

For the purpose of progress payments, fees are apportioned in stages, as follows:

Stage 1: Design/Design Development
25% of $15000 $3750

Stage 2: Documentation for submission to Local Government for building application including shadow diagrams
40% of $15000 $6000

Stage 3: Contract administration (if required)
35% of $15000 $5250

TOTAL: $15000

To date my services have completed Stages 1 and 2. Should you require my assistance during the construction stage, regarding contract administration, that is inspections during the progress of the work, this would be covered by Stage 3.

I trust that the indicative fees meet with your approval and enclose my account for services to-date.

Yours faithfully

Designer

Figure 9.1. Designer's invoice.

Three months later, no payment had been received. On inquiry, it was stated that no payment would be made due to:
- Wrong roof finish specified.
- Ridge height contravened local government regulation.
- Street setback contravened.
- Removal of major trees questionable.

These accusations were made when payment was requested. Because the designer knew that the claims were either false or arguable in terms of local government regulations, she protested the non-payment of fees. It was then left for the client to organise a meeting with the local government, to establish whether the local government would either accept or reject the design. This was not done.

During the next conversation, the dialogue became more personal. The client still would not pay, giving the following reasons:
- The drawings took too long to produce.
- The documentation had been shown to a builder friend who had commented that he could not build from such substandard documentation.

The suggestion was then put to the client couple that they were attempting to avoid the payment of fees due. This was flatly rejected and the conversation concluded.

The next telephone call, two weeks later, was very argumentative with accusations of non payment of fees, and retaliatory accusations of substandard work, and then claims that the initial fee agreement had been for less than the invoice sent. This resulted in the client offering to pay an amount of money ($3000) that it considered had been agreed. This was rejected by the designer. The designer then approached the local magistrates office for legal advice. This service was free and was used because of the perceived high cost of legal representation. The advice was that verbal agreements were not worth taking to court, but that mediation was a possibility.

Comment

The issue at stake was that money had not been paid for the design and documentation of a large prestigious house. The client couple wanted a new house for their family but, as the working relationship developed, the designer became suspicious that the client could not afford such a house, yet was not willing to compromise on the design. The designer wanted to design a home unique to this family and hoped it would further her reputation and bring her more work. These hidden agendas of both parties caused friction, yet neither party would compromise. In the end, both parties felt unsatisfied. Rather than being proactive in resolving the conflict, both parties displayed a reactive manner, and the situation rapidly deteriorated.

It appears the client couple wanted to build a house that they could not afford, and manipulated the situation so that they could receive the design and documentation, but then refuse to pay for it by citing problems. The client did this by:
- Deceiving the designer, by handing over two cheques for $500 each, leading the designer to believe that the client was able to afford such a service.
- Constantly referring to the designer's status as still being only a student and doing it in her spare time.
- Making threats to the designer over the telephone and in person, in an attempt to

undermine the designer's confidence and hoping she would forget the whole thing and accept a token $3000 payment.

The designer, on the other hand, failed in many ways and therefore incited antagonism from the client. The designer did this by:

- Not putting any of the initial discussion and fees in writing.
- Did not appear to listen to the client's initial concerns about the cost of the work, and should have foreseen trouble when the client insisted that the extension should only cost $100k, instead of the estimated $250k it would have allegedly cost.
- Leaving long contact periods between herself and the client, which only increased the anger and hostility.

Options

The options that were available for the designer were:

- To dismiss the incident as a bad debt.
- To negotiate with the client, yet previous attempts had proved fruitless.
- To mediate.
- To litigate, to enforce copyright.

The options that were available for the client were:

- To pay/refuse to pay.
- To mediate/refuse mediation.

The end result

The situation ended up being a win–lose one. The client couple ended up building the house that they wanted for their family, and paying the designer $3000 in total, the designer ending up virtually working for nothing.

Exercise

1. Where were the seeds for the dispute first sown?
2. Given that it was the designer's industry, yet a one-off sortie into the industry for the client, what extra onus does this put on the designer to adopt management practices that avoid disputes?
3. How does the client's 'innocence' assist it in negotiation with the designer?

9.3.5 CASE STUDY — ARCHITECTURAL SERVICES

The following case study centres on three parties — an architectural firm (Hornsby), its client (a school) and the site project manager. As of writing, the case stands unresolved. Different attempts at resolution have been adopted. These have been largely unsuccessful.

The cause of the conflict

An architectural firm, Hornsby, was invited to enter an architectural competition for the design of a school. The firm won the competition and was appointed by the school to carry out the work. An agreed fee was decided upon with the school. The architectural firm engaged, on behalf of its client (the school), a team of sub-consultants (comprising: quantity surveyors, civil, mechanical and structural engineers and landscape architects). It was to subsequently organise the construction through tender and appointment of a contractor. The school had made the decision that it was unable to manage the project, and consequently commissioned a project manager to act on its behalf.

The first point of conflict

The project manager's first act was to write to the architect stating that the fees previously decided upon were too high and approximately twice what the project manager considered in his opinion to be standard. The fees were consistent with the scale recommended by the architect's professional association. This disagreement signalled the first point of conflict.

The architect accordingly made the decision, owing to the aggressive nature and manner of the project manager, to withdraw from the project politely and advised the school of its actions. The architect considered that it was not appropriate to proceed with the project. The client however appealed to the firm to rethink its decision and organised a mediation session. This was carried through with four parties — the head of the school, the project manager, the firm of architects and a mediator. The mediation proceedings went for approximately two hours and were held on neutral territory. It was agreed that the architect would remain with the project at the fees previously agreed upon. There followed, after the mediation, a prolonged exchange of correspondence between the project manager and Hornsby, in an attempt to finalise the terms of engagement (the contract for architectural services).

Meanwhile, work on the project continued. Unfortunately final conditions of engagement were still not agreed. The project manager proposed one set of conditions, part of which was unacceptable to the architect; for example, the 'Fitness for Purpose' clause — warranting that the end-product would be fit for its intended purpose. Eventually these exchanges between the project manager and the architect became filled with tension, and the architect began to gain the impression that the project manager was seeking some revenge. Further internal discussion ensued between the partners of Hornsby as to whether to proceed or withdraw from the project. However, given that the firm had promised the client to proceed, it felt compelled to carry on.

Second point of conflict

The project manager did not reply to the architect's proposed changes in the contract conditions and, despite many further reminders and requests, nothing has yet happened. The work continued on site but without the architect's chosen consultants. The project manager had appointed his own team, several of whom were from subsidiaries of his own company.

Some mistakes were found in the architectural documents — some partially due to the consultants which failed to meet the program, and some due to communication errors between the architect and the client. In the normal course of events, such errors would have been identified during construction and rectified without any drama.

However, because the project manager held total control over the site, the architect was not informed, or given the opportunity to amend the faults in time. Nothing was said or heard until the completion of the project's first stage, when Hornsby submitted its invoice.

The project manager subsequently deducted the costs of the variations from the cost of the architect's fees. An analysis of estimated costs, the value of the errors, was perceived by the architect to be approximately one tenth of what the project manager had deducted.

By this stage, relations on site had deteriorated to the extent that the project manager had threatened to assault the site architect. A director of Hornsby stepped in and requested a meeting with the client and the project manager.

The outcome

The outcome of this meeting was unclear. The architect suggested that the correct way to proceed was for the school to make a claim against the architect's Professional Indemnity Insurance, where the matter could be resolved in a legal fashion. The school promised to reply within a period of two weeks. Three months later however and nothing has been heard of from the client. The architect approached its Insurance Company, because part of the policy allows for the recovery of fees.

Hornsby held a meeting with its lawyers, who advised that litigation should only be used as a last resort. It was decided that one last effort would be made to resolve the matter. The architect consequently made an offer of a 'Monetary Settlement without Prejudice', approximately 10% of what the school claimed the architect owed it. Its final demand was settlement of other outstanding claims for additional fees that had been incurred. This stands as the current state of play.

Exercise

1. Several matters arose during the project. Initially the agreed fee was disputed by the project manager, secondly a written contract was never implemented, thirdly there was an underlying tension between the architect and the project manager, and fourthly a lack of communication led to the prolonged nature of this conflict.

   The architect perceived a conflict to exist between itself and the project manager. It assumed that the project manager was out to seek some form of revenge. How real is the architect's perception? Is 'revenge' a likely motive?

2. The work proceeded on the project without a written agreement in place. A contract would list, amongst other things, the terms of engagement and the methods of dispute resolution to adopt if complications were to arise. What risks do the parties take proceeding without a written agreement?

3. The project manager's and architect's reluctance to communicate contributed to this conflict. Prolonged periods in which no correspondence took place should never have happened. How do you establish communication between two parties who no longer respect each other?

4. The project has reached the stage now where the conflict is not a single problem but rather is composed of multiple layers — fees, design variations, no formal contract, ... What are the possible paths to resolution?

5. When teams work together cohesively and are in alignment about their purpose, their work benefits. However when there is conflict the effects are felt throughout both

parties. Conflicts erode team effectiveness, thwart improvement initiatives, and waste time and effort. What does this case study say in terms of: the cost of taking time to resolve conflicts compared with the cost to manage strife; the resolution of conflicts as early as possible; and the escalation effect on costs and time of prolonging the resolution of disagreements?

9.3.6 CASE STUDY — HIGH-RISE BUILDING

General description

The project consisted of construction in three parts corresponding to levels in the building: basement car parking space, shops and restaurants, and office space.

The owner–developer used its in-house design team for concept design, but engaged a contractor to prepare the detailed design and documentation in accordance with the concept design and specification. The contractor engaged subcontractors through tendering.

The electrical design and installation subcontractor was selected due to its reputation, its previous relationships with the contractor, its past ability to meet completion dates and budgets and maintain quality. The subcontract was documented and, in general, the project ran smoothly. However there were some issues, which arose due to misinterpretations, which led to disputes.

No significant problems occurred in the first few months of the electrical design. However, shortly after, the contractor verbally ordered the subcontractor to: provide extra power supply for the ground floor of the shopping centre; and to rewire the second and third levels of the shopping centre due to a change in the structural scheme and resulting obstructions. The estimated cost of the variations was verbally agreed by the contractor. However, there were no records, no witnesses and no follow-up in writing to confirm this conversation. The agreement was based on trust.

Unfortunately the subcontractor's interpretation of 'extra power supply' and 'rewire' was different to the contractor's. The subcontractor believed that it was to provide an extra power generator for the ground floor, and to move all the existing wiring and the electrical fittings to another location. However, the contractor's interpretation was to provide extra power points for the ground floor shopping centre, so that there were enough power points for organisations to display or promote their products; and rewire meant to relocate all the existing wiring to suit the change to the structural design of second and third levels of the shopping area.

The subcontractor started to bill the contractor for the cost of the variations. However the contractor was unwilling to pay for the work done, as it argued that it had provided sufficient information for the subcontractor to perform the work in the way it had intended it to be done.

Negotiation

In an effort to resolve the dispute, negotiation was started.

Luckily, there was a good personal relationship between the parties. Emotion was not a problem in resolving the dispute. There were no personal grudges that might have affected the resolution.

The parties tried to determine a solution, which suited both parties' interests. The final agreement was that the contractor accepted to pay for half of the variation work done, but the subcontractor was to provide extra power points for the ground floor and rewire the second and third floors of the shopping area.

Following this, a written agreement was signed by both parties. In this document, the parties were careful in specifying the scope of work, the payment and other terms.

Exercise
1. In verbal contracts there is always the difficulty of establishing the exact nature of the agreement if a dispute arises; that is it is an evidentiary problem. How might past dealings between the parties be used in establishing what the terms of the contract might be?
2. How might the parties' expectations and understandings be used in establishing what the terms of the contract might be?
3. Better understanding of a dispute can lead to an earlier resolution, which means a reduction in wasted time, administration costs and other overhead costs on the project. What practices might the parties adopt in order to understand the dispute better?
4. Why do you think the parties to this dispute didn't follow recommended practice of committing the change request to writing in the first instance?

9.3.7 CASE STUDY — SOIL TESTING

This case study refers to a soil testing service used by developers for compliance testing of roads. Most soil testing used to be arranged on the basis of a telephone call.

A large percentage of the business for the soil testing company was carried out for a large developer. Typically, the company would be commissioned by the developer's surveyor/project manager. Invoices would be sent to the surveyor, who would pay based on agreed unit rates.

The parties involved were well known to one another, having worked together for some years. This included the contractors, surveyor, developer, and the soil testing company.

During one project the number of soil tests was significant. This was, in large part, due to numerous retests of failed tests. It transpired that the contractor was not bothering to rework the areas that had failed earlier tests, rather requesting retest after retest until gaining a pass for the area. This practice was discovered by the developer's project manager, when it was noticed that equipment was not on site during a period in which retests were ordered.

On learning of the practice, the soil testing company was advised that the developer was not going to pay for the additional tests, and that these costs should be claimed from the contractor instead. As there was no contractual agreement between the soil testing company and the contractor, the company was unable (and unwilling) to do so.

Exercise
1. The dispute dragged on for a significant period of time, the developer's project manager unwilling to move from his position. When the dispute was eventually settled,

much goodwill had been eroded between all the parties involved, to the degree that the soil testing and other services for projects changed from that point onwards. While still initiated verbally, the testing service was put under the contractor's responsibility. Is this a satisfactory solution? What other solutions could you suggest?

2. The use of verbal contracts, prior to the dispute, was due mainly to the reluctance of the developer's project manager to formalise contracts in writing. In this case the project manager avoided, as far as possible, putting contracts in writing, preferring to manage the contracts by telephone conversations. While this made the management easier from his own perspective, it made it more difficult to manage from the viewpoint of the developer. The new arrangement involved written contracts. What has to be done to make the verbal contract approach work?

9.4 EXERCISES

Exercise 1
It is often not until a disagreement occurs that people realise that their expectations and understandings are different to the other person. How do you address this in order to avoid any disagreement?

Exercise 2
How do you deal with a person who changes his/her stance, presumably unwittingly or unknowingly, but not deliberately? The person is unaware that s/he has a shifting basis for his/her stance. The person genuinely cannot maintain a consistent logic from one time to the next.

Exercise 3
What legal force does extrinsic evidence (evidence outside the contract) have in establishing what the agreement is between the parties?

Exercise 4
Is it likely that a binding form of alternative dispute resolution (ADR), compared with a non-binding form of ADR, will be more/less successful in resolving a dispute in a verbal contract?

Exercise 5
In a dispute involving a verbal contract, what is the likelihood of negotiation succeeding if, at that point in time, personal relationships of the parties had soured? What alternative to negotiation would you recommend in this situation?

CHAPTER 10

Disputes — Case Studies I

10.1 OUTLINE

A number of case studies illustrating dispute occurrence and the handling of disputes are presented from a range of industries and situations. A common culture was present in each case study.

10.2 CASE STUDIES

10.2.1 CASE STUDY — TUNNELLING

Background

The dispute occurred on a major tunnelling project. The tunnelling project involved excavation, followed by lining the tunnel with mild steel concrete-lined pipes, with the annulus between the exterior of the steel liner and the tunnel wall backfilled with concrete.

The agreement for both the excavation and lining of the tunnel, and the fabrication of the steel tunnel liners was contained in a single contract. The reason why the contract was structured in this way was because of the very tight period of time available for project delivery, and it was felt that the liability of the owner to delay claims would be greatly increased if two separate contracts for the excavation and tunnel liner fabrication, respectively, were let.

The contract was awarded to a highly experienced and specialised tunnelling and mining company. This head contractor then awarded a steel liner fabrication subcontract to a specialist steel fabrication contractor. Liner fabrication was scheduled to proceed in parallel with the excavation of the tunnel, with sufficient liners to be delivered to site by the end of the excavation phase to enable an immediate start to the liner laying activity.

The steel liner was made in sections. They were manufactured by rolling and welding steel plates together in an automated process. The ends of the pipes were required to have a spherical spigot-and-socket joint profile that would allow the pipes to be married in the tunnel and the spherical joint arrangement would allow the necessary pipe deflection around curves. The pipe joints were to be fully welded in-situ. Upon completion of the fabrication of the steel liner sections, a welded mesh was spot-welded to the interior of the liners. The steel liner sections were then transferred to a pipe rotating machine which spun the steel liners at high speed whilst a concrete lining material was introduced. The centrifugal force caused by the rotating liner compacted the concrete to a very dense state.

The concrete lining was then steam cured, tolerances checked, and the liner despatched to the tunnel site.

Specified tolerances of the tunnel liners applied to the following:
- Circularity of the barrel, spigot and socket.
- Circumference of the barrel, spigot and socket.
- Internal bore.
- Squareness of liner ends.
- Length of pipes.
- Straightness.
- Lap length of married joint.
- Radius over longitudinal weld.
- Longitudinal profile of spigot and socket.
- Concentricity of concrete lining.
- Concrete lining thickness.

The head contractor mobilised to site soon after the award of contract and the shaft sinking and tunnel excavation work proceeded well. Alas, not so the liner fabrication work.

The fabrication subcontractor had elected to procure the pipe rotating machinery from overseas. Delays were experienced in the manufacture of the machinery, and to compound on this, there was a mix-up in the shipping of the equipment, with the machinery being off-loaded by mistake in transit. The result of this was that the pipe rotating machinery arrived late. In addition, the fabrication subcontractor was experiencing difficulties in forming the spigot and socket profiles within the required tolerances. The profiles were initially stamped in the flat plates, however, when these plates were rolled to the required radius to make up the barrel, the profiles distorted due to work hardening from the stamping process, and a uniform marry-lap of the liners could not be obtained. This was of particular concern, as the confined working space in the tunnel required that the liners marry to the required lap-length and without the need for dogging and wedging.

After the delayed commissioning of the pipe rotating machinery, it was found that the fabrication subcontractor had great difficulty in providing a concrete lining concentric with the axis of the liner. As a result of all these difficulties, the subcontractor had many liners in its fabrication yard, and none of these met the required tolerances.

In the meantime, the head contractor had completed the excavation phase in very good time, but unfortunately had no liners to install. The head contractor was unable to proceed with the work and was incurring costs. As a result of this, the head contractor sued the subcontractor and attempted to join the owner in the action.

The owner's representative was extremely apprehensive about this dispute, and could foresee both parties retreating behind their contractual battlements, the result of which would be the project languishing for a very long period of time.

The head contract utilised a standard form, and there were clauses included in it for a two-stage dispute resolution process: firstly by expert appraisal, and secondly by arbitration.

Resolution of the dispute

As the contract contained alternative dispute resolution provisions, the owner would not agree to being joined in potentially lengthy and costly litigation.

A court hearing was held, and the owner argued that, as there were dispute resolution mechanisms contained in the contract, it was inappropriate to commence litigation until such time as these existing dispute resolution mechanisms were tried. The court agreed and the action before the court was suspended.

The owner's representative convened a series of tripartite meetings with the contractor and subcontractor in an attempt to negotiate a settlement. The first stage of this series of meetings was to attempt to compile an agreed set of facts surrounding the dispute. An agreed set of facts included the number of liners fabricated and the extent of the tolerance violations. The owner's representative realised that the project was in jeopardy and that a solution to get the project moving again must be found as a matter of urgency (subject to the quality and serviceability of the asset not being compromised).

The specification tolerances were re-examined, and the owner's representative conceded that certain specified tolerances appeared contradictory and a clarification/relaxation of certain specification tolerances was granted. The owner's representative conceded that it was impossible to maintain a uniform thickness of concrete lining around the circumference of the liner if the bore was permitted to vary in size.

In view of the above, the owner's representative agreed that the specified thickness of the concrete lining was a nominal thickness only, and a relaxed tolerance for the concentricity of the liner was granted. A relaxation of the tolerance on the spigot and socket profile was also granted; although this was not strictly in accordance with the design requirements, it did not affect adversely the jointing of the liners.

This relaxation of the required tolerances resulted in approximately half of the tunnel liners in stockpile then being acceptable for inclusion in the work. The main contractor then had sufficient liners to recommence work on the project, and the subcontractor was able to consistently produce liners to the relaxed tolerances.

This resolved most of the dispute. There was still the question of which party paid for the delay.

Although there were grounds within the contract for the owner to deny all liability to bear any of the costs arising out of this dispute, a pragmatic assessment was made that to do this was not in the best interests of the owner or the project. Both the head contractor and subcontractor had lost heavily from the delays, a reworked product and abortive litigation, and to leave this situation without a mutually acceptable compromise would be detrimental to the future progress of the work and place the completion of the project in jeopardy.

A commercial decision was therefore made by the owner to offer the contractor payment of delay costs for the period between when notification of specification anomalies occurred and when the relaxation of tolerances was granted. This resulted in the owner accepting liability for half of the delay and extending the time for practical completion. The balance of the delay costs and time (due to late delivery of equipment etc.) was to be borne by the contractor and subcontractor in proportions to be determined by themselves.

This settlement offer was subsequently accepted.

Exercise

1. In the case, the owner's representative was apprehensive about the liner supply situation long before the dispute became manifest. The matter of tardy mobilisation of the pipe spinning equipment was brought up at several progress meetings, and assurances were given by the contractor and subcontractor that everything was all right and that there was sufficient program float to make up the lost time. The liner fabrication subcontractor had an antagonistic demeanour and would not allow the owner's staff to take progress photographs of the fabrication processes. Progress information was scant and reluctantly given. Due to the geographic separation of the fabrication plant from the construction site and the secretive nature of the subcontractor, more stringent supervision by the owner's representative was difficult. The head contractor was experiencing similar difficulties with the subcontractor, and the owner's representative could detect that relationships were becoming strained.

   At the time of the completion of excavation, the owner's representative could perceive that the dispute was very serious indeed, was a severe threat to the timely completion of the work, and had the potential to drag on in litigation for years to come. The subcontractor had made a claim under the subcontract regarding conflicting fabrication tolerances, and the head contractor had passed this claim on to the owner under the provisions of the head contract. Heavy financial losses were being incurred by the head contractor, and a solution to the dispute had to be quickly found.

   Having identified that a dispute was brewing, what actions should the owner's representative have taken at this time?

2. Because there were provisions in the conditions of contract for alternative dispute resolution, the potentially protracted and costly litigation was stopped and the parties brought together in negotiation in an attempt to obtain an amicable settlement of the dispute. The owner successfully argued before the court that the litigation was premature, and that the due process of the contract should be followed prior to any resort to litigation. The parties were then bound to attempt an alternative dispute resolution of the dispute.

   Before the matter could be referred for expert appraisal, the owner's representative was required to make his determination of the claim. Although there was pressure on the owner's representative to make a rapid determination, he resisted this pressure. The owner's representative was of the opinion that a brake was required on the momentum of the events, and that the emotion surrounding the matter had to be stripped away, and the claim determined on a logical examination of the facts.

   A series of tripartite meetings was then held between the owner's representative and senior management of the contractor and subcontractor in an attempt to resolve the issues. The meetings were tense and abrupt at the start, but the owner's representative persevered until an agreed set of facts was obtained.

   What pitfalls should the owner's representative be aware of in firstly slowing the process down, and secondly in trying to resolve the dispute on a person-by-person basis?

3. The owner's representative realised that any outcome that resulted in a win–lose or a lose–lose situation would not be acceptable, as it would result in appeals and the unresolved matters proceeding to expert appraisal, arbitration and possible future

litigation. The cost of this course of action would be substantial to all parties involved and there would be no ultimate winner. A compromise or win–win situation had to be encouraged. It is for this reason that a commercial decision was made to settle the contractor's claim (in part) even though this need not have been done under the provisions of the contract. This concession led to concessions from the other parties, and the dispute was eventually resolved.

What calculation has the owner done here to justify to itself that it has also won? Is this justifiable?

10.2.2 CASE STUDY — SEWAGE TREATMENT PLANT SCREENS

What the project involved

The contract for the screen erection divided the project into three separable portions corresponding to each of the three screen installations. The first two screens were extensions of existing screens, while screen 3 was an additional screen that had to be installed. The aperture of the first two screens had to be altered.

The contract nominated a date for work commencement (or of possession of the site). The starting dates and the date by which each screen had to be completed were outlined in the contract.

As the work proceeded, the contractor fell behind the construction program on screen 1, because of various reasons. The contractor therefore applied for the possession of screen 2 so it could begin work on screen 2, which would allow it to make up some lost time. The existing sewage system could not deal with two screens being off line simultaneously. Therefore, the possession of screen 2 was refused.

The contractor had the hardness of the rock, that had to be excavated, tested. Tests revealed that the rock had a strength twice that nominated in the contract.

During the project, screen 3 was flooded with raw sewage for several days. The flood occurred immediately after the reinforcement had been placed. It took a couple of days for the contractor to dismantle the formwork so that the debris that had attached itself to the reinforcement could be removed.

At the completion of the project, the contractor compiled a claim which contained:
* Latent conditions — for the hard rock.
* Possession of site.
* Liquidated damages.
* Extension of time.

General

The contractor was able to find a 'loophole' in the contract regarding possession of the site. The contract did not specifically state that two screens could not be worked on simultaneously. The dates that were given were very ambiguous and the owner did not show any link between these dates. Therefore the contractor was able to take advantage of this and prepare a claim for possession of the site.

The contract was unclear and ambiguous. If the owner had taken more care in preparing the contract documents and made sure that there were no 'loopholes', then there would have been less exposure to a dispute.

There was ambiguity in the contract about two screens being worked on simultaneously. Under a contra proferentum rule, the ambiguity in this document would be interpreted against the owner, the author of the contract documents, and in favour of the contractor.

Exercise
1. Latent conditions. The contractor submitted a latent conditions claim for the difference in the strength of the rock that had to be excavated. The contractor argued that it should be reimbursed for the 'extra effort' that was needed to excavate the rock, that was double the compressive strength of the rock outlined in the contract. The contractor argued that it could not reasonably expect this difference in rock strength when tendering for the work. The contractor had to carry out additional work that was more involved than expected and additional construction plant were needed in order to excavate the rock at a faster rate.

   Visual inspections at the time of tendering would not have been enough to determine what the strength of the rock was. But should the contractor have gone further than visual inspections?
2. Flood. The contractor also put in a claim for the delay due to the flood. Although this was a natural cause, it was the contractor which had lost time and money due to this event. What is the contractor entitlement, and what is the owner's position? Should the contractor receive its overheads during this delay period?
3. Possession of site. A possession-of-site claim was prepared by the contractor because the contract did not state that two screens could not be worked on at the same time. Although the owner had given starting and completion dates for each screen, there was no link indicated between the three screens, although it could be implied that one screen was to finish before the next one was started. The owner did not want two screens to be off-line at the same time because the system could not efficiently cope. However, this was not stated in the contract. There was ambiguity in the contract. How successful do you think the contractor's claim would be?
4. What part did dividing the contract into separable portions play in exposing the owner to disputes?
5. Liquidated damages. The contractor failed to reach practical completion by the date stated for practical completion due to many factors, such as the extra hardness of the rock and the flooding. The owner had already deducted the liquidated damages, but had to repay the contractor. Comment on the fairness of such an approach to both the owner and the contractor.

10.2.3 CASE STUDY — CONTRACTOR-SUBCONTRACTOR DISPUTE

The case study discusses a project where untidy contractual practices resulted in a dispute between the contractor and subcontractor. The dispute was eventually resolved.

A number of subcontractors were asked to submit tenders for the fabrication and erection of some structural steel. All the selected subcontractors were known to, and had previously done work for, the contractor. All tenders were assessed according to time, cost, and quality, and it became a choice between two subcontractors — subcontractor A or subcontractor B. The subcontract was finally awarded to subcontractor A, the cheaper of the two.

Subcontractor A had problems with meeting quality standards and timely completion of the steel fabrication. Meetings and shop visits were conducted to investigate the problems and then find solutions. The poor quality and delay of fabrication continued. Corrective action needed to be taken to avoid a delay to the project. The contractor conducted meetings with subcontractor A in order to prevent any further delays to the project. Eventually, the contractor decided to terminate the subcontract, claiming breach of contract.

Subcontractor B was approached and was offered the work to complete the fabrication and erection of the structural steel on the project, and that if it accepted it would need to start immediately. Subcontractor B replied by stating that it was willing and able to accept the offer, however it was not clear on how much work had been done by the previous subcontractor and how much rework was required, and therefore it was unable to submit a price immediately. An agreement was reached that subcontractor B would start immediately (the next day) and that it would submit a price within two weeks.

Due to a large work load, subcontractor B had to sub-subcontract most of the work.

Two weeks elapsed and subcontractor B had not submitted its price. Subcontractor B was reminded and it requested further time. It was realised that subcontractor B had a cost plus arrangement with the sub-subcontractor. The contractor became concerned of a potential price 'blowout'. Much written and verbal correspondence was conveyed to subcontractor B regarding the price. The structural steel was complete and subcontractor B had still not submitted its price.

Approximately two weeks after the completion of the structural steel fabrication, a price was received from subcontractor B. The price totalled twice B's original bid. The contractor was expecting a price of the order of B's original bid. An independent quantity surveyor was commissioned to assess the value of the work, and this supported the contractor's expectation.

A meeting was held to discuss the reason for the high price. Subcontractor B presented all its invoices to demonstrate costs, which were proven to be correct and legitimate. However the contractor argued on two points:

- The price was received very late. If this price had been received earlier, corrective action could have been implemented.
- The market price for the work was close to A's and B's original bids.

Based on this argument, the contractor stated that it was willing to accept a price slightly above the original bid.

After several days of meetings and negotiation, the contractor offered to pay subcontractor B slightly above its original bid. Subcontractor B rejected the offer. At this stage neither the contractor nor subcontractor B would negotiate any further. Three options (mediation, arbitration, litigation) were considered in order to resolve the dispute. The parties agreed on mediation.

The mediation was successful. The final decision was for the contractor to pay more but not the full amount claimed by the subcontractor. The resolution was influenced by:

- The subcontract was effectively cost plus in nature.
- Subcontractor B had arranged a cost plus contract with the sub-subcontractor.
- Although subcontractor B incurred extra costs, these costs appeared to represent inefficient practices and/or management and these inefficiencies should not be passed to the contractor.
- Subcontractor B showed no effort to communicate any cost implications to the contractor during the course of the project.

- Both parties realised that if an agreement had not been reached through mediation, then the costs of arbitration (or litigation) would have far outweighed the costs in dispute.

Exercise

1. Subcontractor A was in breach of contract and hence the contract was discharged. In such circumstances, is the contractor obliged to pay subcontractor A for the completed non-defective work? Would a quantum meruit basis apply? Or is the contractor only obliged to pay a total price for the total work?
2. Could the contractor sue subcontractor A for the difference in price between subcontractor A's price and the final price of the work? Would it be worth the contractor's time?
3. In the conditions of contract between the contractor and subcontractor B, negotiation was to be the first method used to resolve a dispute, and if negotiation should fail then an alternative method of dispute resolution should be discussed and agreed to by both parties. The negotiation between the contractor and subcontractor B was unsuccessful. The parties discussed alternative dispute resolution procedures and an agreement was made to undertake mediation. The mediation was successful, and an agreement was reached between the parties. Would you expect forced negotiation to work?
4. In establishing how much is payable by the contractor to subcontractor B for the work of the sub-subcontractor, is the concept of 'reasonable price' relevant, or is it strictly cost plus?
5. The contractor commissioned an independent party to assess the 'market value' of the work. Since the contractor was paying this independent party, how truly independent would you expect it to be?

10.2.4 CASE STUDY — EARTHMOVING PAYMENT DISPUTE

This case study outlines a dispute which took place between Stocking and Gail Contracting. Gail Contracting was a large building company. Stocking's function was to excavate shafts in rock under subcontract to Gail.

Stocking employed people and machinery around the clock. The contract did not include any specific dates for the completion of the excavation. In fact, Stocking was contracted to complete the task in the time it deemed necessary to complete the excavation of the shafts.

For the first six months of the work, progress payments were received by Stocking from Gail Contracting. The excavation continued for a further three months but no further payments were received. One of the directors from Stocking approached Gail's project manager with a progress report which outlined the work which had been performed without payment. Gail Contracting guaranteed further progress payments, however, no payments were received. At this stage, half of the required shafts had been completed, with only partial remuneration.

This had serious implications for Stocking; debts were steadily rising as a result of such a large amount of money being owed to it. Importantly, the machinery necessary for such work was extremely costly to run, as was employing specialised operators extremely costly. The three months with no payment consisted of false promises of payment by Gail Contracting. Potentially Gail Contracting could have bankrupted Stocking. Stocking was placed in a damaging position. The manager of Stocking needed to act very cautiously and carefully in order to ensure the company continued to operate while endeavouring to recover unpaid debts from Gail Contracting.

One of the major drawbacks in approaching Gail Contracting was the fact that Gail Contracting's head office was centred some distance away. The sheer distance and time needed to deal directly with the appropriate management added to the strains of Stocking. To address this, all dealings between Stocking and Gail Contracting took place with Gail's project manager present on the site. Numerous visits by Stocking proved to be unproductive in achieving a resolution to the problem. Negotiation was essential in order to ensure payment of owed money to Stocking and to save Stocking from forced liquidation.

Mediation/negotiation was used in an attempt to resolve the problem. The negotiation focused on the wants and needs of both parties. At this time, Gail Contracting complained about the machinery's constant breakdowns and complications faced daily. Stocking stated that these were false claims and that Gail Contracting was fabricating these claims in an attempt to withhold payment. For this reason, no common ground was established, and so Stocking 'walked out' of the mediation process.

The next stage involved Stocking taking Gail Contracting to court. The court ruled against Stocking. For the survival of the organisation, Stocking called on the support of the owner. Independent assessors from the owner's organisation were called in to look at the work, and found that it was Stocking which had been disadvantaged.

A second negotiation took place. However, this involved the owner defending Stocking, and the owed money was demanded. The matter was settled. Gail Contracting were forced to pay the debts owed rightfully to Stocking.

Exercise
1. The situation, described above, was one involving a large organisation and a smaller organisation. Comment on the wisdom of contracting with a party that is considerably larger than yourself. If you do decide to contract with a considerably larger organisation than yourself, what practices would you consider to be prudent?
2. One party regards the other as not fulfilling its requirements of the contract and feels justified in stopping progress payments, without an understanding or appreciation for the other party. In contrast, the other party feels that it is unjustifiably 'held to ransom' by a larger company, and in order not to lose the work or in fact to ensure payments will continue, it works without remuneration until it cannot continue any further. Given that such a scenario is not uncommon, what practices might you adopt in order that it is not repeated on future projects?
3. The dispute that occurred was one which required litigation, costly legal fees, immense time and much emotion on the part of Stocking. Suggest strategies alternative to taking the legal route.
4. Numerous visits by Stocking to the Gail Contracting offices, requesting payment for work carried out, proved fruitless. Promises of payment had been made, but no money was forwarded. A further attempt to address the problem involved the use of a mediator. Both parties put forward their relevant case histories, with each failing to accept the other's view. As no common ground could be established, an impasse or deadlock occurred. What tactics might be possible to break such an impasse?

5. Stocking sought a resolution to the problem through the courts and this involved time and money, only to lose the case and to feel the decision was unjust. Stocking, seeking justification and remuneration, sought the support of the owner and, through a separate investigation and further negotiation, was able to have the dispute resolved in its favour. This negotiation and investigative process proved to be the only way to resolve the situation that existed. Why would an owner bother to get involved in a dispute between a contractor and its subcontractor?

10.2.5 CASE STUDY — RESIDENTIAL CONSTRUCTION DISPUTES

The case study is based upon the construction of a residential house in an affluent suburb. The complete process of planning, construction and renovation was carefully scrutinised by the owner of the property and its subsequently appointed architect/engineer. There was a substantial amount of money invested in the project and considerable attention was paid by the owner to the building work. The owner was well informed and prepared for this project because of the necessity of getting exactly what it wanted and getting it right.

Dispute resolution process

One such act of preparedness was in regards to the eventual occurrence of any dispute that might arise. One of the steps taken by the architect/engineer was to draw up a careful and systematic framework of resolution procedures, so that for all categories of disputes, the same outcome could be ensured, that is, swift settlement. This clause in the contract, between owner and builder, was to ensure that any disputes would go through the same steps of resolution. Both parties had agreed to these steps.

An outline of the intended dispute resolution process is summarised as follows:

Step 1.
The dispute that arises is brought to the attention of all necessary parties without delay or neglect.

Step 2.
A site meeting is arranged between all affected parties.

Step 3.
All involved participants are brought to an arranged office meeting.

Step 4.
The parties agree on an independent person if the dispute continues; or bring the matter to the attention of a small claims court.

Step 5.
As a last measure, the matter is taken to a higher court.

Dispute between the owner, architect/engineer and builder

The builder completed a stage of work and requested a progress payment from the architect/engineer for the work completed. The architect/engineer checked the work done and estimated that the work completed was not worth as much as the builder claimed. The architect/engineer advised the owner to pay the lesser amount. The builder received the money and put a claim in for the difference by questioning both the owner and architect/engineer. The builder became anxious and needy for the money and threatened to take legal action to gain the difference. The dispute continued between the two parties, each stating that it was right and demanding closure.

At a site meeting, the architect/engineer pointed out that there were several incomplete stages within the progress claim and this he felt was an appropriate reason to withhold funds. The builder countered with the fact that it had completed other areas of the work which therefore necessitated full payment. The builder was not satisfied with the outcome of the meeting and informed the owner that it was going to seek legal advice. The builder felt that it had a strong claim, since it had to pay the subcontractors for their work, and without the full claimed amount the builder would have been substantially inconvenienced. The owner responded by saying that he was going to consult his lawyer regarding the situation as well, and that a full office meeting would be required where both parties and their legal representatives would discuss the issue further.

Each party hired a lawyer to be present for the meeting. Each lawyer was briefed regarding the dispute and after a similar stand off at the site meeting there was no progress made. Each party was convinced that it had the firmer footing, with the builder requiring the necessary compensation and the owner intending to withhold the money until the criteria of progress had been met. The outcome of the office meeting was that both the owner and the builder were advised that they should bring the dispute to the attention of a small claims court (involving few legal or court costs). The intervention of an independent person was deemed inappropriate, because both disputants sought a swift settlement.

The court ruled in favour of the builder. However, an appeal was made to a higher court with a claim from the owner and architect/engineer that the builder had done defective work. This amount was estimated to be far more than the sum difference that the builder was claiming, and so the owner decided that it would pursue the matter further. As the matter stands now, there is a lengthy wait before the court will hear the dispute.

Exercise
1. When analysing dispute situations such as these, what difficulties do external observers face, given that they are detached from the core of the dispute and therefore they lack much of the insight?
2. There seemed to be a large amount of inflexibility on behalf of the disputants. For example, neither party really explored or even took time to acknowledge the other's needs, wants and interests (shorter disruption time for the owner, and the need for the builder to meet its financial obligations). There was a large degree of uncooperative or uncompromising behaviour. Is it feasible that a halfway point (in money terms) could have been agreed upon instead of continuing the dispute? What was operating against reaching a compromise solution?
3. There was the attempt by both parties to create a win–lose scenario. How likely is it that in the appeal to the higher court the outcome will be a lose–lose situation, and costly for both disputants?

Dispute with subcontractors

In another situation, there were two subcontractors in conflict because of a programming clash and unexpected rain.

The floor sander had an arrangement with the contractor to have the inside of the house vacant and available for him to sand, clean, polish and varnish. The time given to him was from 3 pm onwards. There were not to be any other people in the house from that time onwards. All other workers were requested by the contractor to give the floor sander the space and time necessary to complete his work.

When the floor sander wanted to start just before 3 pm, however, there were still people working on the roof. The roofers insisted that they would be able to stay up there and that there was the ladder on the outside if they needed to come down during the time that the sander was working. The sander warned them that once he started they would not be able to come down inside the house and the process could not be interrupted and would take a number of hours. The roofers insisted that there was no need for them to come inside because the equipment they had was usually left out on the roof and they were able to climb down the ladder. This said, the sander started work.

The situation changed at about 6 pm when the weather turned for the worse. As rain started to come down, the roofers realised that they would not be able to finish the work that they had started and that the equipment, some of it electrical, would have to be moved inside. Unable to walk down the ladder in the rain with such heavy equipment, they decided to enter the house. Placing newspaper on the newly polished floor, they walked over the area just worked on by the floor sander. This effectively ruined the floor and a fight almost broke out. The sander decided that it was useless to continue the work that day and that the entire floor needed to be redone. He then proceeded to pack up and head home, telephoning the contractor before doing so to tell him of the situation. The contractor reluctantly agreed to give him the next day to do the work properly and completely, pushing back all other tasks assigned during that day. When the work was completed, the floor sander double billed the owner for the cost of time and materials. This led to a site meeting that included the roofers, the floor sander, the contractor and the owner.

The owner stated that it was not any fault of his and that he was not going to pay. The roofers said that the only reason that they had to go into the house was due to unforeseen circumstances and that they had no way of knowing that it would rain. The floor sander insisted that his work was just to do the floor and complete it on time. He had given the roofers adequate warning about not entering the house. The cost was to be paid by someone else. After hearing all parties, it was decided and agreed upon by all parties, including the roofers, that the roofers were most at fault. They were back charged for costs of all the materials and time before the disruption that they caused. In return for accepting this cost, they would not be held liable for any other costs due to delays from tasks being pushed back by one day.

Informal negotiation was seen as the best way to deal with the majority of conflicts on the building site. Smaller conflicts at the site were solved more often than not by back charging the 'guilty' party. Acceptance of the cost was very common along with negotiation to limit the liability for future costs that may be incurred.

In another similar case, the builder specified a colour of paint to be used on an exterior wall. The subcontractor used a slightly different shade, either by mistake or on purpose, thinking that the owner would not see the difference. This fault was picked up by the architect/engineer and pointed out to the owner. The owner then instructed the subcontractor to repaint the wrongly painted walls. Realising the costs that were going to be involved in repainting the walls, the painter asked the owner to the site to look at the walls. Upon seeing the shade of the walls, the owner agreed that the paint colours were indeed very similar and an honest mistake may have been made. Informal negotiation took place on the site and it was agreed that the owner would bear the cost of the labour and the painter would bare the cost of the paint. In this situation, a win–win situation was reached, as from the painter's point of view, additional labour cost was avoided. From the owner's point of view, savings were made on paint that was almost identical.

In both these cases, an acceptable, cost efficient solution was reached that benefited all the parties involved.

Exercise

4. In both these cases, an informal site meeting with the parties involved led to the negotiation of an acceptable outcome. The structures in place for conflict resolution promoted cost-efficient, swift settlements. Solutions were reached before additional costs were involved. It was also helped by the parties to the dispute being cooperative and willing to talk about different options, realising that to do so would be to everyone's mutual benefit. However, what value would the dispute resolution structure be if the parties were not cooperative?

10.2.6 CASE STUDY — LARGE CLAIM, SMALL CONTRACT SUM

Introduction

Much is spoken and written about the unethical and unscrupulous behaviour of some contractors in the formulation and handling of claims.

However the behaviour of some owners also can be equally questionable; their actions or reactions can initiate or escalate a dispute. Perhaps this style of behaviour may arise not out of dishonesty but from a perception that contractors' claims, particularly expensive ones, are invariably unwarranted and almost certainly unwelcome. Often a monetary claim is outside the owner's budget or contingency provisions.

The contractor, because of what it is doing (that is, claiming whether it be legitimate or not), is judged guilty of an unwarranted act because it is creating a problem for the owner, and it may involve future expense, embarrassment and time, and may well have already caused much 'pain' prior to the actual receipt of the claim.

This case study looks at a claim, large relative to the initial contract sum, where the questionable behaviour of the owner escalated what may well have been a fair claim into a prolonged expensive dispute that would possibly have bankrupted many similar sized contractors.

When the dispute was eventually resolved, the parties were bound to secrecy as to the details of the agreement.

Background

A relatively small civil engineering contractor tendered to excavate a pit.

The design and specification, which formed the basis for the tender, were prepared by a large civil engineering consultancy, which had acted in that capacity on other similar projects. The tender specification included a brief geotechnical report as an annexure. As well as describing the subsoil conditions, the geotechnical report stated that groundwater could be expected at a certain depth and actually stated the anticipated inflow into the excavation; the quantity was minimal.

The contractor, with much experience in such projects, was well reputed, financially very sound and accordingly, because it had submitted the lowest priced tender, was awarded the contract.

Common standard general conditions of contract were used. The owner nominated its engineering manager as 'owner's representative' on site to oversee the work.

Within three weeks of the commencement of work, groundwater was encountered in the excavation with an inflow several thousand times that stated in the geotechnical report.

The contractor immediately notified the owner of the problem; the initial response was that pursuant to the general conditions of contract the contractor had an obligation to fully inform itself of all aspects of the work prior to tendering and hence now had the responsibility of overcoming the problem. However, because of the impact of the problem on such a deep excavation, the owner insisted on the involvement of the consultancy in deciding upon the de-watering scheme.

The contractor revised its construction technique, calculated an additional cost and lodged a variation claim with the owner together with a notification of delay.

The owner rejected the claim, and at the same time took the unusual step of replacing its engineering manager as the owner's representative on site with a person from the consultancy.

The contractor found the owner's rejection of its claim unacceptable. Hence the parties were in dispute. The dispute resolution provisions of the contract then came into effect.

In line with the 'Settlement of Disputes' clause, the following steps were taken:
- The contractor referred the dispute to the owner's representative (who gave a decision in favour of the owner).
- The contractor did not accept that decision and referred the matter to the owner.
- The owner upheld the decision of the owner's representative.

The next recourse open to the contractor in pursuit of the claim was arbitration, but before taking that next step the contractor suggested negotiation to the owner. The owner indicated a preparedness to do so. However after stalling for a month or so, the owner declined to commence negotiation.

The dispute resolution clause stipulated a one month time limit for referral to outside arbitration after the owner's decision. Hence the contractor found itself time barred from arbitration and in a situation where it was forced to continue work. The contractor continued with the work and immediately commenced arbitration proceedings, joining the owner, the consultancy and the geotechnical (sub)consultant.

The atmosphere for the remainder of the construction period was extremely adversarial. The problems on site were exacerbated and the dispute escalated because the de-watering method insisted upon by the owner's representative proved ineffective and it was necessary to revert to the contractor's preferred method.

Extension of time claims were submitted and rejected. The owner continued to make progress payments but, after the contract Date for Practical Completion, noted with each payment that, although liquidated damages had not been deducted, they would be in the future.

The work was eventually completed at a considerable loss to the contractor; as well as being late with the likelihood of an arbitration hearing still some time away.

Several months later, well into the arbitration process of issuing statements, defence, discovery, further and better particulars etc., negotiation between the legal representatives commenced and the claim was eventually settled for less than half of its original value.

The contractor possibly did not recoup its costs but was mindful of the benefit of at least substantial relief of its losses; escalating legal costs; loss of further management time from within its small management structure; the risk associated with an unfavourable award; and very importantly to the contractor, the risk associated with earning a reputation as a 'litigious' contractor in what was a relatively closed industry.

Exercise
1. The contractor pursued the resolution of the dispute through all steps of the contract's dispute resolution clause and observed the correct time limitations and sequence of procedures. The owner appeared to have used the time bar limitation of referral to outside arbitration to delay the resolution of the dispute. Certainly that time bar worked in the owner's favour even if its action in procrastinating over negotiation was without malice. What alternative in the dispute resolution clause to a nominated time period before going to arbitration would you recommend?
2. It would seem that because of his potential conflict of interest, the owner's representative may have been fair but certainly could not have been impartial as was required by the contract conditions. How can you make the role of owner's representative impartial to both parties? Or what alternatives to having a person like an owner's representative are there?
3. The substitution of the owner's representative gave protection for the owner against claims. What liabilities should the owner's representative be aware of?
4. When the dispute between the parties arose, the owner's response could be described as 'active aggressive', a response which was accompanied by the defence of an adopted position.

   What point is there in undertaking negotiation in a situation where a large power imbalance exists or one of the parties obviously has no desire or interest in resolving the dispute?
5. In such cases, does it mean that formalised processes such as arbitration or even litigation will eventually be required to settle the dispute issues?
6. Does settlement of a dispute in this manner result in the total destruction of any longer term working relationship? This idea was not wasted on the contractor; the contractor 'backed away' from arbitration rather than destroy a long-term working relationship (in this case with the industry in question, not only the other parties). Given the attitude and action of the owner toward resolution of the dispute, the contractor was left with no alternative other than capitulation in the resolution of its claim.

7. The inadequacies of a traditional dispute resolution clause such as in this case are apparent when the intransigence of one party escalates the dispute, avoids early resolution and is hence unfair and potentially catastrophic to the other party. Would inclusion of the possibility of using alternative dispute resolution methods, and if this was either inappropriate or one party did not co-operate within this framework, then having an early referral to arbitration, be a better way to go?

10.2.7 CASE STUDY — LATENT CONDITIONS CLAIM

The case study describes a dispute that arose in a bridge construction project and was later resolved by negotiation.

The claim was submitted under the latent conditions clause of the special conditions of contract in which the contractor stated that the rock encountered on site differed materially from that which was shown in the geotechnical report supplied with the tender documents.

To satisfy the clause, the contractor needed to demonstrate the following:
- The condition was physical.
- The condition differed materially from that which would have been ascertained by the contractor if it had fully informed itself of all the tender documents.
- The condition could not reasonably have been anticipated at the time of tender by an experienced and competent contractor.
- The contractor had given written notice to the owner's representative as soon as practicable regarding the latent condition.

The contractor undertook independent laboratory tests in an attempt to prove that the material encountered on site was different. However, the tests showed that it was the same material, but harder than the contractor had anticipated. Because the laboratory results described a rock similar to that given in the geotechnical report, the contractor had a weak case.

It appeared that the contractor had under-estimated the effort required to drill this rock, and consequently provided equipment that proved to be inadequate.

The owner's representative, however, requested the contractor to resubmit its claim, detailing the hardship it had experienced and the time and resources used beyond that which it had provided for in its tender.

After the owner's representative had received the contractor's revised claim, he recommended to the owner that the variation be accepted, as it enabled the contractor to recover costs for the hardship that it had experienced, even though this hardship was caused by oversight of the contractor at the time of tendering.

Exercise

1. One of the more interesting owner's goals, in place for the negotiation process, was the desire for the contractor to make a profit and remain active in the industry, because the number of suitable bridge contractors in the region was low and the list of prequalified contractors needed to be strong to maintain a healthy level of competition and a reliance that projects could be completed by suitably experienced and skilful contractors. How does this place the owner at a disadvantage in any future

negotiations, or could the owner use such conciliatory thinking to its advantage in future negotiations?

2. The approach that the owner adopted in the negotiation process was as follows. Comment on how each might have affected the negotiation process.

    • The response needed to be timely such that the effect on the progress of the other activities in the project could be minimised.

    • The revised claim needed to address the degree that the contractor was out of pocket and not to generate profit that may not have been built into its original price.

    • Both parties needed to believe that they had achieved something out of the settlement. The contractor was pleased that it was compensated for its financial loss, while the owner was content that a strong working relationship had been maintained.

    • An independent expert/arbiter was avoided.

    • The dispute needed to be handled at the lowest management level to minimise the overhead cost component.

3. The basis for approving a revised claim was to enable the contractor to recover its costs, and this kept a quality contractor in business and assisted relationship development that needed to continue on this project and perhaps future ones. A changed attitude was witnessed on site from the contractor and its team after settlement. This could be attributed to the improved communication resulting from both parties exercising open-mindedness and an unrestricted thought process (removal of the weight off one's shoulders). What else could the changed attitude be attributed to?

4. The basis for requesting a revised claim was that the contractor went for too much in its initial claim and, through negotiation, the contractor was asked to honestly assess the impact that the harder rock had on its financial state before the owner would seriously assess the claim. The approach taken by the owner's representative and the owner, in seeking the contractor's true cost and time impact, was an understanding of a contractor's working environment and the difficulties that existed for a contractor in this type of business. Why would the contractor claim high when it knew that it had only a weak case?

5. The benefit that the owner had in this case was that this contractor's pricing structure was lean and at the time of tendering won the work by a fair margin. The owner was then able to support the contractor 'within reason'. Comment whether it is better for the owner to deal with the contractor's price leanness at the time of tender or when a later, inevitable claim is made.

6. The contractor was reassured that it would not experience financial difficulty on this project if it identified what compensation was required to satisfy its needs, rather than taking the initial approach of 'winner takes all'. Why wouldn't the contractor initially approach the resolution in a more win–win style?

7. It was fortunate that the owner was able to satisfy the contractor's needs, with the contractor still producing a quality product at an economical price. Comment on the need to properly assess the credibility of each tenderer's initial bid and then continually monitor the effectiveness of resources such that claims of this sort can be minimised.

8. If a settlement had not been agreed, it is likely that an independent expert would have been engaged before, as a last resort, being referred to arbitration. These options

would have added project costs that, for many reasons, should be avoided. How does the threat of such actions influence the preceding negotiation process?

10.2.8 CASE STUDY — RIGHT VERSUS FAIR

On a pond de-sludging project, a difference in the interpretation of what was right versus what was fair led to a substantial dispute.

The project involved the removal of sludge from a pond, and the subsequent drying, transportation, treatment and compaction of the sludge as landfill via a stabilisation process. The work was advertised for prequalification, and tender documents were offered to five companies, based on an assessment of their past experience and capability of available plant. The tender documents specified a strict project time requirement of three months to be observed; a number of tenderers expressed their concerns that such a deadline was optimistic. The documents also included an assessment (guide only) of the sludge, subject to further investigation and clarification by the tenderers if they deemed it necessary.

An appraisal of the tenders received and associated qualifications revealed prices ranging by a factor of three. The two cheapest tenders were assessed, and when the lowest bidder confirmed that it was a specialist in sludge removal and that the three-month period was easily achievable, it was awarded the contract.

As events turned out, the contractor was good at removing the sludge, however it was inexperienced in the subsequent drying and stabilisation. Large stockpiles of sludge accumulated, and it was found that the sludge developed a crust-like coating of dried pollutants over the surface, thus preventing the underlying material from drying. Whilst the specification called for the material to be spread in thin layers and turned, a calculation late in the project revealed that the area granted to the contractor was insufficient to complete the work within the three months specified.

Three months into the work, the contractor was facing difficulties. The owner's representative stated that as long as the contractor was showing persistence, then liquidated damages would not be applied. In addition, the owner's representative made available a further area, although the contractor would have to pay the additional charge of double-trucking. At meetings at this stage, the contractor expressed that financial difficulties were on the horizon, and that it felt there were some circumstances beyond the scope of the contract for which it should be reimbursed. The owner's representative agreed that he felt some money would be payable, and if the contractor would proceed as per the contract then the owner's representative would give the claim his favourable consideration.

The contractor failed to conform to the specified procedure for claims, even though it was constantly reminded by the owner's representative. The owner's representative even went to the extent (albeit naively) of suggesting someone who could assist in the preparation of the claim. As the project drew to a close, the contractor engaged an aggressive solicitor, who insisted that the owner cease its attempts to contact the contractor to discuss its claim. The contractor appointed a new project manager as a point of contact for the work, and withdrew all other staff from communication.

Approximately six months after the start, the work was completed, with the final progress claim being for many times that suggested reasonable by the owner's representative, however subsequent attempts to contact the contractor were met with aggressive refusals from the contractor's solicitor. The owner was thus forced into contesting the claim. A solicitor was engaged, and he was able to convince the contractor's solicitor to accept mediation in the hope of avoiding arbitration or the courts. An independent expert was also engaged by the owner, and the next few weeks were spent preparing statements and a case strong enough to proceed to court if necessary. All records and documentation from both sides were made available.

The mediation lasted one day. The final demand put to the owner was that the contractor would go to arbitration or court if the owner did not offer at least half the claim. The owner's assessment found that it would be difficult for the contractor to substantiate a claim for more than a quarter of the claim, however the contractor held firm. The situation at that stage was the following. Expenditure on legal fees to date was high. Even if the owner went to court and won, the bankrupt contractor would not be able to afford the additional cost in legal fees. Consequently the owner agreed to give in to the contractor.

Exercise
1. Recourse to the legal system was used as a tool to purposely reject the specification and adopt aggressive tactics. The contract was disregarded in the light of the contractor's refusal to budge. To many people, such events make it difficult to be positive about the legal enforceability of contracts. Or is this discontent wrongly directed?
2. As a criticism of alternative dispute resolution (ADR) techniques, the owner was convinced that with the final option of litigation available to the contractor, the contractor's solicitor could not have been persuaded to act differently. Knowing that it had this option and that the owner had money, allowed the contractor to turn the negotiation from one of a contractual nature to one of a financial nature. In hindsight, given the relative amounts of money of solicitor's fees versus the overall claim, the inability of the contractor to meet the costs should it lose the case became the deciding factor. How crucial then is it to assess the financial position of the other party, because if the other party is facing bankruptcy, you know that you are up against something with 'nothing to lose'?
3. A cynical interpretation of this case, resulting from the contractor's complete disregard for the contract, is that solicitors can interpret contracts any way they please. Solicitors could be thought to be in existence purely to show how their clients' cases are 'special cases' and therefore beyond the bounds of contracts. How valid is such an interpretation?
4. Courts could be expected to lean to what is 'fair and reasonable'. But how do you assess what is fair and reasonable, because opposing parties often have widely diverse interpretations and many points of view are possible?
5. Is there some comfort in knowing that parties cannot totally hide behind a contract?
6. Comment on the approach of the owner's representative, in particular making constant referral to the specification in order to solve problems? Was he merely turning the problem back to the contractor?
7. Is it of little use knowing what is fair and reasonable, if the other party is not going to be able to meet your legal costs should you win?

10.2.9 CASE STUDY — CONSTRUCTION INDUSTRY DISPUTES

This case study considers some causes of disputes in the construction industry.

Situation A

A construction company was awarded a contract for the construction of a wharf. The specification included two methods for repairing damage to the pile coating. The first was more suitable for scratches, while the second was more suitable for major damage such as a pile splice. The owner was carrying out concurrent work at the site and caused damage to some piles. A variation price request was issued to the contractor to carry out repairs. The damaged area to each pile was of the order of a small coin, and neither repair method seemed applicable, and so the contractor requested clarification. The owner's representative undertook some investigation, and on the basis of previous experience with similar products, issued a direction.

During the course of the project, the contractor was awarded another wharf construction contract by the same owner. The specification read similarly to the original wharf construction document, except that the wording on repair methods was ambiguous to the point that one described repair method could be inferred to be either repair method. When the contractor attempted a repair method in the second project, the owner's response was that the specification only should apply. Any dispensations issued on the first project did not carry over. On the subject of ambiguity in wording, however, the owner was reliant upon the contractor's knowledge of the earlier specification to fully understand the intended repair method.

Situation B

The tender drawings, which included truss drawings, had notes indicating that all steelwork was to be hot-dipped galvanised, and the truss components had to be fully sealed. The trusses were long — beyond reasonable galvanising bath capacity — and so the estimators query-tagged the required treatment. Subsequently a tender addendum was issued, and this included a detailed schedule of the protective coating systems. The trusses were listed with a particular paint coating system. This resolved the anomaly to the estimator's satisfaction and the coating system was not queried. The contract was awarded and the construction drawings had the note about sealing of the truss components deleted. Subsequently the owner advised that its intention was for the truss to be galvanised first and then painted, that is the surface treatment schedule did not preclude the possibility of galvanising first, in accordance with the drawing note, and then painting.

Situation C

A contract for the construction of a major road had been awarded by a state transport authority. It was a large balanced cut-to-fill operation with very deep cuts and high fills. The specification allowed for the cut material to be placed as fill in thin layers with individual rocks no more than the layer depth in size. During construction, a large number of clusters of boulders were located. The numbers could not have been anticipated on the basis of the geotechnical information provided. (The geotechnical information provided was quite comprehensive, but did not form a part of the contract; rather was merely included for tenderers to make whatever use they thought appropriate.) The owner refused to accept any change from the specification and so the contractor was required to use rock-breakers just to break up the boulders into useable small rocks. The consequences were: changed and additional equipment, changed and delayed work procedures etc. Finally the owner did permit the use of a much thicker fill layer with individual boulders much larger, but the contractor had incurred and continued to incur extra costs.

Exercise
1. Are disputes an unavoidable part of the construction industry simply by virtue of the nature of the industry? What role do unclear, ambiguous and inconsistent documents play in the dispute landscape?
2. Situation A arose out of an ambiguous document and possibly one which was unrealistic in that it failed to address the typical circumstance which was likely to occur. There was the added unique element of a second contract and the reliance on established precedents which may have been in conflict with the documentation. Suggest ways that the owner may have avoided or dealt with both issues.
3. Situation B was similarly the result of unclear documentation. The drawings were read to infer something different to the specification. This would have been a very simple case to avoid, by having a single comprehensive table of surface treatments. Are contract documents to be read in conjunction, in addition, or in conjunction and addition?
4. Situation C is a classic example of a latent condition claim. Would better site investigation have been cost effective? Which party would be better able to manage the associated risk and better able to mitigate against the consequences?

10.2.10 CASE STUDY — INDUSTRIAL DISPUTE

Background

The owner was a public sector organisation. During a prolonged infrastructure development program, the owner had developed a comprehensive approach to project management. The approach was developed for use generally within the civil engineering and earthmoving industries. It was extended into other industries as more complex and multi-disciplinary projects were undertaken. However, the approach remained largely suited to the earlier mentioned industries.

The owner developed its own general conditions of contract, which generally favoured the owner, but were well understood by the contractors commonly used by the owner.

Project description

The owner undertook a program involving the development of outer-urban sewer treatment plants. Each plant would cater for small populations of people. These projects were multi-disciplinary in nature, drawing on the expertise of civil, mechanical, process, and structural engineers and architects. Contractors and subcontractors would come from the traditional engineering construction, mechanical and electrical, process control, building and plumbing industries.

The particular project of this case study occurred during a peak in construction industry activity. Contractors could pick and choose projects they undertook. The good contractors had full project books, leaving some of the work to less experienced contractors. The timing for this project was critical as a major industrial user was to commence operation at the end of construction.

Tender process

The documentation was standard owner documentation. There was no attempt to customise the documents to reflect the multi-disciplinary nature of the work or the industry condition at the time.

Tenders were received from a limited number of contractors, none of which were contractors regularly used by the owner. The selected tenderer was a medium-sized, family-owned business with most of its experience in the regional power industry. The majority of its work was related to industrial buildings. It wished to expand its experience into the metropolitan market and accordingly submitted a good price that was accepted. The contractor commenced construction, with a twelve-month anticipated time to completion and commissioning.

Performance

The construction work included a number of buildings, the tallest of which was two stories.

One aspect of the owner's documentation was the requirement that site labour be members of the Civil Union (CU). The owner's main day labour forces were CU members. In addition, the contractors normally used by the owner were CU organisations. In the metropolitan area, the CU covered low rise buildings, whilst the High Union (HU) covered high rise buildings.

The contractor employed CU labour on the site. However, on its regional sites it used HU labour. The HU, being an aggressive union, saw this as an opportunity to muscle in on traditional CU territory, and gain coverage of low rise buildings in the metropolitan area.

The contractor was severely compromised in that it had several regional projects using HU labour. In addition, the work had been won with, at best, a slim profit margin. The HU organisers moved in on the site labour and a series of minor industrial disputes resulted. Over a period of several months the situation deteriorated to the point where construction practically ceased.

Industrial dispute resolution

The contractor and the owner referred the matter to industrial (not commercial) arbitration after initially attempting to negotiate the matter on site. The arbitration was ineffective. The contractor paid for dual union membership for all labour. However, the agenda of the HU was unsatisfied.

Work on the site returned to normal production but several months were lost in the process.

Exercise

1. Many improvements could have been made to the documentation to lessen the chances of a dispute becoming disruptive to the delivery of the project. It appears that little if any time was spent in analysing and customising the documentation for the particular project.

   General conditions of contract might have included provisions relating to union coverage considering that the work included a number of buildings. Although industry practice at the time was for such buildings to be built by CU members, the rival HU was keen to expand its area of influence. Indeed, the HU was known to have created similar disputes on sites where a contractor was less compromised than the one mentioned here. Would union issues normally be covered in contract conditions? Suggest how the union issue might have been dealt with in the contract conditions.

2. The tender evaluation might have been more probing in gaining information from the tenderers, in particular their current and recent projects. Would this necessarily have addressed the union issue problem?

3. A careful analysis of the tender price breakdown and referee checks might have highlighted issues requiring deeper investigation. Would this necessarily have addressed the union issue problem?

4. Would it have been more appropriate to break the work into packages that were capable of being handled by smaller, specialised contractors such as commercial builders, civil contractors, and HVAC contractors?

5. Could it have been expected that the contractor should have been aware of the requirement to use CU labour? The contractor had regularly employed HU labour on many other projects. Did relying on the protective umbrella of the owner indicate naivety on the part of a contractor experienced in the construction industry?

6. The contractor had no significant strategy for dealing with the dispute when it arose. What does this say about the contractor's approach to risk management?

7. The initial plumbing subcontractor was a small, local firm with limited resources and expertise for an industrial project. Little was achieved in three weeks and the plumber's services were dispensed with. A second plumbing subcontractor was appointed with greater resources and experience. However its experience was in sandy soil conditions. Rock was encountered in the trenching operations. This coincided with the dispute. Due to the protracted dispute, the second firm went out of business. What does this say about the contractor's approach to subcontracting?

8. Due to the tight price offered for the work, the contractor was employing site labour on minimum rates. These people had little loyalty to the contractor. There was a constant turnover of staff and, due to the amount of work available in the industry at the time, the contractor was only able to engage small numbers at any one time. When the dispute arose, the HU organisers had little trouble in gaining the loyalty of site labour by offering improved employment conditions. What does this say about the contractor's approach to dealing with site labour?

9. It became apparent that the other regional commitments of the contractor were taking the best labour. In particular, the site supervisors and engineers initially on the project were transferred back to regional projects after a few weeks. What does this say about the contractor's commitment to the project at hand?

10. The contractor's staff, at the onset of the dispute, tried to negotiate with the workers and HU organisers. Due to a lack of experience, this initial negotiation was ineffectual and possibly exacerbated the situation. It became more difficult to retrieve the situation from that point, even when experienced staff became available. The HU organisers had won the site labours' loyalty. In negotiation, is a 'little bit of knowledge dangerous', and should negotiation only be attempted if one is knowledgable in its art and skills?

11. There were in effect four parties to the dispute — the owner, the contractor, the HU union and the CU union — all with differing agendas. How might the presence of more than two parties have affected the negotiation?

10.2.11 CASE STUDY — OWNER REPRESENTATIONS

Background

A contractor entered in a design, develop and construct (DD&C) contract with a government organisation for a public building project. An industry standard form of contract was used with substantial special conditions. The contract required the contractor to develop the owner's design provided at the time of tender. This, in part, required the contractor to liaise with users of the building in order to complete the design. To aid both the contractor and the users, there were guidelines for this type of project for both parties to refer to; this was in addition to the normal contract documents such as drawings and specifications. Where the users required a departure from the tender-provided design, the contractor had to identify this departure, price it and assess and advise the potential time impact. The contractor's developed design, as well reflecting the requirements of the users, had to be 'fit for purpose' in the opinion of the owner. Importantly, the design had to comply with all relevant statutory codes and requirements.

Pre-tender

The owner advised that the design had been significantly advanced to the extent that functional and spatial planning was complete. Additionally, the users had had significant input to the tender-provided design to the extent that large-scale drawings had been 'signed off' by the users, and the small-scale drawings were also essentially complete. The owner arranged for a meeting on site to brief all tenderers regarding the project. At this meeting the state of the design was indicated together with the availability of portions of the site for the work. Being an existing facility which was being added to, certain buildings were to be vacated by the owner and subsequently demolished by the contractor during the course of the work. The owner at the time of tender requested the contractors to program the work in a unique manner, and to price the design component of the tender to reflect the advanced state of the design.

Post-tender

Having won the contract under competitive tender, work began in accordance with the contractor's planned sequence of work, the same sequence of work as described in the tender. Within two days of the award of contract and taking possession of the site the contractor convened the first of the user meetings to begin finalising the owner's design. The order in which user group meetings were scheduled related directly to the sequence of the work. This permitted resolution of the structural grid, reduced levels for footings and concrete slab details.

Given the approach and input by the owner at the time of tender, the contractor antici-pated that the design development process would be streamlined. It was anticipated that at users' meetings, the contractor's architect would firstly place a large scale drawing of the relevant area of the building on the table for the purpose of user confirmation that this was the agreed department layout. This would take but a few minutes. Having reaffirmed the large-scale design, it would then be 'signed-off', and the architect would then remove the large-scale drawing from the table and replace it with one of the small-scale drawings and commence discussions on detailed design. Given the representations made by the owner, the detailed development process should entail notional changes only — for example, the minor relocation of a partition wall, additional power points and selection of finishes such as internal colours and floor finishes.

However this was not so. The first user meeting (for one user department) rejected the owner's large-scale drawing supplied at the time of tender as being out-of-date and not the drawing that was signed-off. The contractor undertook a complete redesign of the department. Similar experiences ensued for subsequent departments with varying degrees of departmental re-design being required. The owner's tender-provided design was found to not comply with relevant statutory codes, and a significant redesign to ensure compliance with all relevant codes ensued.

Claim

The outcome of this situation was that a large number of variations were identified by the contractor. The contractor's consultant team claimed additional professional fees due to the revised design input required. Subcontractor's claimed prolongation and increased costs. Including the time-related costs incurred by the contractor, the final claim was in excess of ten million dollars. The owner responded with a suite of defences including: fitness for purpose; time bars under the contract; for the contrac-tor's convenience; implied by the documents; and differing interpretations of contract clauses.

The approach adopted by the parties to resolving the claim was typical of the construction industry. However, the various methods and overlapping of events was somewhat curious. There seemed to be an acknowledgment by the owner that the contractor had a legitimate claim, however the details and quantum remained a sticking point.

During the currency of the work, the contractor advised of the intention to lodge a claim and the owner invited the lodgement of such a claim at the earliest possible time. Since the costs were still being incurred by the contractor, it was not possible to finalise the claim early in the project. The contractor suggested that events giving rise to claims be identified progressively and costs assessed against these items. However this was not the total claim but rather claim-by-instalments. The owner rejected this 'piece-meal' approach and a stalemate resulted.

To an outsider it seemed that it was possible to progressively assess certain discrete items of the contractor's claim on their merits; all the information was to hand. Conversely from the owner's view, it was a poor negotiating position; nothing to make a counter-claim upon, but rather a financial haemorrhage with no end in sight.

Partnering was a feature of the project. However the owner gave the impression that it saw this as more of a problem than a means of helping the project. Previous projects of a similar nature, with 'bad' experiences for the owner's organisation, seemed to prevent the partnering ideals being employed; previous contractors had viewed partnering as a tool for commercial claims, rather than a vehicle for early resolution of disputes.

The contract made assessment of claims by expert determination possible for small claims (an upper limit on value). Hence this avenue was blocked to the parties.

Upon completion, the contractor lodged, for consideration by the owner, a complete claim which was subsequently rejected by the owner on the same grounds with which claims had been rejected during the course of the work. Such grounds were: time bars; fit for purpose; for the contractor's convenience; implied by the documents; failure to establish legal entitlement; and the like. The owner, being a government body, could not act as a commercial identity and truly negotiate, as would a private sector organisation. Certain criteria had to be met to obtain a sign-off on this issue.

Dispute resolution

The parties agreed to a mediation. The mediator selected was from the legal profession and had a reputation for the resolution of difficult disputes. The mediator suggested to the parties at the initial meeting that the dispute might be likened to viewing a coin, with either party on both sides of the coin unable to see the other side. There was a need for each party to view the other side of the coin to gain a balanced point of view. The mediator for his part was viewing the coin from the edge and could as yet see neither side of the coin. A timetable was agreed and the contractor made a series of presentations regarding the issues, always from the perspective that the owner was familiar with the fundamentals of the situation and that quantum was more the issue than the entitlement. For the owner the view was the opposite; the need to establish entitlement was foremost, quantum would follow as a matter of course.

This dispute could not be solved easily. There was a lack of a truly commercial environment due to the owner having restrictions placed upon it by government guidelines and legislation. Part of these restrictions had arisen from newly introduced codes of conduct and the need for transparent and auditable processes. Secondly, part of the contractor's claim was based on consumer protection legislation, and as such a number of untested issues arose — in particular, the commercial status of a public sector organisation; additionally the injured party had to demonstrate that a representation was made, that the injured party relied on that representation and that damages flowed from the representation. Case law in this area was small and as related to such construction matters almost non-existent.

The owner felt comfortable given the onus of legal proof placed upon the contractor. There was also this particular public sector organisation's culture of taking a 'hard-nosed' attitude to claims and disputes as a means of discouraging other contractors from making claims. To settle a matter too early might send the wrong message to the construction community regarding the organisation's attitudes to claims.

Negotiation continued at a very senior level. However the attitudes and positions of the parties remain unchanged and little progress was made. The matter was then formalised and taken to the courts. Given the matters of law involved, waiver and estoppel for certain conditions of the contract, together with action under consumer protection legislation, perhaps it was always going to be a matter for the courts to resolve.

Exercise

1. It is interesting to note the particular public sector organisation has since amended some of its standard conditions and procedures associated with DD&C contracts. Some of the changes may not be for the better, with the right to make a claim as a result of a representation made during the tender period being expressly excluded under the contract. Give your view on this amendment. Would this strategy make contracts more onerous and complicated and prone to dispute?
2. Would true partnering together with the fundamental issue of 'good faith' be a more attractive means of reducing and simplifying disputes? True partnering offers the structure and formality for contracting parties to identify problems, before they progress too far and turn into issues that are too large for project staff to deal with. Do you believe the above matter could have been salvaged at an earlier time employing partnering, prior to the claim figures building to a high level and prior to attracting high legal costs? Does the issue of 'good faith' circumvent some onerous contract conditions behind which a contracting party may hide?
3. Dispute resolution methods need to be appropriate for the degree of complexity of the matter. In the circumstance where there is agreement between the parties that there is entitlement on the part of one of the contracting parties, then direct negotiation as to quantum may be appropriate. This situation may also be appropriate for facilitated negotiation or expert determination. Where there are lesser issues of law, then expert determination may be appropriate. Would you suggest that the ceiling as to the value of claims that may be determined by expert determination needs to be reviewed and the terms of reference for this type of dispute resolution amended? By this is meant that the particular circumstances relevant to the situation become more the determining factor in the selection of the dispute resolution method.
4. In the area of complex law issues such as the matter described above, are arbitration and litigation the only appropriate methods?
5. Unfortunately once a dispute becomes formalised it becomes process driven by the legal profession. Clearly the goal for all contracting parties is not to get into a dispute. All efforts should be expended to avoid this situation in the first place, that is a pro-active approach is required. Alternatively how do you keep a dispute at an informal level?

10.2.12 CASE STUDY — CLEANSER COMMISSIONING

Background

A cleanser was designed, installed and commissioned in a factory by a supplier. The installation and commissioning was long and difficult. The factory, undergoing other work at the time, experienced a number of problems including a boiler breakdown, that meant steam was not available for a few days, delaying the final commissioning of the cleanser.

The two commissioning engineers sent from another town did not return home during this breakdown, because it was unknown when the parts required for the boiler repairs would arrive. The engineers spent several days waiting for steam to become available.

Added to this, the installation had to include some equipment supplied by the factory but of a foreign make belonging to the supplier's opposition. Although the commissioning engineers were familiar with this foreign equipment, they were not aware of all the fine differences between this and their own equipment.

During the final commissioning, water hammer and steam surge problems were experienced. By replacing and fine tuning some valves, this problem was partially overcome. The hammer problem was further compounded by the foreign equipment suddenly discharging product at a force and velocity not expected with that type of unit.

Eventually, all problems were resolved, product of an acceptable quality was being produced and the engineers left the site.

A number of months later, the factory noticed that product was getting into the cooling water circuit, indicating cracks in the metal plates contained in the cleanser. One of the commissioning engineers was sent to investigate, free of charge, to the factory to disassemble the cleanser and ascertain what was going on. Four plates were cracked. These were replaced and the cleanser recommissioned. The observed operation was smooth, and so the engineer returned home.

After the third occurrence of cracked plates, the cleanser supplier received a letter from the factory stating that legal action was about to be undertaken if the supplier did not cover the costs resulting from lost production due to the cracking plates. The supplier refused to accept liability, because it believed that its cleanser was well-engineered and it was the foreign equipment and the boiler causing the problems.

The supplier also stated that it too had lost money on the project due to the extended commissioning and several additional site trips to replace cleanser plates.

Both parties were adamant in their views of the situation.

Resolution

Middle management negotiation over the issue led nowhere. Upper management negotiation went no better. Finally, the supplier received a letter requesting its representation at a meeting to be held at the office of the factory's solicitor.

The meeting was held some two weeks later. The supplier sought legal advice and had legal representation. Each party was represented by three individuals — a solicitor, a senior manager and an engineer involved with the project.

The initial mood of the meeting was hostile. Quips and low level bickering took place between the two parties, although no one got personal. The meeting was going nowhere and after an hour, a twenty minute break was suggested and taken. Each party went its own way and developed strategies for the next session.

When the meeting re-convened, both senior managers acknowledged that the situation was using up precious human and monetary resources and a quick resolution was desirable. The project engineers were then asked to present their versions and interpretations of what had happened. Meaningful negotiation commenced, although the factory still insisted that some compensation was expected for lost production.

The supplier argued that it had supplied parts and labour, free of charge, on numerous occasions, even though it did not feel that it was obliged to do so.

When the technical details, as presented by the project engineers, were discussed and analysed, some engineering solutions to the hammer problems were developed. Input to these solutions came from both parties. Finally, a by-pass of the foreign equipment and an alternative valving system were developed.

With both parties happy with this engineering solution, it was agreed that the factory would drop the compensation claim if the supplier provided materials and labour free of charge.

As such, the dispute was resolved.

Exercise
1. The consumption of resources, prompted the two groups to work together. Once it was recognised that people's time and efforts could be better utilised on other matters, a final resolution was quickly agreed upon, so that human and monetary resources could be freed up. Could the 'consumption of resources' be regarded as a 'common enemy', and used as such to bring about a general resolution of disputes?
2. The dispute was generated by each organisation having a different perception about which party should be responsible for the recurring problem. The situation was that the factory needed the supplier to provide replacement parts, whilst the supplier needed the factory for future business. Tensions grew as the problem continued to recur and production time was lost whilst waiting for parts and labour. Would the inter-dependence between the supplier and the factory exacerbate the dispute or assist its resolution?
3. At the beginning of the dispute, no common ground could be found between the two parties. Hence the almost irrational attitudes of both groups as highlighted by the bickering at the start of the meeting. Although third parties (solicitors) had been brought in, these were not third party neutrals but representatives detailing the legal arguments each side had. How might the biased third parties have interfered or helped with a resolution? How might a third party neutral have helped?
4. Instead of the factory sending the supplier a demand for compensation due to lost production, could the factory have managed the dispute in another, perhaps better, way?
5. Once it was recognised that a resolution was desirable for both parties and all technical issues had been defined by the project engineers, both parties developed a more sympathetic view towards each other. They began to work together and negotiate a solution that was acceptable to both. Neither party gained anything significant from the dispute at the expense of the other party. The dispute was finally resolved and

each party was relatively satisfied with the result. Would you classify the outcome as win–win, lose–lose or win–lose? Explain your choice.

10.2.13 CASE STUDY — EXCAVATION AND CABLE LAYING

A contractor was engaged by an electricity distribution organisation because of the organisation's continuing downsizing and an increase in workload. The contractor was involved in excavation, the laying of ducts, pulling cable, carrying out low voltage joints and temporary reinstatement.

Problems arose between the organisation and the contractor:

- Quality of work. The work carried out by the contractor was not of the standard the organisation expected. This included not maintaining the correct depth of cables and ducts put in the ground, rough reinstatement that was reported as a public hazard, and the condition in which completed work sites had been left.
- Adhering to the work specification. The contract stated that all work must be carried out in compliance with the specific regulations and acts, such as occupational health and safety, and to a level acceptable to the organisation.

  The contractor would be given work involving excavation and laying cable, together with all the information required and sketches on how the work was to be carried out. The contractor was found to be not maintaining the correct allocation in the reserve for electricity mains, and deciding on its own route on how it would get from A to B. No attention was paid to what had been indicated in the sketch drawings; the material and how the cable was to be laid were changed. An example of this was work with a specification designating copper cable to be laid directly in the ground. However it was discovered that the contractor had put the cable in ducts and used aluminium. This had the effect of reducing the original cable rating.
- Excavation. In a large development, local government requests the developer to run the wires underground rather than overhead along the frontage of the properties. In the past, the developer would normally do the excavation and lay ducts, so that the organisation could come along later and pull cable. This was to reduce the developer's costs, because it has the equipment on site and it can do decorative reinstatement before the organisation can carry out the work.

  However, the contract between the organisation and the excavation contractor was unfortunately written in a manner that the contractor would have a monopoly to carry out work for the organisation. With developers doing the excavation of trenches, this violated the terms of the contract. The contractor stated that it would take the organisation to court to recover damages for loss of income.

The problems turned into a stand off, with the contractor threatening the organisation over the excavation issue, and the organisation threatening the contractor over quality of work and non-compliance with specification. Communication with the contractor was minimal and projects were being delayed, causing problems with the organisation's customers. Without a fast remedy the problems would deepen, causing permanent damage to the organisation. No negotiation took place, just each party accusing the other of breaching the contract. There was written correspondence only between the two parties about the breaches of contract.

Exercise

1. The contracting company was made up of ex-employees of the organisation. Would this assist or hinder communication between the contractor and the organisation?

2. It was difficult to have the two parties come together by themselves to resolve the issues. There was a lack of trust and mixed emotions. What sort of third party assisted dispute resolution method would you try, and why? From the organisation's viewpoint, minimal disruption to other projects was a priority. Work was being held up and customers were screaming.

3. The organisation's experiences in using arbitration had been that, at the very least, a binding decision was made and it was quicker than going through the court system for an outcome. Some limited appeals were available. Would this dispute suit arbitration? Why or why not? Is arbitration only suitable where there are technical and practical questions relating to quality and costs which are likely to lead to a long, and possibly complex nuts-and-bolts inquiry?

4. Mediation is a suggested method where a continuing business relationship or a desire for such a continuing relationship exists, and this is what the organisation wanted due to the workload. However at what sacrifice to the organisation would this have been?

5. Would the view of the dispute have changed if, in the quality of work issue, quality was based on industry-accepted practice, or on what was stated in the contract documents?

6. The contractor had a monopoly on the excavation work for the organisation. How might such a contract be phrased — with a broad scope of work? In terms of a schedule of rates? Other?

10.2.14 CASE STUDY — BELOW GROUND CONSTRUCTION

Introduction

A consulting geotechnical engineering company was engaged to undertake the necessary soil investigation. The consultant had already undertaken considerable work in the area. Because of mobilisation constraints, the owner ended up drilling two cored boreholes for the consultant. The consultant took the core and a log prepared by the owner's geologist and prepared a report, which included the factual data from the field investigations plus advice on the excavation.

The owner also asked for a further investigation and report to provide advice in relation to the proposed constructions. The consultant decided to supplement the fieldwork with some seismic refraction investigations, and used the boreholes from recent work to calibrate the velocity profile and interpret the subsurface profile.

The findings of the initial report indicated strong to very strong basalt. It also indicated that the rock mass classification was very poor for a substantial depth. It recommended the use of impact hammers and drill and blast techniques. The second report confirmed the geological framework and confirmed the use of drill and blast techniques in the basalt.

The tender documents, prepared by the owner, included the usual disclaimers regarding the geotechnical information, although the factual data was warranted.

In the tender interview, the (eventually) successful contractor made it clear that it would award a subcontract to a drill and blast specialist. The owner confirmed its position in relation to the warranting of the geotechnical data, and the contractor acknowledged this position.

The contractor was eventually awarded a lump sum contract, and duly engaged the specialist subcontractor to undertake the excavation. The subcontractor soon ran into difficulties excavating the highly fractured rock using drill and blast techniques, and presented a latent condition claim to the contractor; the contractor quickly passed it to the owner. The claim, prepared with the assistance of another geotechnical consultant, stated that the geotechnical investigations overestimated the strength of the rock mass and underestimated the degree of fracturing.

On further investigation, the owner's geotechnical consultant and the owner believed that the excavation conditions matched very closely the drill holes and the discussion of the geology in the report. The owner rejected the claim.

The contractor also employed a consultant, who agreed that in its opinion a latent condition did not exist. It did however, identify some anomalies in the seismic refraction survey report. On balance, the contractor rejected the subcontractor's claim.

After initial conferences aimed at resolving the matter by negotiation, the subcontractor maintained its argument that a latent condition existed. As no further information was brought forward to change the contractor's view (and for that matter the owner's and its consultant's view) the claim was again rejected.

The subcontractor referred the dispute with the contractor to arbitration. Several other 'expert' geotechnical engineers were engaged by the lawyers representing the subcontractor, and a three-tiered case was developed. The first was that a latent condition did exist and the subcontractor had redress under the general conditions of contract. The other tiers of the claim related to actions under consumer protection legislation (deceptive and misleading statements) and the tort of negligence. These last claims aimed at the subcontractor relying upon the geotechnical consulting engineer's opinion that drill and blast was the most appropriate excavation technique, and that there were errors in the interpretation of the seismic refraction investigations.

The matter was heard before an arbitrator some two and a half years after the submission of the original claim. The arbitrator found for the subcontractor on all three tiers and awarded costs with interest.

The contractor and the owner then negotiated a resolution of the matter. The terms of the (head) contract were different to those in the subcontract; the subcontract conditions didn't replicate the clauses relating to the provision of the geotechnical information. The owner engaged yet another consultant who agreed that there was no latent condition, and additionally remarked that the arbitrator's decision was flawed on technical grounds. This expert did agree though that there were errors in the interpretation of the seismic work.

The quantum of the claim increased eight-fold over the period of the dispute.

Exercise
1. One of the major issues in this dispute was that the contract conditions between the
   contractor and the subcontractor, in relation to the limitations and liability aspects

associated with the provision of geotechnical information, were different from the head contract. Could this have been the intention of the contractor (an express intention of the contractor to alter the risk allocation) or a lapse in the drafting of its documents?

2. Consistent and error free documents are important. Does this go as far as including no conflict with conditions in subsidiary contracts?

3. A legal view of contracts includes that all documents should be read together to reveal the true intention of the parties. In what way might a head contract influence the interpretation of a subcontract?

4. In drafting contract documents, disclaimers may be used. There are many practitioners who believe that disclaimers are a practical way of avoiding responsibility for the accuracy of the information provided. Such disclaimers were provided in the contract documents, where the owner only accepted responsibility for the factual data. The arbitrator found that such disclaimers were not useful and the subcontractor could rely on the engineering opinion in the geotechnical reports. Comment on the arbitrator's opinion.

5. At least two of the parties were not happy with the arbitrator's conclusions. Some experts believed the arbitrator's decision was flawed on technical grounds. What does this say about choosing an arbitrator?

6. Generally the owner may not be successful in relying on disclaimers of information provided by the owner and its consultants, where the contractor has relied on the information in the tender documents in framing its price, and some of that information later turns out to be inaccurate, and as a consequence a loss is suffered by the contractor. Does this mean that owners should only provide factual geotechnical reports (which are warranted) which allow the contractor (and its consultants and subcontractors) to make its own interpretation of the engineering characteristics of the ground? What do you think of the practice of allowing a contractor to include a lump sum for additional geotechnical investigations to gather additional data?

7. Disputes need to be managed. In this case, the contractor took a very passive role early and tended to act as a go-between. The dispute was clearly one for the contractor's close involvement. The subcontractor presented its 'latent condition' case at a meeting attended by the owner's representative, the contractor and others from the owner's organisation. The case was considered extremely weak by the owner's representative, especially as the subcontractor could not substantiate that a latent condition existed. Its case was more directed at excavation methodology than excavation conditions described by the logs and the geotechnical model. The contractor forwarded the determination of the owner's representative to the subcontractor. This is when the 'wheels fell off', and the parties left the negotiating table, and the dispute escalated to arbitration. How might this stage of the dispute have been managed better?

8. Arbitration, as practised in this case, was not far removed from court practices, and the lawyers soon took over the proceedings. The dispute moved very quickly from the latent condition front, to claims that the information provided by the owner, and thus the contractor, breached consumer protection legislation (misleading and deceptive conduct) and was also considered negligent (tort of negligence). The most disheartening aspect was the time the process took, with the interim award appearing three years after the initial claim. What does this suggest in terms of dispute management practices?

9. Arbitration need not be a mirror of litigation, with correspondingly long times and high costs to settle disputes. Why has arbitration been allowed to become scandalous like this?

10. In this case, little effort or encouragement was given to the parties to resolving the dispute using ADR techniques. The owner's 'tried' approach did not include ADR. Does this suggest a distrust for ADR or a distrust for anything different? The dispute participants, in hindsight, believe mediation would have more clearly identified the issues surrounding the subcontractor's claim, and have provided the opportunity for someone to view the dispute objectively.

11. Newton's Law for Expert Witnesses ('For every opinion there is an equal and opposite opinion.') certainly applied in this case. Some eight geotechnical consultants were involved, with their support going to whoever paid their fees. Comment on the credibility this gives to the arbitration process.

12. The behaviour of many so called 'experts' is scandalous. They operate as 'hired guns', and their ethics and objectivity disappear when money and payments are involved. Comment on the credibility this gives to the use of expert witnesses.

13. The owner used a lump sum contract and further attempted to transfer the geotechnical-related risk onto the contractor by providing the disclaimer clauses. How much of the dispute could be attributed to risk allocation?

14. The subcontractor initially contended that the ground conditions were different from those that could have been reasonably anticipated from the geotechnical reports, although this view was not supported by the owner. The eventuality though was that the subcontractor relied on the opinion that drilling and blasting was an appropriate methodology, whereas in fact it was not because of the highly fractured nature of the basalt. The subcontractor appeared to be looking for extra money from somewhere. How might knowing this have affected the other parties' approach to dealing with the subcontractor?

15. Would providing for alternative dispute resolution techniques in the contract have influenced the course taken by the dispute in this case. Why?

16. Having gone to arbitration how might the relations between the parties have been affected?

10.2.15 CASE STUDY — TELECOMMUNICATIONS

A telecommunications company, TCC, outsourced a small-sized software development to a private contracting company, SoftComp.

TCC's business group produced a business requirements specification (BRS) detailing a business perspective and deliverables. TCC's information technology (IT) group was approached to develop an application to satisfy this new business need. The intention of this application was to automate and centralise a scheduling system for the company equipment, and supplement the current suite of in-house applications.

The IT group responded with a cost estimate, which was deemed too expensive. It was suggested by management that the project be outsourced, through the IT group, with a fixed price contract.

Several contracting companies tendered for the work. Tenders were based on having no conventions documents (for example 'database naming standards' and 'graphical user interface conventions'), and no technical requirements specification (TRS) upon which tenderers were to adhere for smooth integration with the current in-house applications. The IT group bypassed the TRS because it believed that the BRS would suffice. As well, it did not wish to allocate any person to perform the task because of the cost to the IT group. Ultimately the BRS was used as the TRS, to the dissatisfaction of the responsible IT project manager.

The contract was awarded to the tenderer SoftComp, on the basis of lowest cost and a generous 'warranty' period.

SoftComp visited the premises of TCC numerous times. Discussions were held with the technical people of the IT group with regard to the current databases, and the impact the new application would have. It was found necessary to modify databases for the new application.

Two months later, SoftComp delivered the application for user-acceptance testing. The IT group produced a user-acceptance test plan and a user-acceptance test guide.

Two weeks of vigorous testing produced an enormous amount of 'test exceptions' (situations/scenarios where the application was deemed not to behave as required by the specification). A significant portion of these were 'graphical user interface' conventions, that were not written into the BRS. SoftComp agreed, using its time and expense, to conform to these conventions as long as no penalty (time or cost) was placed upon it for this work.

For the remaining test exceptions, SoftComp stated that the exceptions fell into two categories:

- Possible software errors — SoftComp argued that the application behaved as specified, and that the testers should retest the application to confirm their results. Upon re-testing, it was found that only one-third of the remaining test exceptions were actually valid. SoftComp agreed to correct the software errors, and for each error a time estimate was given.
- Software enhancements — SoftComp stated that any software enhancements would be costed and, if approved by TCC, work would be performed and charged for.

TCC's business group maintained its stance that the application did not work as specified; the so-called software enhancements were exceptions in its eyes. This was an attempt by the business group to obtain substantially more than was originally tendered for. The IT group intervened and investigated the situation. Many meetings were held with SoftComp to come to an amicable solution.

Meanwhile, the in-house development team investigated the source code of the application and data produced by the application, and concluded that SoftComp's application conformed to the BRS — with the current data in the database. The reason behind the business group's disgruntling was that the database contents were pivoted at the date of installation of the company equipment, not the last known equipment service date. Consequently, all new scheduled equipment service visits were sooner than expected and it would cause a enormous number of service visits to be scheduled but not be able to be performed in the time allocated, if this application was to be released. The IT group asked the relevant people for the pivot point, and was told that the installation date would be satisfactory. However, it seems that this information was not conveyed to the testers, who thought the pivot point was the last known service date.

Several meetings between the IT group and the business group regarding the application followed. It was agreed that the IT group populate the database with the information that the testers were expecting and that a subset of tests be re-performed.

After re-testing, only a handful of software enhancements were required. Nevertheless, one particular enhancement was radically different to the specification for which the application was written. The business group stated that it had made an error in its judgment and, with the enhancement, that is the way it had originally intended the application to execute, and the enhancement was required before it would allow the application to be delivered to the users. The group also stated that the cost to re-develop the application would not hinder it in financing the project, believing the enhancement would only require minor development time.

SoftComp costed the software enhancements; the cost nearly matched the original tender cost. The business group was startled with the amount, and argued that the cost of the major enhancement was excessive. The group was told that a significant portion of the application would be re-written due to this enhancement. The business group thought that all the work could be done in one day. A degree of such 'overhead' work was included in the original cost estimate by the IT group, based on past experience with the business group.

After several meetings and 'warm' discussions, the business group agreed to the cost, and SoftComp was allowed to commence work on the application.

The application was delivered and re-tested. All parties were satisfied that the application executed as per the BRS.

The application was 'signed-off' and the business group gave permission to the IT group to deliver the application to the users. However, the business group stipulated to the IT group that the delivery mechanism for the application be a particular in-house application. Consequently, the IT group allocated a resource to modify both applications so that the business stipulation would be satisfied. Several days of development work followed. Such work was included in the original cost estimate by the IT group, and not by the tenderer SoftComp.

The applications were delivered as one 'module' to all users by the local area network (LAN) administrators of the IT group. Such work was included in the original cost estimate by the IT group and not by the tenderer SoftComp.

Exercise
1. The BRS, which was the basis of the contract, was poorly defined. The contract was just 'paperwork' to try to guarantee that the application was delivered and accepted prior to payment. The contract work came close to ending prematurely because the business group thought the application was a near-total failure due to the enormous number of test exceptions. Also, the poor specification was used by the contractor to its advantage when the substantial amount of re-work was needed. Comment on the link between contract documentation and disputes.
2. The contract was a lump sum contract. The successful contractor was to accept all over-budget expenses (which can be quite frequent and significant in the software engineering/information technology arena). Comment on the applicability of using such a contract, given that the scope of work was not well defined and the documentation was not complete. What risks associated with subsequent increased costs and delays would TCC be expected to bear?
3. Should the procurement of the development of the application have been totally in-house? Do the hidden costs of project management, the time required to arrange and attend discussions and meetings (including mediation meetings) between the

TCC and SoftComp, and the poorly defined requirements from TCC's business group point to using in-house over outsourcing? Do the advantages of doing the work in-house (direct control), such as flexibility in timing and funding and the ability to make changes and react quickly by redeployment of resources, outweigh the advantages offered by outsourcing? Resources were available in-house in this case because the cost estimate was prepared by those resources.

4. The tenderer, SoftComp, was aware of TCC's internal price transfer for software developments costing and used this to its advantage. SoftComp was able to produce a bid substantially lower than that of the IT group. Can you infer anything about whether there was true competitive tendering or not, and also whether the bid was indicative of the market situation?

5. SoftComp deliberately set a bid level in order to get the work, even at a loss, and to enter the field of contracting. The company had no recognised expert in the application development environment. Comment on the risk to TCC of accepting the lowest tender price and discovering the contractor's true level of competence.

6. In this case a great deal of dispute handling was required. The disputes were not over the contract clauses themselves, but rather the interpretation of the BRS by both SoftComp and TCC's business group (the authors of the document). The dispute resolution in this situation was quite amicable. The resolution was in the form of mediation. The mediator assigned was a person, with technical and business expertise, from the IT group and was able to guide both parties to a solution. How might such a person's apparent bias defeat the purpose of the role of an impartial mediator? Why in this case did it work to both parties' advantage?

10.2.16 CASE STUDY — INTERACTION BETWEEN CONTRACTORS

The project involved the construction of a box culvert. The project also included the provision of a gross pollution trap (GPT) near the outlet of the new culvert.

The owner issued separate contracts for the culvert and GPT construction to different contractors. The conflict issues that arose between these two contract areas were:

*   As working space was limited, there were conflicts over which contractor had possession of which part of the site.
*   During the project there were a number of local storms and each contractor alleged that the other's management of stormwater flows caused it delays.
*   The contractors argued as to which was responsible for installation and maintenance of common sediment traps.
*   There were arguments over which contractor was responsible for restoration of various parts of the site.

Whilst initially the two contractors and the owner were on good terms, and seemed to be willing to help each other out, the above conflicts eventually resulted in a breakdown of communications and a progressive withdrawal of interaction.

There were also numerous allegations that the 'other contractor' was not meeting its contractual obligations, and there were increasing problems for the owner coordinating the whole project and activities of the contractors.

The relationship between the two contractors broke down, requiring the owner to act as a go-between. Each contractor became only interested in satisfying its own goals, with no concern for the overall project.

The conflicts contributed to the project running over time and slightly over budget. There were also a number of claims for contract variations which had to be resolved.

Resolution

No formal conflict management techniques were used to resolve the conflicts that occurred during this project.

When the issue of possession of site came up, it was generally left to the contractors to sort the matter out themselves, as each contract document gave possession of the site to each of the contractors. The contract documents did not indicate any way for sharing the workspace, but did describe that any work completed by others needed to be protected.

Early in the project, the contractors were still communicating with each other. Through on-site discussions between themselves, they negotiated suitable arrangements which both contractors seemed to be pleased with. However, as the project progressed, it was clear that the relationship between the two contractors was declining.

Every time there was a storm, there was conflict between the two contractors over which should manage the stormwater flows. The contract documents did not detail any responsibilities in this regard. Both contractors refused to accept any responsibility for the stormwater and generally blamed each other and the owner when there was any damage.

After stormwater significantly damaged the contractor's GPT work, the owner's project supervisor instructed the other contractor to construct temporary retention structures and pump the stormwater around the site. This was an extra cost to the project, but it resolved the matter.

Conflict between all parties occurred with the installation and maintenance of a common sediment trap. Soil erosion and sediment control were part of each contractor's responsibilities, as indicated in the contract documents. The first contractor on site installed the sediment trap, but refused to maintain the trap when the second contractor started work and created a large amount of sediment runoff.

This was resolved by the owner instructing the second contractor to maintain the trap, because it was creating the most sediment. The second contractor was also instructed to remove the trap as part of its contractual obligations. The owner had to act as a conciliator in this instance to resolve the problem and determine what was a reasonable share of responsibilities between the two contractors. Both contractors accepted the owner's determination as being reasonably fair.

Restoration of gardens and the road pavement was also another source for conflict between the two contractors. The contract documents failed to define which party was exactly responsible for what restorations. As above, the owner had to determine what was fair for each of the contractors and instructed each accordingly. However, this was not accepted by one of the contractors, indicating that it would submit a variation claim for the work.

Exercise
1. The management of conflict situations on this project was far less than ideal. There are many ways in which the conflict situations could have been managed better.

How could the plans and specification for the work have been prepared with more care and thought? How could you have clearly defined which contractor was responsible for soil erosion and sediment control and all restorations?

2. What value would there have been in the owner meeting with the contractors before the start of work to establish common project goals? How would this have affected each party's motivation? Would it have identified and resolved possible conflict situations before they arose? Would it have created a win–win situation for each contractor from the outset?

3. Could the whole project have been planned and scheduled to improve the coordination of activities between the two contractors? How might the timing of the work and possession of the site have been planned better, in consultation with the contractors, in order to avoid some of the conflicts?

4. How might communication and co-operation between all parties have been improved? Would regular meetings have helped? How could the meetings have been used to reinforce the common project goals, coordinate work and predict possible sources of conflict?

5. How might meetings have been better used to carefully define what the problem was and brainstorm possible solutions?

6. During the project, the relationship and communication between the two contractors broke down, requiring the owner to act as a go-between. Each contractor became only interested in satisfying its own goals with no concern for the overall project. If conflict management had been used from the outset of the project, would the relationship between the stakeholders have been significantly improved? Would this have resulted in the project being completed without any time delays or without variations? Would the stakeholders have left the project feeling that the project was successful?

7. On projects generally, where multiple contractors are engaged, self interest is often at heart, and each has little concern for the other contractors. How might such an issue be tackled in general terms?

CHAPTER 11

Disputes — Case Studies II

11.1 OUTLINE

A number of case studies illustrating disputes and dispute resolution practices are presented from a range of industries and situations. More than one culture was present in each case study. The intent of the case studies is to illustrate a range of practices.

11.2 CASE STUDIES

11.2.1 CASE STUDY — PROCESS TANKS

The project was considered to be critical for a number of reasons. From the Asian-based owner's point of view, the current operation of one of its plants was due for expansion, both in order to stop other organisations capitalising on a rapid increase in the local market, but also to provide cash flow to fund further expansion in the area. From the foreign contractor's viewpoint, the income from the project would help avoid a potential cash flow crisis, provide capital for the development of an Asian-based manufacturing plant and secure a niche for the organisation in the local market.

For these reasons there was a severe time constraint placed on the early stages of the project. The owner wanted to move quickly to gain its market advantage and the contractor wanted to please the owner and get its cash flow moving.

The project was initiated following negotiation between senior managers of both organisations resulting in a 'letter of undertaking', which set out that the companies would be committed to the contract, but not setting out any terms for the agreement. The project managers and contract executives from both sides were not involved in these discussions. A number of verbal agreements for the project scope were made during this negotiation.

The task for the contractor was then to get the project moving as quickly as possible. Design and procurement were fast-tracked along with the requests to the owner for the contract to be established. Goods were forwarded from overseas suppliers before the establishment of a 'letter of credit' or the receipt of the contract documents.

When the contract documents were received and analysed there were a number of problems for the contractor. While the scope was clear in terms of the number and sizes of vessels to be supplied, other agreements that were made in the negotiation were ambiguously stated. One point of confusion between the parties was the responsibility for customs clearance into the country.

The contract was based on standard international conditions of contract for electrical and mechanical work, where the contractor was responsible for all transport costs associated with the movement of goods and materials to the site. The owner was responsible for the clearance of goods into the country.

When the goods arrived at the docks the contractor had provided the documentation necessary (to its understanding) to the owner for customs clearance of the goods. The owner believed that it had responded by supplying the documentation required for the contractor to move the goods through customs.

It had been agreed during negotiation that the owner would be responsible for clearance. The contractor believed that this entailed providing to the contractor a clearance certificate to be presented to the local authorities. At the same time the contractor was waiting for the owner to go to the docks and shepherd the goods through customs, as this was its understanding and experience of customs clearance, which in part prohibit a party other than the consignee (owner) from clearing the goods. Both parties were capably managing the aspects of the contract that were their responsibilities, according to their understanding.

What was missing was a consensus of what these responsibilities were. If these had been unambiguously defined in the 'scope of supply' the resulting problems would not have occurred.

With hundreds of tonnes sitting on the docks and both parties expecting the other to expedite clearance, the 'free time' normally allowed for clearance was expended without any effective action, and demurrage fees began accruing daily.

With these fees accruing, both parties steadfastly refused to take further action, considering that this would involve accepting liability for these costs, and so they accrued further.

Eventually, the independent contract administrator, as defined by the conditions of contract, issued an instruction for both parties to make all efforts to expedite clearance with the contractor to submit a claim for the administrator to adjudicate. The goods were then cleared and forwarded for fabrication.

The contractor submitted a large claim for additional costs such as demurrage, underutilised labour and time spent on clearing the goods. Additionally a claim for extension of time for completion was made for over half of the original schedule. The claim is yet to be settled.

The rapid initiation of the project and lack of involvement of project and contract professionals here combined to provide a project of ambiguous scope. All the reasons for the hurried start, the quick start-up of the plant for the owner, securing cash flow and capitalising the manufacturing facility for the contractor were either nullified or put in jeopardy by the delays and costs that stemmed from the confused scope. If the scope had been clearly defined by the project and contract professionals available, this problem could have been avoided.

Exercise

1. In hindsight, such a conclusion is reasonable. But how do you foresee such problems? How do you identify every foreseeable thing, every possible work item?
2. What role does communication play in contracts management? What role did it play here?

11.2.2 CASE STUDY — LOCAL VERSUS IMPORTED MATERIALS

Outline

An apartment construction project consisted of constructing three main parts — a main high rise apartment and office centre building; a shopping centre; and a park and recreation centre.

The owner of this project was the PCF Group, a property developer. The players involved in this project were the owner, a design team, one main contractor, four subcontractors and two independent contractors. The design team consisted of consultants which were responsible for architectural design, structural design, mechanical–electrical design, quality control and quantity survey. Two contractors were called independent contractors because they were not bound to the main contractor. These two contractors were directly appointed by and were contractually linked to the owner.

One of the independent contractors was IMB. IMB was trusted by PCF to work on this project because IMB had a good performance reputation. It always met completion times and maintained quality work. The types of projects which had been done by IMB before this project were the construction of factories, buildings, dams, bridges etc.

In this project, IMB had to erect the steel frames of the main building. Work started in July and finished in October. All preliminary work on the site was done without any problem, while prefabricated materials were delivered from IMB's workshop. The project was divided into four stages. According to the contract, inspection was to be carried out by a quantity surveyor (QS) and a quality controller (QC) before the start and at the end of each stage.

No significant problem occurred in the first three stages of the steel erection process. However, in September the QC claimed that materials used by IMB did not follow the specification. The QC told IMB to replace most of the steel with imported steel. This upset IMB, and also interrupted the work. At that time, the fourth stage was underway and it was expected to be completed one month later.

In the specification it was not stated that the steel must be imported. In this case, IMB used the local material which could meet the strength requirements of the specification. However, IMB acknowledged that it was impossible to find local steel with dimensions which were exactly the same as the specification. For example, most lengths of steel columns were less than that specified, requiring joints in different locations.

The disagreement became serious because both IMB and QC defended their stands. IMB contended that all materials had been inspected by the QS and QC before the construction was carried out. Nevertheless the QC asked the project manager to stop IMB from continuing via a letter written in September. This interrupted the execution of the whole project.

Seven days after the QC's letter was received, negotiation to resolve the dispute was offered by the project manager between IMB and the QC. In the negotiation, the project manager represented the owner. One of the considerations taken by the project manager to offer negotiation was that the QC's argument was not strongly supported by the specification. In all other matters, IMB had performed very well on site. The other contractor companies said that the dispute possibly arose because of the personality of the QC. He always forced contractors to accomplish their work as quickly as possible. As well, he was always dissatisfied. In his view, mistakes were always on the contractors' side.

Though a hard negotiation, finally an agreement was achieved to resolve the dispute. The final resolutions were that IMB accepted half of the delay as a time extension, and the QC accepted local materials for all the work.

Exercise

1. The dispute occurred as a result of different interpretations of the contract by both IMB and the QC, and was also influenced by personality factors.

   Why would you have a specification that precluded local materials unless local materials were not desired? Or was it an oversight in the specification?

2. Do you believe personality is an issue here or has the thrust of the dispute been transferred from the problem to the people?

3. An agreement was reached within seven days after the QC's complaint letter was submitted. Negotiation provided a fast resolution method.

   How might the dispute have been resolved at lesser cost?

11.2.3 CASE STUDY — CONSULTANT AND SUB-CONSULTANT

General description

A bank planned to build a luxurious and smart (fully automatic operation) high rise building as its head office. The bank established a new firm, Knab, which had the responsibility to build and operate the building. This firm acted on behalf of the bank.

Knab hired consultants to provide the concept and detailed design. After the total design was completed, Knab engaged a contractor through tendering. A traditional delivery method, together with a lump sum payment was used.

One of the consultants that Knab hired was Meds, which had as a scope of work to provide mechanical and electrical design (concept and detailed) and supervision in the field. Meds had a direct contractual link to the owner (Knab). Meds did not want to do the work itself, but engaged a sub-consultant, Angry, to do all of its work. Angry did not have any contract with Knab.

Knab would pay Meds for mechanical and electrical design and supervision. Then Meds would pay Angry after making some deduction for Meds' own profit.

However, Angry perceived the payment from Meds to be unsatisfactory. Angry asked Knab to pay an additional fee directly to Angry, and not through Meds. This request was refused, although Knab knew that all of Meds' work was being done by Angry. This was a problem between Angry and Meds, and nothing to do with Knab, because Knab had no contract with Angry.

Angry tried to persuade Knab to hold the additional fee until the dispute between Angry and Meds was resolved. Again, that request was ignored by Knab because there was no reason for Knab to delay the payment to Meds. Moreover, the owner of Meds had a close relationship with one of the highest level managers at the bank.

Angry became angrier. Angry did not want to take this dispute to court because its position was weak and the cost of going to court was quite high. Further, Angry was only asking for a small amount of money.

As a last resort Angry threatened Knab with its connection. The owner of Angry had very good relationships with government officials that oversaw the banking industry. Angry told Knab that if its demand was not reconsidered, it would use its connection power.

Exercise
1. Meds subcontracted all of its work to Angry. What is your view on such a practice? How does the owner prevent it from occurring?
2. It is unsure whether there was any contract between Meds and Angry. If there was a contract between them and there was a formula to calculate the fee, why did the dispute still arise?
3. The only control on the consultant was its professionalism. Ethics and professionalism may have been lacking. One example of this was the consultant's specification of a particular brand of product. How do you deal with ethics and professionalism issues?
4. Connection and relationship to the decision-makers can override what is in a contract. Angry had something more powerful than just a contract, and that was connection. This is not an uncommon event. Why then bother with contracts?

11.2.4 CASE STUDY — HYDRO POWER CONSTRUCTION

The project was funded by international money. The main construction in this project involved a concrete dam, an underground powerhouse, a tunnel spillway and a log pass facility. The owner of this project was a specially set up government corporation, the engineering was done by a local company, and the contractor was a joint venture company with a foreign construction company as its major shareholder.

In the contract between the contractor and the owner, the contractor took responsibility for purchasing and installing all the monitoring instruments in the dam. After being installed for four weeks, the instruments would be transferred to the engineering company with all the documents and four weeks of records.

Most instruments were ordered from InstMon and Rivim. When the first group of instruments arrived at site, a disagreement between the engineering company and the contractor arose. The engineering company asked the contractor to do accuracy calibration tests for each instrument before installation and to use site calibration results for determining equipment factors. And, calibration test results should be issued to the engineering company as a part of an instrument document. However, the contractor's laboratory did not have equipment to do such accurate calibration tests, and the contractor thought, because all the instruments came from highly reputed manufacturers, each instrument would have had accurate calibration testing before shipment. It was thought not necessary to repeat the calibration test at the work site.

Concerning calibration, the contract read: 'calibration should be done according to the requirements of manufacturers'. If the contractor accepted the engineering company's requirements, calibration equipment had to be purchased without recoupment.

One thermometer from InstMon was found to be out of order, and test results showed the factors provided by the manufacturer were wrong. Another contractor, doing other work on the same project, used the same brand instruments, but it carried out calibration tests. Therefore, the engineering company strongly argued that the contractor should make accurate calibration tests and re-measure instrument factors on site.

In order to settle the dispute, a representative from InstMon was asked to come to site, and meetings were arranged to discuss these issues. After three days of negotiation, the two sides achieved an agreement: accurate calibration was not necessary, but the contractor should purchase fundamental facilities to make rough tests to ensure that the equipment was not damaged after its long distance shipment.

Exercise

1. The project was one of a few projects in which the owner used an international contractor. The absence of an understanding of each other's different culture and habits caused a lot of disputes in which the owner and contractor had different interpretations of specified requirements in the contract. The situation described above was just one of those cases.

   Is this really a cultural issue? Or improper specification issue? Is culture being blamed for poor contracts management skills?

2. Before this project, few foreign instrument companies had entered the local market. Most monitoring instruments were produced by local manufacturers. After shipment, the calibration of the locally-produced instruments might have changed slightly, and so calibration tests were commonly carried out on site.

   However, this was a different situation. From past experience, the contractor trusted products from highly reputed manufacturers; it just did simple tests to ensure that the instruments were not damaged after shipment. Calibration equipment was expensive, and so the contractor did not want to spend money on this.

   InstMon used different technology compared with the local counterparts. Long distance shipment had little impact on the product, and it was thought that it was not necessary to make calibration tests on site.

   Again, is culture being blamed for, this time, different standards of equipment between countries?

3. Although the two parties held different opinions, calibration was not a big issue. Both parties aimed for a long-term relationship and tried to settle the dispute in a friendly manner. A representative of the manufacturer acted as mediator during the negotiation process, and provided the technical support for decision making. The dispute was settled efficiently without a high cost.

   How valid is the assumption of long-term work for the contractor? And hence why would not a profit motive be used by the contractor?

4. The construction contract between the owner and the international contractor was based on a standard international contract.

   The owner did not have the necessary experience and it employed an overseas consultant company to help it understand international practices in the construction field, and to manage the contract issues. It believed that it had proper protection of its interests. It remained passive when handling disputes with the experienced international contractor.

   Does having an international consultant shift the cultural barrier from between owner and contractor to between owner and consultant? Is this a gain?

5. InstMon did not have as high a reputation within the country as the contractor had expected. Before this project, a similar nearby project had used InstMon's instru-

ments and monitoring system, but after construction, when the asset was transferred to the owner, much of the monitoring system was found to be out of order; the reasons have not been established. The owner did not approve using all InstMon products in the dam and ordered some instruments from Rivim.

The engineering company accepted the resolution on calibration, with concerns about the instruments embedded in concrete. The resolution was accepted because the engineering company wanted to retain a good relationship with the contractor. On the other hand, with limited knowledge about international arbitration, it also did not have confidence that if the dispute escalated, it could find support in the contract.

How do you evaluate the chance of success if you decide to submit to international arbitration? How many people are expert advisers in international arbitration?

11.2.5 CASE STUDY — SPECIFICATION DISPUTE

Introduction

A contract was established for the construction of a clinic in a remote area for a provincial health office. The work included the installation of a water system. The contractor made arrangements with a subcontractor for the installation of the water system. In a year's time, the clinic was finished but was not accepted by the provincial health office because the water system installed produced non-potable water. The contractor claimed that the contract did not specify the type of water to be produced. On the other hand, the provincial health office claimed that it was implied that the clinic would be utilising potable water. To end the dispute, the matter was referred to the regional health office. An amicable settlement was reached wherein the provincial health office would shoulder the expenses for the materials to be used to cure the water if the labour was provided free by the contractor (or subcontractor).

Dispute resolution

The centre of the dispute mentioned here was the water produced by the system installed by the subcontractor. The contract did not specify the type of water produced but it might be reasoned that the clinic would utilise potable water. The area was remote and with the type of land, it was very difficult to find potable water

The method used to resolve the dispute was by mediation whereby a third party (the regional health office) assisted the two parties to end the dispute. The regional health office took an active, inquisitorial role to identify and understand the matter in dispute. The contractor acted on behalf of itself and the subcontractor.

Exercise
1. The source of the dispute was due to the interpretation of the contract. Both parties had valid arguments. The remote area led to great difficulty in obtaining potable water. A swift resolution of the dispute was achieved without going to arbitration or litigation.

Is it an implied requirement that water for a health office would be potable? What use would the health office have for non-potable water?

2. The use of mediation was appropriate because both parties wished to resolve the matter and make their own agreement. The authorised decision makers of both parties knew the details of the dispute and were available to attend meetings to resolve the dispute. In the end, both parties required a written opinion regarding the dispute to provide auditable evidence that an agreed resolution was properly based on the merits of the matter.

How influential is the need for something auditable in the selection of a dispute resolution method?

11.2.6 CASE STUDY — ROAD REHABILITATION

Background

The project involved the rehabilitation of a by-pass road which connected two major highways. The consultant (engineering and supervision) selected for the project was one of the best local consulting firms, and possessed good experience and had been involved in many construction/reconstruction projects in the country. The main contractor of the project was a local construction company with a similar background, and was selected from among the well-known local and foreign construction companies registered with the owner.

The project was to bring back to life the whole section of the by-pass road which was destroyed during war, and the neglect of maintenance thereafter. Previously, this road served as a main link between regions, connecting commercial zones. The rehabilitation of this road was considered viable in order to avoid the inflow of heavy vehicles onto other already congested roads.

Scope of work

The scope of work was divided into two parts relating to minor and heavy states of damage.

(A) Minor — Rehabilitate most of the existing roadway, including:
  (A1) shoulder repair
  (A2) pot-hole repair to the existing asphalt pavement
  (A3) reconstruction of culverts
(B) Major — Reconstruction of the remaining heavily damaged section:
  (B1) road embankment reconstruction
  (B2) asphaltic cover to the roadway
  (B3) entire drainage system reconstruction
  (B4) rebuilding of bridges

Delivery

The delivery method was design-and-construct, where the contractor was responsible for producing an end-product that met the requirements of the owner (as given in a brief from the consultant). The type of contract payment was a combination of schedule of rates and lump sum; the final quantities for a significant part of the work could not be accurately determined in advance. The reconstruction period was nine months, including mobilisation and demobilisation. The contract documents were locally produced and particular to this project, but with agreement from all concerned parties — the owner, the consultant and the contractor.

Aware that there may be further unforeseen political developments in the country, and because of budget limitations, special provisions were included in the contract concerning variations — a variation had to be authorised by the consultant and endorsed by the owner's representative, before the contractor could start on that variation. At the briefing before the start of work, the consultant was alerted to the procedures for communication with the owner's representative before issuing any order to the contractor to perform a variation.

Force majeure

Security measures were taken by the contractor to prevent guerilla disruption of the construction, as well as against the common practice of theft. During the dry season, the contractor doubled its number of security guards and as a consequence had increased costs. Further, during the excavation of side ditches along the road, two unexploded artillery shells were found, and this prompted the contractor to employ a mine clearance unit from the military to clear the site. Additional artillery shells as well as other explosives were later found alongside the roadway and at the bridge sites.

The contractor informed the consultant and the owner's representative by letter, but failed to wait for a decision and formal instruction from the consultant and the owner's representative, before acting.

Change in scope of work

During the project, at about the one third complete stage, there was a government cabinet reshuffle following a general election. As a result of the reshuffle, a new minister and several new senior officials were appointed. During a site visit by the new minister and the senior officials, in the presence of the owner's representative and the consultant, the contractor was asked to resurface the road in (A) to match (B); originally (A) involved only pot-hole repair to the existing pavement. Further, the contractor was requested to change the design of the superstructure of the bridges from steel to concrete.

Preparation for the amendment

Considering problems that might occur because of the limited budget, and following the site visit by the minister, the owner's representative issued a letter to the consultant requesting the preparation of an estimate for consideration and to amend the original contract as early as possible. However, some weeks passed before this amendment was finalised, endorsed by the owner's representative, and issued to the contractor; bureaucracy and administrative matters within the government were the main reasons for the delay.

Additional work

As the schedule for completion was tight, and because of the fear of work disruption by early seasonal rain, the contractor decided to do all the extra work in conjunction with the original work straightaway, not waiting for the completion of the amendment and estimate by the consultant. Later, in its claims for payment, the contractor gave quantities and rates, which in both the opinion of the owner's representative and the consultant were much higher than they were expected to be. As a result, an urgent meeting between the owner's representative, the consultant and the contractor was called to discuss the matter. At the time, the parties did not come to an agreement on rates, but agreed to provisional payment based on the old rates, to keep the project running. By this time, the project had proceeded to about half complete.

Negotiation

Negotiation on rate claims by the contractor included additional overhead costs covering security and the clearance of explosives, the asphalt overlay on work (A), and the additional costs of the bridges. No agreement was reached. The contractor wanted to keep the rates as high as possible, emphasising its risks and difficulties in performing the task with security and the disruption of the work by unseasonal rains, despite time allowances having been given by the owner. In its argument against claims by the contractor, the owner's representative mentioned a lack of a binding agreement for the extra work; the work was done prior to any consent from the owner's representative on new quantities and rates, and there was no consultant's progress certification.

Solution to the dispute

Soon after the completion of the project, the owner and the contractor reached an agreement to have the consultant review its quantity and rate estimate for the extra work. Finally, an amicable solution was reached whereby they agreed to accept the new quantities and rates provided by the consultant.

Exercise

The answers to the following will depend of course on the actual conditions of contract, but answer in this light.

1. Could the decision to increase security and to call in the explosive clearance unit by the contractor lead to any liability on the part of the owner? The contractor reacted on these matters alone, without waiting for a response from the consultant or owner's representative. How would such an approach affect the contractor's entitlement for extra time and costs? What the contractor did was only to inform (by letter) the consultant. There was no formal agreement binding the contractor and the owner. Is this the contractor's risk and should the contractor bear all the additional costs?

   [The owner, nevertheless, did agree to grant a time extension and extra payments to the contractor based on the consultant's estimates, following the completion of the project.]

2. The contractor decided to carry out the extra work straightaway based on a verbal instruction from the consultant, and without the endorsement of the owner's representative. Should the contractor bear all consequences arising thereafter? Should the contractor refuse to carry out extra work unless the owner promises to pay for it at agreed rates?

3. Does the fact that the contractor has carried out the work entitle it to claim for payment? Must the contractor show the existence of some agreement before claiming payment? Is the contractor entitled to payment only if it can prove that the work was properly ordered under a provision in the original contract?

4. Is the owner liable only for costs that it knows or agrees to?

5. Consider the failure of the contractor to abide by the contract requirements in terms of extra work. Does the consultant have no authority to supervise the construction of additional work or to certify on progress of the work and for claims of payment?

6. The variations amounted to a sizeable proportion of the original contract price. The contract required the consultant to agree on rates with the contractor, and the contractor to give notice if it intended to claim extra payment for the variation.

    Does the contractor have the ability to offer new rates and quantities in its claims or has it to keep the old rates, unless agreed by the consultant and the owner?

7. What are the contractual implications of the verbal order for extra work issued by the owner's senior officials?

8. What are the obligations of the consultant in dealing with variations in a timely manner?

9. How independent is the consultant in the process of developing new estimates and rates for the extra work? Is there a conflict of interest here? Should someone more independent than the consultant have been engaged to develop the estimates?

10. Both parties did not pay much attention to their contractual obligations; they reacted only when it was deemed necessary and ignored their obligations at other times. This practice is not uncommon in circumstances where there are no proper or unified contract formats, or where there are inadequate and ambiguous contract clauses. What implications does this have in trying to resolve any disputes that arise?

11.2.7 CASE STUDY — AIRPORT REHABILITATION

Introduction

The project was an emergency rehabilitation of an airport. The consultant (engineering and supervision) selected for the project was a local company due to its experience and participation in earlier airport repair work. The main contractor was a foreign construction company, which was selected through tendering from local and foreign construction companies registered to tender for the project.

The case study describes a dispute that arose during the airport rehabilitation, between the owner and the contractor relating to additional work verbally requested by the owner. The dispute was amicably resolved after the completion of the project.

The aim of the project was to bring the airport to a reasonable state, after war, by minimum disturbance to current aircraft flights. A major upgrading of the airport was expected to follow in the near future.

The scope of work involved:
- Asphalting an existing runway.
- Asphalting an existing taxiway and constructing a new taxiway.
- Reconstruction of an existing drainage system.
- Upgrading an existing aircraft parking area.

The delivery method was detail design-and-construct, and payment was a combination of schedule of rates and lump sum. The construction period was three months. The contract documents were locally produced and particular to this project, but satisfactory to all parties concerned — the owner, the consultant, and the contractor.

Because of budget limitations, the consultant was especially informed of communication procedures with the owner, for any variation which might be beyond that contemplated by the contract, before issuing any order to the contractor to commence the work. The contract also contained a provision that, before the contractor commenced any variation, the variation had to be authorised by the consultant and endorsed by the owner.

Changing scope of work

The airport was still operating while the rehabilitation was going on. While in transit at the airport, a senior government official asked the contractor to widen the existing taxiway.

Additional work

The concept of the extra work (enlarging the taxiway width and extending the aircraft parking area) was explained by the senior official to the contractor; its specification was to be the same as in the original contract. On this basis, the contractor started the extra work straightaway, believing that it was its responsibility to do the detail design.

Preparation of an amendment

The owner issued a letter to the consultant requesting its preparation of an estimate for consideration, and then requested an amendment to the contract (concept design, performance specification and other related material required) regarding the additional work. Having completed the amendment, the consultant issued it to the contractor (but withheld the cost estimate), with the owner's endorsement, requesting the contractor's new rates for the additional work. By this time almost one third of the additional work had been completed.

The contractor's rates for the additional work

The contractor provided new rates, which from both the owner's and the consultant's points of view, were a lot higher than expected and higher than the consultant's estimate. Negotiation was required. By this time the additional work had progressed to above half complete.

Negotiation for a reasonable price

The negotiation for a reasonable price for the extra work involved all three parties, but failed to reach any agreement. The contractor's argument centred on the intensive nature of the construction, despite a time extension being allowed for the additional work, and frequent work interruption by aircraft flights. By this time the additional work had been completed approximately two thirds.

As no agreement for the additional work had been reached, the consultant had no authority to supervise the additional work, but only observed.

Claim for the extra work

The owner rejected the contractor's monthly claim for the payment of the extra work. The rejection was based upon the lack of a binding agreement for the extra work, having no agreement on the new rates, and no consultant progress certification.

Dispute resolved

The rates for the new work items of the additional work were amicably agreed by both parties based on the consultant's estimate, after the completion of the project, and the contractor received the payment then.

Exercise
1. Is the contractor entitled to charge for additional work and materials based on the verbal instruction from the senior official?
2. Does the contractor carry a risk in doing the extra work without the consultant's written notice?
3. How might reasonable rates be obtained for the extra work? Estimates of net labour and net material prices together with overhead and profit? The view of an expert? Measurement of work done and materials supplied? Other?
4. One consequence of the failed negotiation over reasonable rates for the extra work was that the consultant had no authority to supervise the additional work, which contained many items hidden from public view. However, even at the start of the negotiation, more than half of the extra work had already been completed. The contract required each work stage to be certified by the consultant before the next one was started. How do you resolve matters relating to work done before there is an agreement?
5. Why might the contractor have an expectation to recover costs for the additional work even though written orders weren't obtained?
6. Is there an implied promise to pay by the owner?
7. The dispute was not about the extra work, but rather the applicable rates for the extra work. The owner took advantage of this to retain the progress payments.

   What are the consequences in this case of not using a standard form of contract which contains precise provisions for variations, and their communication?
8. For such projects, how should the contractor make itself aware of the hierarchical decision making system in the public sector, public sector budget limitations, and public sector delays in decision making?

9. Do you believe the contractor has adopted some particular strategy here for its benefit, or is it expected contractor behaviour?

11.2.8 CASE STUDY — COMMERCIAL COMPLEX

This case study is recounted from the contractor's viewpoint. The delivery method was traditional. The owner engaged a design consultant (architectural and structural design) and a contractor (on a schedule of rates basis). The design consultant was responsible for technical matters and certifying the contractor's progress claims.

Occurrence

Changes were made in the design load to cater for the installation of a water tank and an airconditioning plant with increased capacity. As a result, the depth of isolated footings increased. There were a number of footings affected by this change. An instruction was given to the project engineer on site by the design consultant to excavate deeper for the footings. The revised drawing was to be supplied in due course. This change order was issued orally and the contractor performed according to the instruction given.

At the time the instruction was given, the excavation for most of the footings had been completed and pouring plain bedding concrete had been completed in some of the footings. Some sets of steel plate formwork (as specified) for pedestals and columns were already on site. Steel reinforcement had been cut and bent for all the footings, columns, and pedestals up to ground level.

Local practice

Local practice, because of cheap labour, was for manual excavation with pick and shovel, with an excavator or backhoe only being economical where major earthworks were involved. Steel reinforcement was ordered in full lengths in different diameters and cut and bent on site according to the drawings supplied by the structural design consultant. The pouring of concrete up to a manageable height in the building was carried out manually by using wheel barrows or concrete pans, and by erecting scaffolding or vertical lifts. Concrete pumps were expensive compared with using this method.

Effect of design changes

The following items were affected by the design changes:
• The excavation depth for the footings increased.
• The height of isolated footings increased.
• The height of the pedestals increased.
• The height of columns below ground level increased.
The labour force was diverted to excavate further the footings already excavated.

The bedding concrete, complete in some footings, had to be removed to allow further excavation.

In order to have concrete pours no greater than the heights specified, new sets of column and pedestal steel formwork were ordered, and these were delivered after about a week.

Extra reinforcement was cut and bent. Steel already cut and bent was not used because the design consultant would not approve the lapping of steel.

Extra money and time were claimed for these items in the first progress claim. Some money was claimed under the schedule of rates (where it fitted within the schedule), some as lump sums.

The design consultant, who was responsible for processing claims, wouldn't consider the claim for extras at that stage, but promised to pay extras some time later after discussing them with the owner.

The contractor continued to work in good faith. The project was completed and the design consultant never settled for the extras, and ultimately refused to pay for the extras. This matter went to court and settled for all items except that claimed for the reinforcement because the design consultant denied not approving the lapping of steel.

Exercise

1. The change order was a verbal agreement between the design consultant and contractor. The contractor performed the work without issuing a notice that it expected extra costs. The contract provisions required any changes to be in writing and the cost agreed before execution, and signed by both the owner and contractor. None of this was done. The owner was not aware of the fact until the end of the project. The original contract was between the owner and contractor, while the change was ordered by the design consultant. Also the items claimed under the variations were covered by the general scope of work. How strong/weak is the contractor's case in such circumstances?
2. How can the contractor prove that the executed work is a true variation?
3. What practices would you suggest be adopted and what documentation would you suggest be kept in future similar situations by the contractor?
4. Without the necessary supporting documentation, would the contractor be better advised to pursue the matter through negotiation, mediation, arbitration or litigation?
5. The design consultant acted as the owner's agent. Could the contractor be led to believe, therefore, that an instruction from the design consultant was equivalent to one from the owner? Can the owner be committed by its agent's action and therefore have liability to pay?
6. The design consultant knew that the contractor expected extra money and time because the contractor claimed extras with the first progress claim. The change was authorised by the design consultant through the issuing of revised drawings; the design consultant had no option other than to increase the depth of the footings because the design load had increased. Do such a set of circumstances waive the requirement for or replace the need for a change order in writing?
7. In considering an oral change order, what influence do the answers to the following questions have on the successful/unsuccessful outcome of a claim?
   - Was the disputed work covered by a changes clause?
   - Was it a valid contract alteration?
   - Did the owner have knowledge?

- Was additional compensation (consideration) expected or promised?
- Was it ordered by an authorised agent?
- Were contractual requirements waived by words, action or inaction of the owner?
8. Recommended practice might be:
   - Reduce oral directives to writing; detailed written confirmation.
   - Take minutes of meetings.
   - Maintain a file for changes.
   - Clearly establish lines of authority for approvals.

   Suggest other practices that you would recommend.
9. How did the conflict of interest, whereby the consultant was both a designer and contract administrator, affect the progress of the dispute?

11.2.9 CASE STUDY — CLAIM REJECTION

The project was to construct a power station designed to supply electrical power for a pulp-making factory. A schedule of rates contract and a traditional delivery method were used. The owner engaged a local civil engineering consultant to assist as the owner's representative on site.

The work was executed under a tight fast-track schedule. There was much interfacing work — civil, structural, piping, concreting and electrical work.

To speed up the civil and structural work, the owner offered the contractor an incentive for completion of the structural work (which was divided into parts) ahead of schedule, amounting to 1% of the relevant part contract price per day. On the other hand, fines would be imposed on the contractor for work delays, amounting to between 0.5% up and 0.7% per relevant part contract price per day.

The work was very challenging to the contractor's team, because it had to work hard to keep to schedule while having to overcome obstructions. The surroundings were unfriendly (unfriendly landscape and malarial disease threat) and material supplies were remotely located. There were no proper roadways; instead river paths were used for materials transport.

After a long and hard struggle, the project was successfully completed ahead of schedule.

Unfortunately, the owner denied the contractor's claim for the incentive payment. The owner argued that instead of receiving an incentive, fines would have to be paid by the contractor. The owner's reason was that the contractor had failed to complete several of the project parts on time, as gauged by the original schedule. In fact, the original schedule had been revised and adjusted because of actions of the owner — it had delayed delivering electrical instruments to the site, and this resulted in the delay of the contractor's concreting work. This delay, however, had been anticipated by the contractor, which had notified the owner and its engineer representative on site regarding recovery for the delay. To overcome the delay, the contractor had proposed a revised construction schedule to the owner's engineer representative. The revised schedule was accepted and agreed by the owner's engineer representative. A copy of the revision was sent to the owner. Within the contract, the owner's engineer representative had full authority to direct, to suggest and to approve any proposal submitted by the contractor.

The owner denied the contractor's claim for an incentive payment, since it argued that it had never approved such a schedule revision, although its representative on site had. In the final monthly meeting held just before the handover, the owner once again confirmed its denial of the claim, although a bundle of documents that supported the contractor's claim had been delivered to the owner a few days before the final meeting took place.

The contractor regarded the owner's denial of the claim as equivalent to a contractual breach. A week later, the contractor sent, for the second time, a letter claiming an incentive payment. In this second letter the contractor expressed its intention to take further action through arbitration and report the case to a public servant of higher authority.

On receiving this letter the owner replied, inviting the contractor to a meeting to negotiate the claim. The case was amicably settled in a friendly atmosphere of negotiation. The contractor's claim was accepted and approved by the owner, without any objection.

Exercise
1. The contract required the parties to attempt an amicable settlement before going to arbitration. With regard to negotiation and arbitration, the contract required at least 56 days to elapse before arbitration could be commenced.
   Why 56 days?
2. To prepare for the final negotiation, the contractor believed that it had to implement a 'smart strategy', by which 'the enemy' would be cornered, had no other choices to defend itself, or even to strike back, but to surrender. It required a special effort to checkmate 'the enemy'. Why would the contractor think of going for such a win–lose approach in these circumstances? Or does such an approach reflect common contractor thinking?

11.2.10 CASE STUDY — BUILDING CLAIM

Background

Both the owner and contractor were local companies. It was a turn-key, fast-track delivery, using a lump sum contract, for a new five-storey office building, and the design and refurbishment of an existing office building.

The owner terminated the contract due to continual slippage of target milestone dates and concern that the owner's requirements would not be achieved.

The contractor claimed time and costs against the owner for extensive delays to the design and construction work. The method of dispute resolution selected was arbitration.

An expert was appointed by the owner and briefed by a legal adviser to review documents and prepare an expert report on the performance of the contractor in the carrying out of its obligations under the contract. The expert later attended the hearing as an expert witness.

The brief included a review of the contractor's programming and management of the work, especially design and documentation.

Documents reviewed

The documents reviewed in the process of preparing the expert report included:
- The contract.
- Statement of claim.
- Defence and counterclaim.
- Programs.
- Cost reports.
- Scope changes.
- Variation listing.
- Progress reports (design and construction).
- Conceptual design drawings.
- Amended agreement.
- Correspondence.
- Minutes of monthly progress meetings and coordination meetings.
- Organigrams and quality plans submitted by the contractor.
- Schedule of drawings.
- Witness statements.
- Reply and defence to counterclaim.
- Further and better particulars of the owner's statement of claim.

Technicality of claim

An analysis of the technical aspects of the claim was undertaken through developing an as-built program and assessment of the critical activities, based on an examination of the documents referred to above. A comprehensive drawing schedule was also prepared from examination of the title blocks and revision blocks of all the hundreds of drawings that were in the possession of the owner.

The expert's conclusion

Briefly, the expert's view was that the base cause of the termination of the contract stemmed from the initial failure by the contractor to manage and control the design process during the early stages of the project. The failure to comply with the specified program, and reporting and cost control procedures caused substantial time overruns, and as a consequence additional costs in design.

Time and cost of arbitration

The claim, due to its complexity and the number of documents to be reviewed, involved extensive research, time and cost by the expert. As well, there were the costs for the expert to attend the hearing as an expert witness.

Other experts were also engaged to prepare reports and attend the hearing as expert witnesses for the owner. These reports included a structural engineering and building services report, and an assessment of architectural issues.

The arbitration hearing took place more than a year after contract termination, after being delayed on many occasions.

Exercise

1. In this claim, the arbitration proceedings were formal and costly for both parties involved. Alternative dispute resolution clauses were contained in the contract. These however failed to help settle the dispute. Throughout the project it was apparent from the examined documents, particularly the correspondence and witness statements, that an adversarial attitude existed between the parties. This was exacerbated by the termination of the contract by the owner for lack of performance by the contractor, to the point that both parties did not wish to discuss the dispute issues. The dispute was therefore referred to arbitration.

    The advantages of arbitration are that it can be quick, cheap and simple. However, the disadvantages are that it can be abused with delay and can be expensive in long-running disputes. The disadvantages of arbitration were evident in this case. Arbitration in this instance was a complex, expensive and time consuming procedure for both parties. Why does arbitration commonly head towards the disadvantages end of the spectrum, rather than the advantages end?

2. The expert's report was clear, concise and well laid out. A methodology was included in consideration of readers with less specialist knowledge. All documents referred to were noted as well as all information which was accessed. A clear conclusion was presented based on the analysis undertaken. Comment on the introduction of experts into the case versus having no experts at all, given that experts are usually paid for by one of the parties. Would having a neutral expert be a better way to go?

3. Rarely is the preparation of expert reports for arbitration and litigation simple and straightforward. The preparation of such documents entails extensive research and reference to numerous documents. Frequently, the work has to be discussed with others involved in the project. Developing an as-built program, along with the preparation of the report is a time consuming exercise. The final expert report may be a bulky document containing the findings of the analysis, conclusions and the annexed back-up information. Comment on the usefulness of conclusions drawn from as-built programs.

4. The arbitration hearing took place over a year after the expert's initial brief. How can time delays be avoided in dispute resolution, and also not be used as a tactic by one of the parties?

11.2.11 CASE STUDY — CIVIL ENGINEERING WORK

It is generally agreed to be in the best interests of both parties to a contract to minimise the number and magnitude of claims. Nevertheless such a practice is not always adopted. Effort must then be made to ensure that claims resolution is timely and equitable. Without this, disputes will inevitably arise.

This case study illustrates the impact that poor claims administration had upon a project. The case study is based on a civil engineering project, with a construction period of two to three years.

A number of claims were submitted by the contractor. Some of these were quite early in the project, as is common with contracts involving civil foundation work. On this project there were many bored piles, and the ground conditions were quite variable in reclaimed land.

The contractor's initial claims generally involved additional costs and time resulting from unforeseen ground conditions. The contract made provision for such circumstances, and the contractor was entitled to compensation. However, through a combination of inflated claims by the contractor and consequent heavy handed retaliation from the owner's representative, what should have been a routine contractual matter evolved into ongoing disputes that delayed the settlement of all claims for over a year.

The owner's representative tended to be more concerned with refuting claims than judging them fairly in accordance with the contract. Payment of claims, or even agreement to payment, was unnecessarily delayed. Extensions of time were also held back. The tendency was to leave the resolution of all claims and disputes until the completion of the work in the belief that bargaining power would be reduced if additional payments and extensions of time were granted, regardless of their entitlement.

In retaliation, and perhaps from ill feeling and mistrust, the contractor started submitting trivial and invalid claims. The reasoning behind such a response was difficult to support. Its actions only served to fuel the fires of the dispute, and undermined trust in its other legitimate claims.

Conflict between the owner's representative and the contractor's manager led to unnecessary delays and costs. A great deal of time was spent in a paper war that not only affected the claims in question, but also unnecessarily complicated the progress of the work, progress payments and relations on site.

The case study illustrates that quick resolution of claims and disputes is needed to prevent unnecessary costs and delays. As disputes go on in time, people lose sight of the original issues and personalities can become involved. Perhaps more importantly, quick resolution is essential for maintaining good and cooperative working relationships on site, allowing the work to proceed as efficiently as possible and minimising further disputes brought about by mistrust and ill feeling.

Exercise

1. Management of the claims process in this example could be considered the antithesis of good practice. Both the contractor's manager and the owner's representative were found wanting of a reasoned and professional approach to their responsibilities.

   The owner's representative is not a party to the contract, and has a responsibility to be independent and fair in the administration of the contract. But being directly employed by the owner, how can this conflict of interest be reconciled?

2. Antill (Contracting and Construction Engineer, Vol 33, No 11, 1979), writing of the owner's representative, states that ... *his proper role is that of a 'preventer of disputes' in contradistinction to that of a refuter of claims.* In this case, the owner's representative acted more in the interests of minimising claims, either for the benefit of the owner or to prevent any poor reflection upon the contract documentation or control of the work, rather than striving to fairly resolve the claims in accordance with the contract. What other motives may have been driving the approach of the owner's representative?

3. What do you think the submission of inflated or bogus claims by the contractor did for its own cause? How was the contractor's credibility affected? Would it have been better served by cooperating with the owner's representative and acting sincerely? Or was there some underlying tactic involved (for example, to use later on as concessions to give away in return for concessions from the owner)?

4. Both the owner's representative and contractor's manager allowed themselves and their staff to be led by emotional behaviour, and driven by personality clashes, power struggles and stubbornness. The resulting lack of cooperation, motivation and sincerity on the project was evident, as was its impact upon the progress of the work. How do you remove the people from the problem?
5. It is believed that the owner's staff had a fear of claims, because they had been responsible for preparing the tender documents. As a result, the emphasis wrongly tended to be on refuting claims rather than acting out their duty of impartial administration. How do you remove this conflict that the owner's staff may have?
6. The poorly managed dispute created a pool of further unresolved issues, frustration and resentment, and resulted in subsequent and escalated conflict. But once this action-reaction process starts, how do you stop it from escalating further?

11.2.12 CASE STUDY — SUPPLY DISPUTE

The dispute

During the defects liability period in the construction of a prestige building, a dispute arose between the contractor and the owner regarding the failure of several dimmable lighting systems throughout the building.

A few weeks after practical completion, the owner notified the contractor of a defect on one of the dimmable lighting circuits. Several fluorescent lamps on the circuit had become inoperable and under the terms of the contract the contractor had been instructed to rectify the fault at no cost to the owner. An inspection of the inoperable fittings revealed two problems:

- A number of the fluorescent lamps had failed prematurely.
- A number of individual electronic ballasts had burnt out.

The contractor replaced the defective components. However, the owner received a letter from the contractor shortly thereafter stating that, in its opinion, 'the faults were caused by power surges from the electricity supply authority's mains and/or by the incompatibility of the dimming system electronics and the electronic ballasts.' The contractor also indicated it did not believe the problem with the dimmable lighting was a defect and intended to seek costs from the owner.

The owner disagreed with the contractor, stating that the contractor had not provided any evidence of the alleged power surges or any documentation in relation to the incompatibility of the specified components. In the owner's opinion, the contractor had supplied defective equipment and had not, as the contract stated, supplied 'a fully integrated, workable system capable of performance within the design limits specified.'

Over the next few months the remaining dimmable lighting systems throughout the building failed. The owner issued defect notices on the contractor. The contractor was unwilling to rectify the defects, claiming the problem was outside its control. Meanwhile sections of the building remained without adequate lighting.

Fortunately, however, both parties realised that such a prestigious, high profile building could not be allowed to languish in the middle of a contractual dispute and approached the owner's representative. The owner's representative suggested that an agreement be made to undertake interim measures to provide adequate lighting until the dispute was resolved. The cost of the interim work would initially be borne by the contractor, but would be subject to the outcome of the dispute. Both parties agreed.

Negotiation

The owner's representative called the owner, the contractor and the lighting subcontractor to a meeting to discuss the best method of resolving the dispute. The outcomes of this meeting were as follows:

- It was agreed by all parties that arbitration and the possibility of litigation would cost everyone time and money and was to be avoided if at all possible.
- All parties agreed that it would be counter-productive to apportion blame at this stage and decided the best course of action was to work through the problem together. It was decided that an investigation would be undertaken to ascertain the exact technical nature of the faults. This included:

  a) A thorough assessment of the specification for the dimmable lighting system including checking the conformance of the lighting components.

  b) Testing of the electricity supply authority's mains to identify the severity and frequency of voltage spikes, prolonged periods of high and low voltage and the number of power outages in a given period.

  c) The manufacturer of the components was be consulted in regard to the configuration of the system and the performance of its component parts.

  d) Impartial technical consultants were to provide a report on the dimmable lighting system.

Over the course of the investigation several meetings were held between the owner and the contractor to review the situation and set target dates. Not all of these meetings were amicable. Parties agreed to disagree until clarifying information became available, and target dates slipped when the information was not provided on time. The manufacturer of the dimming system sent a representative to the site to make its own assessment. Several technical consultants also visited the site and made recommendations.

Results of the investigation

- The contractor and its subcontractor had supplied and installed the dimmable lighting system as specified in the contract. The statement in the contract that the contractor was to 'supply a fully integrated, workable system capable of performance within the design limits specified' was considered (by all parties) to be somewhat unreasonable given that the lighting components were expressly specified in the contract documents.
- The owner's design consultant had specified the component configuration for the dimmable lighting system — presumably on advice from the manufacturer.
- The impartial technical consultants were able to confirm that some of the components of the system were incompatible. The manufacturer's representative later concurred, but indicated the design consultant had incorrectly specified the components.
- Testing of the electricity supply indicated that it was 'dirty' and had the potential to cause damage to electrical equipment and components. However, it was not considered to be the main cause of the failures by the impartial technical consultants.

Resolution

Based on these findings the owner's representative advised the owner that in his opinion neither the contractor, nor its subcontractor were at fault and that the matter may have to be taken up with the design consultant and/or the manufacturer. The owner agreed and gave an undertaking to carry the cost of the interim and permanent rectification work. The owner also indicated it may seek damages against the design consultant and/or manufacturer.

Exercise

1. The owner was of the opinion that the contractor had not fulfilled its obligations under the contract and was legally bound to rectify the (apparent) defects at no cost to the owner. Conversely, the contractor was of the opinion that it had complied with the specification and was not legally bound to rectify defects that it believed were caused by events outside its control.

   Might this have indicated (among other things) that the parties were interpreting the contract documents differently and that the contract documents were likely to play a central role in the outcome?

2. The general conditions of contract defined the method by which disputes were to be resolved. Accordingly, the owner's representative advised both parties of their obligations under the general conditions. In considering where this might lead them, both parties recognised that the dispute had the potential to cost them time and money. They needed to select a method of resolution that would not waste time and resources and would help to reduce the financial impact that might be imposed on either party.

   At this point there were two basic options:

   • Attempt to negotiate a resolution through an objective assessment of the problem without formally advising the owner's representative of the dispute.
   • Formally advise the owner's representative of the dispute in writing, thus committing both parties to the method of resolution defined in the general conditions of contract, namely arbitration or litigation.

   The first option had some clear advantages:

   • It gave the parties access to several methods of alternative dispute resolution. If these methods failed they could then resort to the second option. It was recognised that it may take some time to obtain the information required to resolve the dispute. The owner's representative therefore advised both parties that they could set their own (agreed) time constraints by choosing not to formally advise him in writing.
   • The owner's representative would not have to give a decision in writing. Rather, he could act as a facilitator in negotiations by offering informed opinions. This allowed the parties to be masters of their own dispute resolution.
   • Parties outside the contract (that is, an independent arbitrator and legal consultants) may not need to be involved and additional costs would be minimised.

   By comparison, the second option had some distinct disadvantages. Had the parties advised the owner's representative of the dispute in writing he would have been bound by the contract to make a decision within 28 days of receipt. He may well have had to make this decision without having all of the relevant information available. If either party was dissatisfied with the decision of the owner's representative, or the owner's representative was unable to make a decision, then the dispute may have ended in costly litigation.

   Comment on the rationale of the thinking behind the method chosen for the dispute resolution.

3. The course taken by the parties to investigate and objectively assess the dispute utilised some of the elements of different ADR methods:

    *   The company used by the subcontractor for procurement from the manufacturer independently collected technical specifications of the components and acted in a quasi-conciliatory role between the subcontractor and the manufacturer.
    *   The owner's representative acted as a mediator between the owner and the contractor and played a central role in coordinating the process and helping the parties reach agreement on several technical issues in order to resolve the dispute.
    *   Independent technical consultants were used to carry out an appraisal on the system configuration and technical specifications of the components. They provided an informed opinion that would assist in facilitating a negotiated settlement between the parties.

    The parties were able to resolve the dispute by this means. Comment on this method in terms of cost effectiveness and time efficiency.

4. How important is it that the parties recognised that they would have to co-operate with one another to stop the dispute escalating?

5. How acceptable to the owner was the outcome, given that the owner had to bear the initial cost of rectification work, but did seem to have a valid claim to seek damages against the design consultant and/or the manufacturer? Would a damages claim against the design consultant or manufacturer be worthwhile, or could the owner expect to carry all the cost even though it was not at fault? Why is it that many designers seem to escape being out-of-pocket in such circumstances, and the dispute doesn't go beyond that between the contractor and the owner?

6. What is to be gained by trying to rigidly enforce statements in contracts that are clearly unfair or unreasonable? In this case the statement 'supply a fully integrated, workable system capable of performance within the design limits specified' falls into this category, particularly when the components of the system were expressly specified.

7. Are such statements used when the writer of a specification is unsure of what is required to satisfy the owner's needs or there is some design problem that has not been properly addressed? Do such statements have the potential to invoke disputes rather than satisfy a need? Can this type of statement also be the conduit for personality clashes between parties with attendant flow-on disputation?

11.2.13 CASE STUDY — REFURBISHMENT PROJECT

Background

Within this case study refurbishment project, had better project and contract management procedures been adhered to, these would have negated some, if not all, of the resulting issues and disputes.

   A new project manager for the owner took over the project part way through. The owner's general approach to contract management up to that point had not been systematic. Dealing with both expatriate management and the on-site foreign national staff of the contractors posed some problems and peculiar issues. The project was not considered extraordinarily different from past experiences of the project manager. The project was, at that stage, well behind schedule, and was suffering from poor relationships between the contractors, project personnel and facility tenants.

The refurbishment was undertaken in conjunction with normal tenant business proceeding on site, a usually difficult, disruptive and potentially conflict-producing formula, and was subject to changes in personnel to both parties to the contract. All of this led to the need for some 'damage control' to be exercised by all parties involved — owner, contractors and local tenanted staff (who at this stage were becoming increasingly antagonistic and understandably uncooperative).

There were a number of unresolved issues that had to be addressed, in order of priority, and these required some political intuition and in most cases, straight talking. It was important to get some form of communications strategy in place, to help placate ongoing issues and problems that were running over and impacting on the project schedule. It was also important in this case, that some of the warring factions were isolated from each other, as there was too much history on this project to make any effective headway in trying to 'make good past differences'. Fortunately there was enough scope in the project to enable this to happen.

The following lists some of the issues that affected each of the parties involved and describes them in relation to good contracts management procedures. Looking back on the project, the resolutions were moderately successful.

Project issues on commencement of new person assuming overall project management role

Owner

- Slippage in the schedule. The actual performance was approximately three months behind that planned. This was a manageable aspect of the project, however the different values, desires, needs and habits of the parties did not match and this therefore resulted in lost time.
- Changes in the project staff. This highlighted poor project planning and people management. Poor documentation procedures left a difficult and tortuous paper trail by which to determine the project status. The project consultants were located overseas.
- Less than favourable opinion held of the project manager, by contractors, subcontractors and tenants, to manage and deliver the project.
- Failure to address dispute resolution in a timely manner. The contract clauses did not stipulate 'dispute resolution' as such, but only referred to 'arbitration' in the event that there were any differences that arose between the owner and the contractor. This was a major failing in the contract and inevitably led to protracted and time consuming action by both parties over relatively minor issues and amounts — for losses resulting from the withdrawal of work that was neither properly documented at the time or agreed to at the progress meetings.
- Poor on-site meeting practices, and poor information on material delivery time frames from suppliers and site access issues. Although information was not delib-

erately withheld, the owner did not positively contribute as strongly as would, or could be expected in order that contractors and tenants were better informed of actual information relating to the project.

Main contractor and subcontractors

- Changes in staff employed on the project. This was a result of poor people management and scheduling.
- One subcontractor had been dismissed from site for failure to deliver within contract requirements. (With a badly underpriced tender, throughout the project the subcontractor continually tried to 'shortcut' delivery in order to try and compensate for losses.) Poor attention to tendering by subcontractors was apparent after it was revealed that the subcontractor concerned had failed to allow for a number of key aspects of the subcontract. The main contractor should have been aware of this, based on the submitted price, and either questioned the tender price at the time of submission or followed this up much earlier in the project.
- Contractor's on-site staff subjected to verbal abuses from owner's representative. (This was often in the presence of other contractor employees and tenants.) This issue, which was extremely detrimental to the project, should have been addressed by all parties.
- Poor on-site meeting practices, and poor information on material delivery time frames from suppliers and site access issues from the owner's representative. The contract conditions failed to stipulate proper meeting procedures.
- Failure to address dispute resolution in a timely manner. Again, this could be attributed to poor contract construction.

Tenants

- Continually missed time frames.
- Poor communication processes at all levels — senior, middle and frontline staff.
- Disregard for occupational health and safety issues by contractors' staff, and this directly impacted on site safety for all personnel on site.
- Disruptive and noisy work practices. There was no attention or regard for the tenants, even after being communicated to the project manager.

Exercise

1. Previous experience of the project manager in this type of situation has shown that a major cause of this level of issues and disputes arising from within a single contract is that of a lack of proper communication practices. However, in this case, the problems were also the result of a failure to properly address a few basic management fundamentals that are required of any project. Classify the instances given in the case study into specific (poor) management practice categories.

11.2.14 CASE STUDY — DEFECTS AND PAYMENTS

Background

The case centres on two aspects of contract documentation for the construction of accommodation buildings for a public sector owner:
- Defects liability clause.
- The payment of subcontractors by the contractor.

The contract was a lump sum contract. Security, retention money and performance undertakings were required to ensure the proper performance of the work. Security was required in the form of a government bond and a retention of ten percent of the total contract sum was to be held for a period of twelve months after practical completion.

There were a number of nominated subcontractors that the contractor had been directed to use for the supply of specific items and work associated with the project. The nominated subcontractors were contracted to, and under the control of the main contractor for the duration of the project and all payments made to the subcontractors were via the main contractor.

Problems

Whilst there were numerous minor contractual problems encountered during the life of the project, a more serious situation arose after practical completion in finalising the payment to the main contractor as the main contractor proceeded to wind up its operations in the country. No final payments were made to any of the nominated subcontractors by the main contractor and, as a consequence, the subcontractors involved approached the owner for assistance in lobbying the contractor for money owed.

In addition to the issue of non-payment to subcontractors, the owner was beginning to wonder whether the main contractor would honour its undertakings with regards to the rectification of defects as a result of faulty materials and workmanship. No doubt if any defects were attributable to any of the subcontractors, there would be little incentive on behalf of those subcontractors to carry out any remedial work if they had not been paid for such work in the first instance.

This was in fact what eventuated as a result of the main contractor going into liquidation as far as its overseas operations were concerned. As a result, several of the subcontractors were forced to close their businesses and file for bankruptcy protection from their creditors, and so it went on down the line.

Within the first six months after practical completion, a substantial list of defects was compiled for the contractor to fix. As the contractor's in-country operations had been terminated, there was no point of contact with the contractor. Eventually contact was made via the contractor's head operations overseas, and whilst numerous undertakings were made with regards to both the payment of subcontractors and the rectification of defects, neither was forthcoming.

In accordance with the conditions of contract, the contractor was notified of the intention by the owner to have the defects rectified by another at the contractor's expense. The defect correction was paid for initially by the owner, the intention being that at the end of the defects liability period, the owner would seek to recoup this payment from the retention held. The owner also sought a ruling from an expert for the right to utilise any retention money remaining in order to pay the subcontractors, although this would not fully compensate the subcontractors.

Exercise
1. Standard conditions of contract, that dealt with both defects liability and the payment of subcontractors, were used. And so it could be expected that the contractor was aware of the situation regarding both issues, and no misinterpretation was likely. Is this a reasonable expectation or assumption?
2. The contractor presumably realised, that as a result of winding up its operations in the country, if there were any issues that were required to be rectified as a part of defects liability, then the contractor would be in no position to do this work. There are two reasons why this might be so:
    *   The contractor would have no workers in the country to do the work.
    *   The contractor had not finalised payments to the subcontractors (and so they would most likely be unwilling to do the work).
    What other reasons might the contractor have for not attending to the defects?
3. In this case the owner had nominated the subcontractors that it wanted, and the contractor accepted assignment of the subcontractors as part of the contract. This included a provision for the contractor to make payments to the subcontractors on behalf of the owner. Is this the key to the problems that were experienced? If the more usual arrangement for nominated subcontractors was utilised, that is the owner makes direct payments to the nominated subcontractors as directed by the contractor, would the impact on the project have been reduced or even eliminated? The subcontractors would have been paid for their part of the project. Then, irrespective of whether or not the contractor wound up its operations, would the subcontractors more likely have been willing to remedy any defects? In this particular case, the majority of the defects involved work originally performed by the nominated subcontractors.
4. How do you deal with the situation that some owners may prefer to nominate certain service providers or contractors to carry out specific sections of a project, but at the same time, not want to become involved with issues such as the payment of those subcontractors? Would this be the preferred option in most cases? Should, in this particular instance, the owner have directed payments to the subcontractors through the owner's on-site project manager with the knowledge of the main contractor? Would this have avoided the majority of the problems associated with the dispute?
5. The final outcome of this dispute involved most of the subcontractors being paid from the retention money held by the owner. This took place upon the expiration of the defects liability period and after all defects had been rectified (utilising others) to the satisfaction of the owner. Also, a recommendation was made never to again utilise the services of the contractor in question. Comment on the reasonableness of this outcome from the point of view of the owner, the contractor and the subcontractors.
6. As a subcontractor, what practices would you adopt if it was suggested that:
    *   You be a nominated subcontractor, or
    *   The subcontract was on a 'pay if paid' basis (that is, payment only occurs if the contractor has been paid)?
    (Note, in some localities, legislation governs practices on these issues.)

7. As a contractor, what contracts management practices would you adopt if nominated subcontractors were suggested?

11.2.15 CASE STUDY — ELECTRICAL AND MECHANICAL WORK

Introduction

A bank decided to open a new branch in a shopping mall. Contractors were selected through tendering to carry out the renovation, electrical and mechanical work (for example, lighting and air-conditioning), plumbing, computer system installation etc. The project was scheduled to occur over only a few months.

Serem was awarded the electrical and mechanical contract. Dolite was chosen by Serem to be its electrical subcontractor through competitive quotations. The dispute described in this case study was between Serem and Dolite.

Situation

Dolite finished what it was supposed to do, but was not paid according to its agreement with Serem. Meanwhile a consultant carried out inspection, testing and commissioning. A report of defective work was subsequently distributed to Serem so that it could carry out corrective work.

The defective work that Dolite was responsible for related to:
- The transformers for all the ceiling downlights were to be laid on ceiling channels, not on the ceiling boards.
- All exposed cable had to be covered by flexible conduit, not just those cables that could be obviously seen.

Serem wrote a letter to Dolite regarding the defective work, instructing it to correct and finish the work by a given date. Dolite refused to do so. Dolite insisted on Serem settling any outstanding money owing to it before it did the work. The given date passed, and Dolite had not done anything about the defective work.

A number of letters were written to Dolite on the same issue but all these letters were ignored. The relationship between Serem and Dolite had become soured. It was made worse when Serem wrote a final warning letter to Dolite, telling it that if it did not carry out the work as instructed, Serem would appoint another subcontractor to do so, and any cost of doing the work was to the account of Dolite. The letter produced no result; Dolite still maintained its position. Serem's attitude hardened and refused to pay because of Dolite's attitude.

Serem then appointed another subcontractor (Nodark Pte Ltd) to carry out the defective work. Serem paid Nodark at an extremely high price to do the work, and this amount was deducted from Dolite's account. Finally the work was completed. Serem paid both Dolite and Nodark accordingly. Dolite received its payment, but was unsatisfied with the amount that was deducted, and thought of suing Serem.

Serem was confident with its position, and had documents to support its stance. Serem's argument was that it was compelled to appoint another subcontractor to perform the work, in spite of a higher price, because it had to maintain its good reputation with the owner.

Nodark's reasons for its high price to carry out the defective work were:
- Access was difficult because much had been covered by the renovation work.
- It was dangerous work because the ceiling was extremely high.
- All other work, including the painting and renovation work, was almost completed; Nodark would be responsible if it damaged the well-scrubbed ceiling and other renovation work, while carrying out its work.
- Not many people want to do such work because it is both dangerous and risky; as well there were many poor electrical wiring connections done by Dolite, and Nodark had to perform the trouble shooting.

The court case lasted for about a year. While everyone thought that Serem would not lose the case, the judge argued differently. The reason given was that Serem had no right to withhold the payment to Dolite. If Serem was dissatisfied with Dolite, it should have paid Dolite in full first, before suing for discontinuing the work.

Exercise
1. The contract documents were very brief, and there was no method provided for resolving disputes. Although the dispute did not cause much disruption to the project, it caused both Serem and Dolite to lose some reputation and cost time and money to resolve. Does the presence of a dispute resolution clause imply careful and efficient handling of disputes?
2. Litigation was used as the method of dispute resolution. Is the world becoming more litigious? Both parties realised that it was definitely not a wise move from a business point of view. Dolite went straight to litigation without considering any other form of dispute resolution. Why might this have been the case? When people are easily offended, and once a relationship has become soured, is legal action the natural outcome?
3. Should courts only be used for more complex disputes? With an expected waiting time of a year, disruption and loss of reputation to the disputants, why wasn't another method used instead of litigation? The relationship between Serem and Dolite was destroyed, and this prevented them from working together again.
4. The outcome of litigation can be unexpected, and its fairness is arguable. Despite all the arguments in favour of Serem, the judgment went against Serem. Past experience had indicated that subcontractors discontinuing their work would not be viewed favourably by the courts. Serem had not always been prompt in paying its subcontractors. How might the past performance of Serem influence the judgment? The amount that Serem deducted from Dolite was unusually high. Although Nodark had produced reasons for this, how might this have affected the judgment?
5. In many cases, litigation is not the issue of how right your stance is; it depends on the interpretation of the judge of the evidence presented. It also depends on how good the lawyers are able to present the evidence, and what is considered the normal practice of industry. How might Serem's overconfidence have affected the outcome?
6. A common cause of disputes generally is one party withholding payment in response to (believed) defective work. Both actions may constitute breaches of contract. Given that this is a common cause of disputes, suggest general practices that will avoid its occurrence.

11.2.16 CASE STUDY — PROFESSIONAL SERVICES

Introduction

This case study details the resolution of a contractual dispute associated with a project involving an international standard, marine-based, theme park and resort hotel.

The overseas consultant was contracted to provide a full range of professional services through the project. These services included architectural, engineering, environmental, animal procurement, financial, facility operation and management.

The owner was a large local company, with its core business being the construction of industrial buildings. The proposal to build the theme park and resort hotel was its first incursion into tourism-based development.

An in-house management team was established by the owner for the purposes of administering the professional services agreement with the consultant. The owner's management team was answerable to the principal of the owner's company.

Relevant details of the situation

The professional services agreement stipulated that progress payments for the services provided were to be made at milestones in the development process. The quantum of each milestone payment to the consultant was a stipulated percentage of the total fee. The total fee was based on a fixed percentage of an approved but variable development budget.

The attainment of the specified project milestones was in some cases easily determined, for example on obtaining authority approval, however the agreement also provided for milestones based on itemised schedules of the work. Problems arose in the determination of project milestones where a schedule of work was used to define a milestone, because such schedules were generally ill-defined and not appropriate to the project specifics.

An attempt was made by the consultant during the project to clearly define the services to be provided in order to achieve the payment milestones. This definition was forwarded to the owner for its approval. No written response was received. However, in the consultant's opinion the owner verbally indicated its agreement to the proposal.

When the consultant considered that it had reached a particular milestone, it advised the owner, and presentations of the work undertaken were made by each professional discipline to the relevant members of the owner's management team and finally to the owner's principal. This process was prolonged by the owner's repeated requests for further information, with several trips between countries needed to be made by each member of the consultant team. After a period of some six months, it was the consultant's understanding that the owner's technical queries were satisfactorily addressed and an increase in the project development budget was agreed. A progress payment claim for the achievement of the relevant milestone was therefore lodged.

Some two months after the lodgment of the claim, an inquiry as to the status of the payment was responded to by the owner with a request for further technical information and an instruction to reduce the project development budget to the pre-milestone presentation level.

Negotiation by correspondence and in person with the owner's management team proved unsuccessful. A notice of the consultant's intention to stop work was issued. No response to the notice was received and work by the consultant on the project effectively ceased.

Dispute resolution

The impact of the dispute on the consultant was severe because without the milestone payment it was unable to pay its staff or its sub-consultants. This financial problem was exaggerated by the fact that it had entered into an agreement with a financial agency to provide funding on a monthly payment basis between the milestone payments. Such funding periods between milestone payments were up to twelve months, with large borrowings involved. The delay in receipt of the milestone payment resulted in the loan amount attracting extra interest.

The owner was also inconvenienced because its management team, responsible for administering the agreement, was unable to respond to requests from the owner's principal, local authorities etc. Requests for information from the owner were constantly received by the consultant. However they were not responded to.

Eventually, as a result of the above impacts, dialogue between the consultant and the owner's management team was reinstated. Again, the discussions did not resolve the dispute, with the consultant becoming frustrated by the 'mixed' messages received from the owner and its reluctance to process the milestone progress payment claim.

A direct appeal to the owner's principal was made by the consultant's principal. A meeting between the two was held and the dispute was resolved amicably. The result of the meeting was:

- A project development budget was fixed.
- The value of the milestone progress payment was calculated and payment approved.
- A future monthly cash flow was agreed.

Exercise

1. The situation described above is many faceted and complex. As with most situations that result in dispute, it is difficult to attribute the dispute to a single cause. What do you consider to be major contributors to the dispute?
2. The fundamentals of good contracts management were not observed in the tendering and formation phases of the agreement.

   Each party was guilty of not fully understanding the project and therefore an agreement, not specifically applicable to the project requirements, was formed. Consequently, the administration of the agreement proved to be difficult, with each party interpreting the unclear agreement to best suit itself.

   Further, proper administration was not insisted on by either party with alterations to the agreement, namely adjusting the scope of services, necessary to achieve the particular milestone, not being documented.

   Would you say this is a common fault with technical consultants, where the attention is focused on the things the consultants are good at, namely the technical matters, at the expense of things they may not like, namely administrative matters?

3. With little or no experience with dealing with international companies, both parties found communication, negotiation and in particular, understanding the motivation of the other party, difficult. This lack of understanding was not considered by either party to be a significant problem because it was the stated intention of the principals of both parties that the agreement was to be performed in a spirit of goodwill and cooperation.

The first major test of this goodwill, being the agreement on the achievement of the particular milestone, resulted in a dispute. A lack of understanding by both parties of their cultural differences contributed to this situation.

How might cultural differences be dealt with within an agreement, or can cultural differences only be treated outside any formal agreement?

4. The duration of the project had a major impact on the consultant through its commitment of resources and overhead costs, and through its financial arrangements, namely borrowing money between milestone payments, resulting in significant interest costs.

The owner however, was seemingly unconstrained by time. It was apparent that the management team was more motivated by obtaining a 'perfect' product from the consultant, no matter how long it took. This predisposition with not making mistakes, or ensuring that all alternatives had been considered, was symptomatic of the style of the owner's principal. The principal delegated the responsibility for the administration of the agreement to the management team. However, the authority to make the important decisions remained with him.

How do you then conduct a negotiation where to one party time is important, but to the other party time is of no consequence? Or should any agreement incorporate time matters that are to be observed by both parties?

5. The consultant's lack of appreciation of the responsibility and authority structure within the owner's company was compounded by its lack of understanding of the local culture, and in particular the indirect form of communication. The owner's management team, unable to make a decision on the achievement of the particular milestone and faced with compelling information from the consultant that the milestone had been achieved, chose to provide an indirect response on the acceptance of achievement of the milestone. This, it appears was a face saving measure, important in the local culture. What might be a way of communicating between two cultures such that the real meaning of communications is transferred?

6. The resolution of the dispute was negotiated face-to-face between the principals in a spirit of goodwill with both parties striving to understand each other's motivation. The stated intention of the original agreement was honoured by both parties reflecting a solid commitment to a successful relationship. Why might a dispute be resolved at principal level but not be able to be resolved at lower organisational levels?

7. What other lessons can be learnt from this contractual dispute?

11.2.17 CASE STUDY — FUNDING DELAYS

Outline

IQM was formed as a private venture, in order to undertake hydro power work and supply electricity to the national supply grid. IQM was a consortium of foreign and local companies. The foreign company saw this as an opportunity to gain a foothold in a country with a huge potential for future hydro power development. The local partner company had carried out several small hydro-electric development projects with overseas aid funds.

IQM was to deliver the project as a BOOT (build, own, operate, transfer) scheme.

Whilst a formal partnering agreement was not used, the relationship between the parties involved in the project adopted the principles of partnering. The parties saw the potential of working together on future projects. Within the constraints of commercial realities, the foreign partner intended to develop the capabilities of the local partner through training and technology transfer.

An application for loan funding for the project had been made to international finance bodies. The international banks imposed lengthy review and legal requirements which delayed commencement of the construction beyond the planned start date. The government was very cooperative and the negotiation between the government and IQM was at an advanced stage with a power purchase agreement already in place. At this point, approval from the banks was believed to be imminent. In an attempt to avoid delaying the project completion date, the joint venture partners agreed to commence construction of the critical work by investing their own capital. This also provided ongoing work for the local partner which had just completed a smaller hydro power project, enabling it to retain its experienced workforce.

The project required the construction of an access road, river diversion, intake structures, tunnels, an underground power cavern and a transmission line. Construction was expected to be completed in seven years. In order to assist an early start to the project, the government agreed to bring forward funding for the access road which had been planned as part of a road network expansion program. The major critical path activity was the tunnel. Therefore, while the road was being constructed, excavation of the tunnel was begun by IQM.

Having had considerable experience with similar projects in remote sites, the local contractor made good progress. However the same couldn't be said for the bank's protracted approval procedures. Meanwhile, the local contractor's equity resources were being exhausted, and so site work was scaled down to a narrow focus on only the most critical activities. Eighteen months after commencing construction, the foreign partner withdrew from the project pending bank approval of finance. With no alternative source of funds, this action forced the suspension of the project and resulted in significant standdown costs with closing the site and laying off most of the workforce.

A dispute then arose between the joint venture partners regarding the additional costs resulting from the closedown.

How was the dispute resolved?

After a six month suspension of the work, the banks finally completed their review procedures and released the loan funds. Construction recommenced almost immediately with commissioning eighteen months later than initially planned.

The dispute between the local and foreign parties took months to resolve. A mediator was named in the agreement document. He was a senior, retired foreign engineer who had worked in the country for a number of years. His background suited him for the role and enabled him to effectively broker a resolution. The outcome involved the local contractor being promised reimbursement of actual standdown costs, but only out of future profits from power sales once the project was completed, and being advanced funds in the short term to meet its cash flow requirements in remobilising the work.

The strain on the goodwill between the fledgling joint venture parties came close to a breakdown. The character of the relationship appeared to have been permanently affected.

Exercise
1. The local partner could not understand why the foreign partner took such drastic action without any consultation, leaving the local partner in a very destitute situation having invested all it had in the project. The local company perceived the foreign partner as having unlimited financial resources. Culturally, abandoning the relationship in an hour of need when it had the resources to support the local partner, caused offence and seriously damaged the relationship. The local value system which placed greater importance on relationships rather than commercial considerations led the local contractor to believe that its new foreign partner was only intent on exploiting the relationship for its own benefit. What obligation is there for the local partner to be familiar with foreign business culture?
2. From the foreign company's perspective, having analysed the project in the cold light of financial risk, it was time to draw the line and stop pouring resources into a 'black hole'. After all, it wasn't a development agency. Its action was perfectly reasonable in its view. Its intention was not to terminate the joint venture, simply to suspend the project until loan funding was available. The reaction of the local partner, which expressed such strong criticism for its action, came as a complete surprise. What obligation is there for the foreign partner to be familiar with local business culture?
3. The local partner placed a high priority on maintaining continuity of employment for its staff, since company loyalty and lifelong employment were strongly held values. For this reason, the local company was prepared to commit much more in order to keep the project going. The local company had a social conscience knowing that there was high unemployment in the country and no social security system. This was not understood by the foreign partner which adopted an attitude that human resources are as disposable as any other resource.

   Both groups acted in the situation in accordance with their cultural norms with little or no insight into what motivated the behaviour of the other party. A case of culture shock ensued and the new relationship was severely tested. Could such a dispute have arisen between parties of the same culture but different value systems? What can be learnt from the above case?
4. Should the possibility of delay in financial settlement have been foreseen? Given the pioneering nature of the project, was it likely that the financial institutions would be more thorough in their procedures and the approval process more protracted? Should provisions have been made in the agreement covering this eventuality?
5. Should the joint venture partners have undertaken some training to alert them to the different value system of the other partner? Would this have helped in understanding the reaction of the other party?
6. Despite the apparent mutual benefit of gaining a head start for the project by starting construction ahead of the funding approval, the joint venture underestimated the risks involved. Why do you think it was that the possibility of the loan application being

unsuccessful was never seriously considered? Under what circumstances would you commence construction before funding had been approved?

7. In spite of the amicable resolution of this dispute, the partners resolved to ensure that a much more adequate dispute resolution procedure was included in all future agreements. This procedure would involve a staged procedure for the escalation of disputes, beginning with assisted negotiation and progressing to mediation and other alternative dispute resolution methods if required. Is this in keeping with the partnering ethos of the relationship?

11.2.18 CASE STUDY — DESIGN INFORMATION

Outline

The project involved design-and-construct delivery of a mass transit railway depot. The depot itself formed one part of a new railway line and stations being built by the owner. The work was undertaken by a foreign international contractor. The design was undertaken on its behalf by an international consultant with a well-established office in the country. However, the majority of the design work was performed in the consultant's offices overseas.

During the course of the design and construction, a dispute arose about responsibility issues, design information, and late owner approvals, and the impact this was having on construction progress.

The causes for this problem existed from the very start. Although the contractor had an established contract with the owner, it was several months before a document was signed between the contractor and its consultant. This was primarily due to the negotiation of fees, and issues over the scope. This resulted in the design not starting until six months after the owner–contractor contract had been signed. Meanwhile the owner–contractor contract milestones remained unchanged.

As the design progressed, there was a continuous problem of getting final design details of the many system wide contractors (SWCs) that were employed by the owner to design and install system wide features, including trackwork, signalling and electrification, of the overall project.

This lack of information affected the ability to finalise the design, and therefore also the approvals that were required before construction could start.

The contractor was holding the consultant responsible for not finalising designs and approvals, and hence delaying construction and critical payment milestones. The consultant was also suffering losses due to the delays, which it considered to be out of its control, and was looking to claim these costs from the contractor.

How was the dispute resolved?

The contractor–consultant contract conditions were ambiguous in their allocation of responsibility for the performance of the SWCs. It was agreed that the consultant was responsible for coordinating with the SWCs, but the contract did not specifically name responsibility for their performance.

The dispute continued throughout the construction period of three years. Progress payments were withheld from the consultant. Considerable correspondence was exchanged between the contractor and the consultant, both maintaining their positions that losses had been incurred through the fault of the other.

It was not until construction was nearing completion that more productive negotiation took place. The process started with each of the parties preparing its case, bringing together the considerable amount of information that had been generated. As some of the original personnel had left the project and the organisations, this was a time-consuming process.

Negotiation only involved a small group of the senior project management staff and executives from the two parties. They had not been involved in the more detailed and technical aspects of the dispute, and therefore had to be fully briefed. Other disputes were also to be discussed at meetings, and so similar effort was required for those matters as well.

A number of meetings were held over a one week period. Discussions were fairly general in nature, and generally consisted of each party establishing and maintaining its claim that losses had been incurred, and these were out of its control and in its view the fault of the other.

A final meeting was held the following week, where it was agreed that the consultant would be paid outstanding fees plus approximately half of the claimed amount. Additional fees would also be paid subject to a claim that the contractor was negotiating with the owner.

Exercise
1. The dispute was elevated to senior levels in the contractor's and consultant's companies. What are the advantages and disadvantages of only having senior staff, who are unfamiliar with the origins and details, resolve the dispute?
2. The resolution of the dispute did not occur until well after it had occurred. This had its effect on relations during the project, and contributed to a low degree of satisfaction for those involved on a day-to-day basis. Additional duties were being undertaken by the consultant in good faith, but there was concern as to whether there would be fair payment. The toll of low staff morale was also reflected in the high rate of staff turnover, which was costly and affected continuity and background knowledge. What would have been the barriers to a more timely resolution?
3. Given the long duration of the dispute, and the detailed technical issues involved, the negotiation process itself was relatively short and simplified. How might you explain such a paradoxical event?
4. Both parties preferred that relations were not damaged by the negotiating process. Overall the project had gone well. The contractor did recognise that in many areas the consultant had saved it a considerable amount of money by optimising designs, particularly in some of the repetitive elements. How might such thinking have contributed to the speed of the resolution?
5. Both would have benefited from a successful outcome of claims against the owner. How might such thinking have contributed to settling their differences quickly and concentrating on a combined front in that direction?
6. The contractor could see advantages in keeping a tight hold on the consultant, maintaining the threat of delay costs that far outweighed the consultant's more modest

claims for additional fees. This encouraged the consultant to strive for a final and approved design as soon as possible, putting the other issues to the side. Give your view on the use of such a tactic, which contributed to a delay in the settlement, particularly when the designer and contractor are meant to be working as a team.

7. The owner approval processes put the contractor at some risk. There was little recourse for contractors which believed they had been treated unfairly, because the owner (a public sector authority) wielded considerable power. Not being fully familiar with the processes, or the people, put the contractor and consultant in a position weaker than they were used to when operating in their own countries. What approaches are possible to address the issue of a lack of local knowledge (practices and culture) for foreign companies?

8. The concerns to the contractor included extended construction periods, building from unapproved drawings so as expedite the work, and liquidated damages payable to the owner. To transfer the associated risk, some responsibility for the delays was attributed to the consultant. Although the consultant had a sound case against these claims, it was a big enough stick to encourage considerable effort from the consultant to get the design approved as soon as possible. Are tactical reasons for disputes as important as their negotiation and resolution?

11.2.19 CASE STUDY — CLAIMS AND ROAD CONSTRUCTION

Background

A joint venture, involving a foreign construction company and a local company, won a contract for the construction of a public road. The road was funded by an international loan bank, designed by a foreign consulting engineering firm, and the contract supervision and administration was handled by a foreign consulting engineering firm. Standard international conditions of contract were used.

The contractual dispute arose as a result of a series of claims submitted by the contractor and being rejected by the owner's engineer. The rejections resulted from the claims being poorly presented, from both factual and contractual aspects, and/or because the owner's engineer was acting within policy guidelines which had been set by the government (owner). The rejections were not taken lightly. The contractor retained a compatriot claims consultant to prepare the claims, which, if not resolved amicably, would be taken to arbitration for a decision.

The joint venture relationship existed in name only at this stage, and the local company had signed over the responsibility for making and pursuing any claims, and finalising the work, to the foreign partner.

The claim covered the following issues.

Possession of site

The conditions of contract were definite in that the owner had to provide possession of site to the contractor to allow construction to proceed. Unfortunately, the handing over of the site was not cleanly done, in that many of the landowners had not received their full compensation. Many confrontations occurred between the local people and the contractor's workforce. The local people correctly surmised that action would be taken by the contractor with the local government if problems were caused with delaying tactics and lockouts, and thus this would assist them to obtain compensation for their land and trees. Because the majority of the workforce itself was local, there was a natural sympathy towards the landowners, and although little physical hindrance or barriers were used, the workforce respected the 'no-go' lines pointed out by the landowners. Some of these delays became quite protracted.

Geotechnical

At the time of contract award, the proposed approach for doing the work was for starting at one end and developing hard rock quarries along the way based on a geotechnical report. This approach could not be followed, because the geotechnical report was in error. As it turned out, there was only one hard rock source for the complete length of the road, and it was located at about half way along the road. The hardness of the rock was also much greater than detailed in the report. Thus, the approach of relocating the site office close to the road head by moving several times, as each planned quarry was opened up, did not eventuate. The type of crusher mobilised had to be replaced. The haul length for the quarry product was much greater than anticipated, with greater maintenance costs on vehicles.

Access

Just as the mobilisation was commencing with the shipping of plant and equipment from overseas, a severe storm struck and destroyed an access road, and the nearby port facilities were seriously damaged. This caused delays to the start and continuing delays to the project because the access road was never reconstructed in time.

Survey

Errors were detected in the original base survey, and therefore all the earthwork quantities, both cut and fill, given in the bill of quantities were incorrect. The major impact of this was a delay — new work could not start until a new agreement or a new base survey was made. Then, at the end of the construction, all quantities had to be re-measured, a procedure agreed to by both parties, but one that caused a further delay to the issue of the contract completion certificate.

In accordance with the conditions of contract, the contractor gave notice of its intention to seek resolution of the claims by arbitration, after these were rejected by the owner's engineer.

Resolution of the dispute

Several months after the construction had been completed, the arbitration process had progressed to the stage where both parties were attempting to agree on a short list of arbitrators with international experience. On the contractor's side, the claims documentation had been prepared, and legal procedures were well advanced. The foreign company's regional manager paid a visit. The company had other projects underway in the country.

The project manager for the road construction project was finalising the detailed joint quantity check, providing assistance to the claims consultant, and supervising the removal of all plant, buildings, facilities etc., either by sale or shipping out of the country.

Even though the claims quantum was quite large in percentage terms, no bitterness existed between the parties involved in the project. Therefore, it was relatively easy to arrange for an owner's official and the owner's engineer (the foreign consultant contract administrator) to have dinner with the contractor's regional manager and project manager. The owner's official wished to know if the contractor's regional manager could negotiate on behalf of the contractor. When the basics were agreed, the official placed an opening offer on the table.

The contract value (not including the claims which were the subject of the dispute) was in the last stages of calculation and agreement with the finalisation of the survey quantities. The owner, by not accepting any of the contractor's claims, considered the final value of the project to be approximately half the contractor's figure, including the value of all the outstanding claims.

The official stated that the government would be prepared to pay a bit more, and he asked what would be the contractor's lowest figure. He was given a slightly reduced sum. Although both parties had come some way to the centre, there was still a gap, which seemed insurmountable. Both parties agreed to meet the next morning and to explore the matter further.

The meeting on the next day was a much more formal affair, with a stenographer present to keep an accurate record of the negotiation, but with the same personnel as had attended the night before. The meeting was convened at a neutral venue. After an initial opening statement from both sides, the actual meeting of the minds took place very quickly, and a compromise figure was accepted before the planned lunch break. Following a recess for lunch, in which the agreement was prepared and the regional manager spent on the telephone to his head office convincing his own board members of the wisdom of the settlement, the agreement was signed by both parties.

The dispute had been resolved.

Exercise

1. One contractor's general view is that it is doubtful if any conscientious project manager would set out to deliberately contrive a claim. It is considered that there is enough to do without the added burden associated with issues that may not produce income. However, when it is obvious that matters outside the control of the contractor are having a severe impact on the project, claims for the recovery of costs would be mandatory. What might be an owner's view?

2. All the claims, except for that dealing with the possession of site, were ones that could be encountered in any country. The reaction to land being resumed for construction and compensation not being adequate or late in being paid was peculiar to the local culture. There was also an element of danger involved. In this case, firearms and knives were frequently seen, but not used. However, given that the basis of most of the claims might be the same in other countries, how might you expect their handling to differ between different cultures?

3. Initially, the claims were prepared and submitted by those on site without assistance or input from the contractor's overseas head office. However, it was felt by the contractor that external (to the project) assistance would be necessary to properly document and prepare the claims. This led to the retention of a compatriot claims

consultant. This type of specialist was not available locally in any case. There were further difficulties encountered by virtue of the distance between the project site and the claims consultant, and the communication between the project manager on site and the overseas consultant was critical to good claims preparation. How might this physical separation affect the preparation of claims?

4. If this case had gone through the full arbitration process, how might you select an arbitrator with international experience, and a 'neutral' country in which to hold the arbitration?

5. Because the project was funded by an international loan bank, and because the foreign contractor did the majority of its international work on projects funded from this type of source, action was taken by the contractor to approach the bank to ensure that the disputed claims would be financed. Comment on the irregularity and pragmatism of such an approach.

6. An aspect that was critical to the final negotiation with the local government was that the contractor did not wish to jeopardise its relationships with the government. The contractor had been working in the country for over fifty years, and wished to continue. There had to be allowances therefore in the dispute resolution process for 'face saving' by the government, and in fact the final agreement did not fully cover the extent of expenditure incurred by the contractor. This loss was seen as part of this face saving exercise. What determines how far the contractor should go in making concessions, when presumably they are eating into its profit?

7. There was also conjecture on the reasons as to why the owner had acted to resolve the issue when it did. It may have felt that the claims would succeed in arbitration, and therefore that an early resolution would be advantageous. There may also have been pressure applied from the international loan bank to resolve the dispute as quickly as possible. Whatever the reason, is it unexpected to see disputes settled so quickly?

8. The predominant lesson from the case study is the importance of land to people. The threat of violence can be unnerving for all. What other practices, besides covering land issues thoroughly in the contract documentation, in terms of responsibilities, actions on default of compensation, and the bureaucratic system that is to be followed for the resumption of land for infrastructure projects, would you recommend?

9. In this case, there was no bitterness evident from either side, and that made the final negotiation much easier. If there had been any feelings of rancour, then it is doubtful if the dinner, which led to the resolution of the dispute, could have been arranged. How unusual do you regard this that emotions were not involved?

10. How might the interface between foreign and local cultures have affected the rise of the dispute and its resolution?

11. Negotiation, in this case, was over money alone. This can be difficult. Why might non-financial issues not have been introduced into the negotiation?

11.2.20 CASE STUDY — DESIGN-AND-CONSTRUCT ROAD DELIVERY

Introduction

This is a case study of a contract dispute between a public sector owner and a joint venture contractor on a project involving the fast-tracked design and construction of a major road.

The work started on a fast-track basis (with motives of prestige, personal gain and political gain), but was suspended pending a government inquiry and was not reinstated. The contractor attempted negotiating costs for the termination. The negotiation was unsuccessful and the contractor sued the owner.

Contract details

The owner approached the joint venture partners with a proposal for the design and construction of the road. Both of these partner companies were reputable companies in their respective fields and the owner approached them directly as a result of a desire to fast-track the project.

The road reserve was steep terrain and the original brief was for tunnel construction. Based on this design brief, a conforming offer was submitted together with a number of alternatives offering cost savings. One of the alternatives (a conventional road cutting) was accepted.

Standard international conditions of contract were used, with negotiated amendments to better suit the design-and-construct process. Construction started in the year following the owner's original approach. A stopwork order was issued soon after construction started and the work was discontinued.

Description of the dispute

The government of the day changed soon after contract award and claims were made by the new government that the previous government had not followed a reasonable or fair contract award process. Suggestions that government officials had abused their positions and possibly took bribes and misused funds, were made.

The new government was looking to strengthen its political position, and thus made corruption accusations against the previous government. The road might also have been viewed as an accomplishment of the previous government rather than the new government. The new government was against the contract and the media voiced similar opinions.

After work had ceased, the government set up an inquiry on the award of the contract. The inquiry found in favour of the previous government's actions and no misappropriation charges were made. Under the contract, the contractor could have expected to have been paid for a suspension of the work, and for the work to resume.

The work however did not resume and the government cancelled the contract with no costs awarded. All attempted negotiation by the contractor with the owner failed. Facing significant losses, the contractor then pursued the matter through the courts.

The resolution process

The dispute proceeded quickly to court. The court found in favour of the contractor, and awarded costs and damages to the contractor.

The contractor however decided that completion of the project would be far more beneficial to the company, and the owner was approached to reconsider the project, in lieu of the compensation costs awarded by the court. The owner accepted the contractor's offer, and design and construction continued for the project.

Exercise

1. The government of the day's term was nearing an end with a new election imminent, and it wished to use the award of the contract for political promotion. This happens in all countries — political parties lay claim to projects started (more often than those built) whilst they were in government, for political credits. In this country, groundbreaking ceremonies appeared common, and much was made of the awarding of contracts. As a contractor, how do you deal with the risk associated with politics?

2. The government in power, when the contract was awarded, may have had a genuine interest in rectifying a transport problem. However (unproven) accusations of funds misappropriation and abuse of position were made. Is this something for contractors to be concerned about, or can they operate ignorant of such goings on?

3. Many countries now are concerned with probity issues in the awarding of contracts, and force the public sector to be transparent in its dealings with tenderers. What practices change for a tenderer operating in a transparent environment compared to a non-transparent environment?

4. It would be very unlikely that public sector owners in other countries would not negotiate with a contractor to avoid court proceedings. Court proceedings would generally be a last resort. Does this say something about contract administrators on this project or this country generally? People can be promoted to public service positions by their association to people in power. A brotherhood system of employing friends can mean that many inexperienced people are occupying positions they aren't capable of handling. As a contractor, how do you deal with the risk associated with dealing with inexperienced contract administrators?

5. The owner wasn't aware of the position it had created by the decision to terminate the contract. The decision possibly arose directly from a very senior politician with little advice or questioning from public servants, such was the power distance relationship that existed. As a contractor, how do you deal with the risk associated with dealing with a lack of delegation to lower organisational levels, and heavy political involvement in business?

11.2.21 CASE STUDY — PETRO-CHEMICAL WASTE

The project

The project described in this case study was part of a larger venture involving the construction of an industrial complex consisting of various manufacturing facilities (steel, petro-refineries, machinery etc.). To enable the industrial complex to operate, a large amount of service infrastructure was required.

The infrastructure included power stations, water desalination plants, water treatment and storage facilities, service distribution facilities and pipelines, waste collection and disposal facilities and an industrial standard road network. This infrastructure was spread over a wide area.

Project organisation

The local government provided the finance for the project. In addition, its representatives wished to be involved on a day-to-day basis in the decision making with the appointed project manager.

The project manager was a foreign project management/construction management firm. Its particular expertise was in the petrochemical industry development and maintenance area. In delivering this particular project, a large number of consultant firms were used to undertake the investigation, design and documentation for the various packages of the project.

For the water and wastewater packages a foreign consultancy firm was selected to undertake the investigation, design and documentation. This firm's experience was in the civil engineering area, chiefly water and wastewater. It also happened that the firm had a good relationship with the local government and had worked previously with the government representatives assigned to this project on a number of other projects in the region. The consultancy firm was from a different country to that of the project management firm.

The dispute

The consultancy firm was briefed on its part of the project and set about undertaking the investigation, design and documentation activities. These activities were undertaken in its country of base, whilst the project manager's staff were based in another country as well as the country of the project. The designs were prepared to standards that normally applied to the water and wastewater industry internationally. As the design phase progressed, the government representatives viewed the documentation in the consultant's office on a regular basis. The project manager's representatives were infrequent visitors to the consultant's office. The government representatives assessed the documentation at the 30% design stage.

At the 60% design stage, the project manager's design review team travelled to the consultant's office to review and accept the documentation prepared to that stage. During this review the team realised that its accepted practice was for the documentation to be prepared to international petro-chemical standards. This was on the basis that most of the construction contractors for this project were involved in the petro-chemical industry rather than the civil engineering field. The project manager directed that all documentation be redone to the petro-chemical industry standards, at no cost to the owner, for the 60% design stage. The alternative given to the consultant firm was to be relieved of the work with no payment. The consultant firm protested strongly at this direction and appealed to the government representatives for consideration. The likely cost of redoing the documentation would exceed the current fee arrangement, due to the consultant having to engage additional staff with petro-chemical experience. An impasse had been reached.

Dispute resolution

At this point, all parties studied the documentation covering the arrangement for the provision of consultancy services. The agreement documentation had not been fully completed by all parties. It was decided that the cases of both the project manager and the consultant firm would be presented to the government representatives in a conciliation style arrangement. Even in this there was a lack of clarity in the acceptance or otherwise of any decision.

Following this, the conciliators decided that the design process would proceed, but with an additional hold point. This was termed the 75% stage, an additional stage before the originally agreed 90% stage. The documentation to be produced for the 75% stage was expected to demonstrate that the consultant could change the design to meet the petro-chemical industry standards. If acceptable, authorisation was to be given to proceed to the 90% stage.

The 75% design review was undertaken and the documentation still failed to meet the expected requirements. Further conciliation took place, with the outcome that the consultant would continue the design to a 90% stage with a 90% fee payment. The documentation would then be handed over to another consultant to complete to 100% design stage.

Exercise
1. The three parties to the dispute had different cultures that impacted on the commencement of the project, the dispute and its resolution. The project management firm comprised task-driven engineers with little attention given to communicating their expectations. The consultant firm had a strong belief in its technical expertise and relied on its relationship with the government representatives. Its communication, due to the absence of representatives of the project manager, was mainly with the government staff. The local staff held previous relationships to be important and respected the consultant firm's abilities, whilst expecting the project manager to deliver a project on budget. What does such an arrangement need in order for it to work?
2. It appears that there was very little recognition of the different cultures of the parties to the project at the commencement of the project. Why might this have been the case?
3. At the stage of consultant selection and again at the commencement of the design, significantly more effort needed to be given to communication, particularly with respect to design standards. The agreement between the parties should also have been finalised and understood by all parties. What does this in particular say about specifications in an international context where several standards may exist around the world?
4. How do you go about clearly setting out the expectations of parties to a project? What forms of documentation and face-to-face contact would be involved?
5. What regular communication practices would you have recommended be established and used throughout this project, particularly considering the parties were long distances apart?
6. Good contracts management practice, irrespective of whether the project is an international one or not, is to get the documentation correct up front. What do you think contributed to this not being done in this case?
7. The relationship between the project manager and the consultant was meant to be very businesslike. However the familiarity between the consultant and the government representatives clouded this and produced a misunderstanding at an advanced

stage of the design. Is it necessary that relationships between all parties to a project be on the same footing? Should all relationships between parties to a project be on a business basis? Can you have personal relationships in business situations? If not, what practices should be adopted?

8. The appointment of the government representatives, as conciliators, did not produce an independent and decisive resolution of the dispute. The government representatives were sympathetic to the consultants, but the project manager's staff were not aware of this due to their lack of visits to the consultant's office to familiarise themselves with the design process and the consultant firm. In hindsight, who might have been a better conciliator, or mediator?

9. Is it necessary that an acceptable dispute resolution process be agreed at the commencement of a project, or can you wait and see what type of dispute arises and then fashion a process to match?

10. The outcomes of this project were less than optimum with sub-standard documentation, damaged relationships between parties, delays in the project delivery, extra costs for design and documentation, excess consultant staff, and damaged reputations for all parties. What lessons can you learn from this case study?

11.2.22 CASE STUDY — SCOPE OF ROAD CONSTRUCTION

The case study looks at the management of a road construction project with an international contractor and a public sector owner.

The construction contract was awarded to a relatively small company with minimal local experience. The ownership and management staff were foreign. The company's major road work and contracts management experience had been gained overseas.

There were a number of disputes that were difficult to resolve. These highlighted some cultural differences that affected the resolution of the disputes. Some of these disputes and issues are discussed below.

The delivery method was design-and-construct. Major difficulties, which were the source of most of the disputes, involved the contractor's lack of understanding of the scope of the work.

The contractor saw the contract documentation as only a starting point for the project, not defining the actual scope of the work. This perception was further exacerbated by the use of design-and-construct delivery.

The contractor had difficulty in dealing with the owner due to preconceived ideas on how a contract should be administered, based on experiences with government work in its own country. There were also various problems with the contractor attempting to make deals outside the contractual requirements. Locally, it was considered inappropriate and unlawful for a public sector body to partake in such deals, as it was accountable to the public. Such corruption was considered unacceptable, but was the contractor's usual practice in its own country.

Difficulties arose during the design phase when the contractor did not meet the architectural/aesthetic requirements of the contract. And as a result of a strong stance taken by the owner, disputes resulted. These disputes were centred around the documented scope of work, and the owner's desire to obtain the best possible product for the tendered price.

The contract documents had, what appeared to be, an adequate dispute resolution process:

- The owner's representative considers and determines the issues/claim.
- If the contractor does not concur/accept the determination, the owner's representative can review and re-determine the issue.
- If the contractor still does not concur/accept the determination, the contractor can request that the owner's representative review the determination with consideration of additional information.
- If the contractor still does not accept the determination, the matter can be referred to arbitration.

However the contractor was adamant that it wanted the matters regarding the scope of work negotiated directly with the owner's senior management (on a man-to-man basis), skipping the owner's representative review stages.

To expedite the resolution of the issues, the owner's management agreed to discuss the matters directly with the contractor, but first required the contractor to forward a formal written submission supporting the contractor's claims. This did not suit the contractor, as such a submission required the contractor to justify its claim based on the contract.

The contractor's business manager had a high level of optimism with respect to his negotiating skills ability. He was a flexible and skilled bargainer but gave low priority to the contractual/technical arguments.

During the negotiation, the contractor relied on negotiation skills, and did not address the documented requirements of the contract. The owner took a strong stance on the documented requirements of the contract, and this conflicted with the business manager's way of undertaking negotiations.

The result was a stalemate, until the contractor finally yielded in part to the owner's position, after legal advice on the issue.

Commitments made during the negotiation were not followed up by the contractor, resulting in further disputes between the contractor and the owner.

Exercise

1. A more amicable solution could have been found earlier in the project through open negotiation, and this would have benefited both parties to the project. Why might this not have happened?
2. The contractor was operating in a foreign contract environment. The owner's representative approached the contract in a manner that would have appeared unyielding and aggravating to the contractor. Should the contractor have recognised local traits of contracts management and negotiation? Conversely should the owner's representative have considered the cultural traits of the contractor, particularly that of a more open negotiation forum?
3. Culture establishes the way people interact with each other. How important is it then for project personnel to understand the impact of different cultures on relationships within the project. Different cultures inevitably lead to different management styles. Once project personnel identify cultural differences, can strategies be implemented to minimise or address problems that could result from these differences?
4. In an international project how do you acknowledge the existence of cultural differences, try to understand the reason for the differences, and develop strategies to

address the differences before they become issues that may be detrimental to the project?

5. Would the practice of prequalification have eliminated such a contractor from doing business with this particular owner? Under what circumstances would it have not eliminated the contractor from tendering?

6. Would a structured, well-documented approach to contracts administration on behalf of the owner have eliminated any of the problems experienced on this project? Or was this not the issue here?

11.2.23 CASE STUDY — VALVE MANUFACTURE

Description

A manufacturing company won an order for a large valve that represented, for both manufacturer and owner, a first experience with this type of product. The specification was copied almost one-for-one from that for another type of valve, and as a result was somewhat deficient in some critical areas. The owner approved a detailed general arrangement drawing of the valve, and this formed part of the technical documentation in the contract.

Past experience with this owner was that it had the habit of first approving a valve design, but then requesting modifications to the original design, while refusing to accept that it carried a cost.

Later, as the work progressed on this valve, the owner requested some extras.

Some of the extras were quite minor in terms of cost, while others were rather costly. The owner claimed that while they were not shown on the general arrangement it approved when placing the order, the approval review of the design revealed that the extras were required to ensure a safe design. On every occasion that the owner requested an extra, the owner was notified that it was beyond the scope of supply, it was informed of its cost and it was informed that, on account of the tight delivery schedule required by the owner, the request was going to be accommodated without waiting for approval about its cost. The owner never replied to any form of communication or to reminders about this matter being unsettled.

Finally the manufacturing company received a fax from the owner, where it informed that the issue of unpaid extras would be discussed by its employee on the occasion of his visit to the manufacturing company's premises to assist the operational tests of the valve.

As the construction of the valve proceeded, it became apparent that the delivery date promised could not be complied with. The manufacturing company first requested an extension of the delivery date, which was accepted by the owner. A few weeks later another time extension was requested, and it was again accepted but with strong protests. The owner had made arrangements with many subcontractors to carry out work during the shutdown for the installation of the valve and the latest delay was going to interfere with the subcontractors' work schedules. The owner felt that the delay was going to cost it a lot of money.

Clearly the matter was controversial. The manufacturer believed it was in the right in its claim for the payment of extras; it desperately needed the money it was claiming yet the owner was a very valuable one and it was important to avoid blemishing what was regarded as a good commercial relationship.

A negotiator, with knowledge of the owner's language, was appointed by the manufacturer. This person was initially involved with the mechanical design of the valve and later as project engineer.

From a telephone conversation with the owner it was clear that the owner regarded the extra payment claimed as excessive and it was obvious, as a minimum, that the owner was going to ask for evidence that the payments were justified.

The negotiation about the disputed claim for payment began two days after the arrival of the owner's employee to the manufacturer's premises. The main purpose of his visit was to assist the operational tests and it was important to convey to the owner that the manufacturer appreciated this. The owner's employee and the manufacturer's negotiator knew each other from other business meetings and no personal familiarisation was necessary for this negotiation.

The first negotiating session began with the manufacturer's negotiator giving a presentation about the issue and listing all the facts. In particular it was highlighted how the manufacturer had always promptly notified the owner about the cost of the extras and about the fact that these were beyond the scope of supply. It was pointed out that the owner had never acknowledged any of the company's faxes.

Every item claimed was duly explained even with reference to the specification, and the justification of their cost was explained to the owner's employee.

While the manufacturer's cause was being outlined, the owner's employee frequently interrupted with verbal interjections and excessive body language that was unnerving.

After the opening presentation and the review of all contentious points, the owner's employee stated that although he had full power to settle the issue himself, he would not do so and that it was his intention to clarify the facts and let his head office make a final decision. He also stated that he found rather provocative the fact that, while the manufacturer was going to deliver the valve with considerable delay causing the owner financial loss and inconvenience, it had the cheek to demand payment for extras. The manufacturer's negotiator replied that the negotiating session was called to discuss only the issue of the payment for extras, not to review the project as a whole.

Predictably, the owner's employee objected to the actual amount of some of the payments, and asked to see the invoices for the extra materials.

The session was adjourned to the following day. After the session the manufacturer's negotiator reported the results of the negotiating session to the technical manager, who pressured the negotiator to try to settle the issue at once by convincing the owner's employee to make a decision himself. When it was suggested that the only way the deadlock might be broken was by offering a discount on the amount claimed, as a gesture of goodwill, the suggestion was scorned.

In the second session, the invoices were tabled and the owner was convinced that there was no overcharge in the payments claimed. The issue switched to the technical interpretation of the most expensive extras requested by the owner. It was stated, without proof, to the owner that the request was technically unjustified.

Just as the owner's employee appeared to be yielding, without notification the technical manager entered the room and offered the discount that he found unacceptable the previous day. In response, the owner's employee, rather than modifying his position, stiffened and refused to discuss the issue further, stating that the matter would be agreed upon his return to his head office.

After some telephone haggling the issue finally settled for an amount about one fifth lower than that initially claimed.

Exercise

1. Do you believe negotiation, without any third party involvement, was the most appropriate way of resolving this dispute? Comment in terms of time and costs.

2. The manufacturer's negotiator did not get a full mandate to negotiate; he was merely assigned a task that was considered distasteful. This was highlighted by the technical manager's interference, diminishing the negotiator's authority in the eyes of the other party. What does having a negotiator without the full support of his company, do to a negotiation?

3. Prior to the negotiation it was not established that the owner's employee had the necessary authority to negotiate. Had this been done, it would have become clear that the negotiation had no chance of being concluded before the owner's employee reported to his head office. Can you conduct negotiations successfully if either or both parties do not have the necessary authority to conclude a deal?

4. Since the most controversial item, additional reinforcing, was not in the original general arrangement drawing approved by the owner, the associated work should have been preceded by structural analysis calculations indicating its necessity. In fact the analysis was carried out at a later stage and showed that what was requested by the owner weakened, rather than reinforced, the structure. Had this been done earlier, might the owner have seen that it need not have adopted a competing style but rather a compromising one?

5. Should the concession of a discount have been made by the negotiator himself rather than a person of higher authority, and traded for something like a speedier settlement of the issue? Should the technical manager have played a supporting role to the company's appointed negotiator, rather than interfering with his work or dictating to him?

6. Since the manufacturer had a well established business relationship with the owner, should provision have been made in the contract for dispute resolution? In this case, the solution was on an ad hoc basis. How acceptable is that?

11.2.24 CASE STUDY — BUILDING CLAIM

Background

This case study is written from the point of view of the organisation selected by an international bank to project manage the construction of a medium-rise building.

The project was truly international. The architect, quantity surveyor, owner's project manager, and the head contractor were all from different countries.

The bank's core business was providing funds for development work. To receive funding from the bank, all proposals had to adhere strictly to in-house procurement and development guidelines.

This strict adherence to policy resulted in the owner having to accept the lowest priced conforming tender, although the bid was much lower than the next competitive bid and lower than the quantity surveyor's estimate. This anomaly was thoroughly investigated and the contractor extensively questioned to ensure the bid was valid. Against the recommendations of the project manager, the owner awarded the contract to the lowest bidder.

The contract was let as a lump sum contract with full documentation provided. The contract had limited grounds for any extension of time claims and, as part of the bill of quantities, a provisional rate for delay costs was required from all tenderers.

The contract also included a unique dispute resolution procedure, which was required because of the involvement of the bank. The bank was not governed by the laws of the host country and as such was outside the jurisdiction of the host country's legal system. Therefore, to alleviate the concerns of the tenderers, a specific resolution clause was written into the contract and tied the owner into a process that was seen to be fair, reasonable and most importantly workable. The owner gave an undertaking to honour any decision from this procedure.

The dispute resolution procedure was two staged. Firstly, the contractor could challenge any determination by the owner's representative. If challenged, the contractor had to submit, in report form, its case to a third party appointed by the owner. The intent of this person's role was to assess the contractor's arguments and determine a solution that was fair and reasonable to both the contractor and the owner.

If this third party's determination was disputed, the matter could be referred to international arbitration.

The structure of the resolution procedure was intended to allow any disputes to be settled quickly and efficiently. The advantage of this procedure was that the owner could make a decision based on commercial grounds. This option wasn't available to the owner's representative. This procedure also gave the owner the opportunity to review the issues and remain informed of all aspects of the contract work. It was assumed that the third party would be nominated from the owner's senior management team and would have the authority to make a determination with the best interests of the whole project in mind, not simply the contract work.

The dispute

The dispute to be discussed was about the circumstances and issues leading to, and the resolution of, a substantial final claim submitted by the contractor.

The claim was in part, retaliation for the owner's representative enforcing liquidated damages on the contractor for not reaching practical completion by the required date, and the subsequent loss of the performance bond.

The claim was based on a myriad of issues that the contractor used as justification for completing the project late. The claim was for a very large amount.

Given the project completion was many months late, it was well within the duties of the owner's representative to claim liquidated damages on behalf of the owner and, in addition, claim the contractor's performance bond.

The contractor believed that it was being treated unfairly and complained to the owner. To support its case, the contractor submitted a one hundred page document, disputing the determination of the owner's representative, and outlining issues it believed entitled it to the additional money.

The report consisted of statements referring to various meetings and discussions with different parties. Particular reference was made to statements and meetings with the owner where the owner had implied that the contractor was entitled to additional money. This was particularly difficult to dismiss, because the owner at meetings had made remarks and representations which could only be interpreted in that way. The phrase 'If you do a quality job, we will make sure you are smiling at the end of the day' was well known to all. It was very difficult to read this any way, other than the owner doing a deal which would be to the contractor's satisfaction.

Included in the claim were legitimate issues related to fluctuations in exchange rates. A lot of plant and specialised finishes were purchased outside the country, and this affected the contractor's costings.

The claim was possibly an attempt by the contractor to recoup losses encountered because of poor management of the contractor's risks. As a fixed price contract, there was little ground for the owner's representative to accept any of the claim.

The resolution

Assessing the final claim resulted in the project management team remaining on the project for a further twelve months, painstakingly reviewing all correspondence associated with the contractor's claim.

In initial discussions with the contractor, the owner stood its ground and rejected the contractor's full claim. The project manager's assessment suggested the contractor might be entitled to a small amount.

This attitude escalated the dispute and senior people from both organisations became involved. At that point, the contractor revised its claim significantly down. After further lengthy discussions, the owner agreed to a small additional sum, no liquidated damages and the return of the performance bond.

This resolution could only be considered an outstanding success for the owner, as the final cost of the project including the final resolution, was still under the original estimate. Although the project was late in finishing, the owner didn't incur any great financial loss because existing premises were rent-free. Therefore, recouping liquidated damages was not of primary concern. More important was ensuring that the building was completed on budget and to the right quality.

Comments

The dispute was a function of many things — cultural differences, poor management, competitive tendering and a lack of experience.

The dispute was over delay costs; the single biggest contributing factor to the delay was the fact that the contractor, upon winning the work, obtained new quotes for the subcontractor packages in an effort to force down the subcontractors' prices and cut costs. This process took up valuable time and resulted in the head contractor being almost immediately behind an already-tight construction program.

Cultural differences in all aspects of the project contributed to the project's problems.

Firstly, it was unusual for the quantity surveyor to have the responsibility for the preparation and coordination of the contract documentation. Therefore, managing and controlling the project was somewhat more difficult, from the viewpoint of the owner's project manager.

The attitude of the architect didn't help. Despite numerous design meetings with the architect to discuss the best way to document the project, the architect would submit documentation completed to its own standards and in its own way. Therefore the documentation produced was difficult for the contractor to fully understand. This was a contributing factor in the contractor's claim.

Another issue, which contributed to the contractor's problems and ultimately the dispute, was the architect's practice of designing the building in imperial measurements and then converting to metric. Although this would seem to be a minor issue, the straight conversion often had the contractor building to odd dimensions. This led to numerous occasions of misunderstandings and incorrect interpretations of reduced levels and offsets. In some instances, the contractor had to redo work because setout dimensions were misunderstood.

A more tangible issue was the lack of importance the contractor's management placed on the contract. The contractor placed more reliance on verbal undertakings and performance than on the contract. Therefore care had to be exercised in any discussion or meetings to ensure the right message was being portrayed and that the emphasis remained on the contract. The contractor's attitude to the requirements of the contract contributed significantly to both the dispute and its resolution.

The contractor's focus on performance often placed both parties in dispute. The contractor believed that it was more important for the owner to see the building taking shape, than to meet the required quality goals. This could possibly be attributed to the language barrier, as it was often more difficult to interpret a foreign specification than to discuss in broken English what the owner's views and problems were. This situation was also reflected in the language of the final claim.

Poor interpretation of the contract, and a general indifference to quality issues, resulted in the contractor having to pull down and rebuild portions of work in order to meet specified quality goals. The owner placed significant emphasis on having a quality building.

The attitude of the owner's third party appointee also contributed to the dispute. It was hoped that this person would assess both arguments and work with the parties to find an amicable solution. Instead this person viewed his role as a quasi-owner's representative and made his determinations based on the contract. This resulted in most cases being determined against the contractor. This frustrated the contractor immensely.

It was unfortunate that this third party didn't make his decisions with a commercial focus, as decisions made created a very tense working relationship. In some cases, this person even overruled the extension of time determinations of the owner's representative. In short, the attitude taken by this person made the dispute resolution process unworkable, and this in turn resulted in the large final claim.

Exercise

1. It was expected that the dispute resolution procedure would mean most claims would be resolved with no further action being required. It was hoped that the owner's third party appointee would be a senior manager with enough authority to take a commercial overview of the project and therefore broker deals with the contractor to resolve any dispute. Unfortunately this did not occur. In hindsight, what qualifications should have been required of this third party person?

2. Attitudes of the contractor possibly contributed the most to the dispute. Some contractors have a reputation for making money from claimsmanship tactics. Given the low bid of the contractor, was it to be expected that claimsmanship would arise?

3. The claims procedure didn't work as well as all would have liked. This was a direct result of the attitude taken by the owner's third party in assessing any dispute. Ultimately this meant the contractor disputed nearly all determinations. Suggest a possibly better mechanism for dealing with claims, one that is fair to both parties and is manageable.

4. The owner's third party avoided confrontation at all costs and would avoid any provocation in face-to-face meetings. He would then pursue the issue in correspondence, or through directions to the project manager. This made it very difficult to manage the contractor. How does a contractor counter such an approach?

5. Undoubtedly an international contractual dispute has unique obstacles to negotiate and manage. Dispute resolution clauses, and to a greater extent contract documentation, will be interpreted differently by people of different cultures. Whose responsibility is it to ensure that all parties clearly understand their rights and obligations under the contract and that they continue to operate under those guidelines?

11.2.25 CASE STUDY — CONTRACT TERMINATION

This case study discusses a dispute between a head contractor and a subcontractor on a liquefied natural gas plant extension. The project was undertaken by a joint venture (JV, of two companies from two foreign countries, acting as designers and constructors). This JV split the design responsibilities between the two parties. This led to major problems. Head offices were located in two foreign countries. This, combined with a difference in language, made communications difficult.

Based on the design of an existing plant one company attempted to adapt this design to the site, and to amend the plant cooling system.

Concurrently the other foreign company was preparing specifications and bills of quantities. This was a fast-track project, and hence the concurrent nature of events. Subcontract packages for specialised painting were issued several times and subsequently withdrawn over an extended period, due to errors and a changing design and specification. The design prepared in one country and the specifications and bills of quantities prepared in the other were not compatible. The project incurred delay and additional costs as a result. The project commenced later than planned, and the project was on a fixed completion date, with no extensions of time, combined with extremely large liquidated damages.

Subsequently, items for the project were fabricated in many countries. These items had been ordered and fabricated to a design that was progressively developed. The mechanical contractor prepared shop drawings and fabricated pipework using different generations of the design drawings. Needless to say, there were problems with components not fitting together when assembly on site commenced. Items of structural steel arrived as deck cargo by ship. Salt water and spray contaminated the steelwork. Thus, every item had to be washed down, grit blasted and repainted. The specification failed to state handling and transport requirements. This is an example of unanticipated additional work that was required to be completed at short notice.

The foreign subcontractor for specialised painting and insulation was required to grit blast and apply a multi-coat paint system to all pipework in a controlled factory environment. The only painting to be carried out on site was touch-up for minor damage that occurred during site erection. The plant piping was insulated in situ, using a specialised foam insulation, manufactured on site. There were additional processing units built. They were constructed sequentially, with trade subcontractors moving from one processing unit to the next upon work completion.

With the delays experienced, project start up was late and the overall project duration was reduced. This program compression meant that the duration for individual processing units was reduced dramatically and there was now considerable overlap between processing units. This overlap required people to work on more than one processing unit at certain periods of time. As the project entered the stage of pipework erection for the first processing unit it became apparent that the design had significant errors, resulting in pipes not fitting together. Fittings such as valves were missing; branches had to be cut in to pipe already erected. The result of these problems on the subcontractor's work was that pipes that were finish-painted were extensively damaged by cutting and welding or by welding 'slag'. Finished insulation that was signed-off as completed was removed to permit installation of missing valves and fittings. Variation work such as remedial painting to structural steel, as mentioned above, added labour.

The head contractor directed that the damage to the subcontractor's painting was to be repaired at subcontractor's cost, as it was no more than 'touch-up' painting. Touch up painting was allowed in the subcontract tender price. Approved insulation that was signed off had its completed status reversed and the repair of insulation was deemed to be part of the subcontractor's contracted scope of work. Variation claims were submitted and rejected by the head contractor.

The subcontract had a clause, which permitted 'termination for convenience'; it was not contingent upon agreement between the contracting parties, rather it was at the head contractor's discretion. The subcontract was terminated at a point when approximately half of the contracted scope of work was completed. The subcontractor was prevented from entering the site to recover scaffolding and equipment, rather the replacement subcontractor used this equipment. Typically this type of clause might be applied when circumstances outside of the control of the contracting parties prevent completion of the contract. It was not intended as a mechanism to permit one of the contracting parties to gain a commercial advantage by terminating the contract. The basis of the resulting dispute was that the termination of the contract was wrongful and this gave rise to a quantum meruit claim.

The subcontract required that any claims for additional payments be lodged within a defined period from the date of termination of the contract.

Resolving the dispute

On the subcontractor's side, a task force was assembled and based locally. This group priced all variations, extensions of time, acceleration and work inefficiency claims. Further, all evidentiary material and records were assembled and copies placed in ring files.

A formal claim document, including the legal basis of the claim was prepared and forwarded to the head contractor.

Attempts were made to negotiate a settlement. However this failed, because the parties were too far apart on a settlement figure. In accordance with the relevant conditions of the contract, the dispute was formalised. The dispute resolution provisions of the contract contemplated an international arbitration under local law and be held in the country. The subcontractor's legal representation was a joint venture of a firm from the subcontractor's country and a local law firm. All preparation work was undertaken in the country, with a period of one year taken, prior to proceedings commencing. The argument put forward was that it was a wrongful termination of the contract and flowing from this were damages in the form of a quantum meruit claim. The arbitrators comprised a panel of three persons from various countries. Sittings were held both locally and in a neutral country.

Exercise
1. The dispute process became a project management exercise in its own right. It required time and cost planning to meet the desired outcomes — the claims and evidentiary material — by the contractual deadline. The period of time immediately following the termination of the contract was frenetic. Preparation of the claim was conducted in offices located adjacent to the site. Some existing project staff were retained for tasks associated with the claim. Some staff were retained for the duration of the dispute on the basis of knowledge and to act as witnesses. Additional staff were obtained for costing the work. The claim group was split into small subgroups, some pricing variations, others preparing write-ups on events or the conduct of the parties. The senior members of the group worked on the costing of loss of efficiency and acceleration claims, together with research on agreed issues for the lawyers. Separately, a team of support staff photocopied and paginated all document files and records. This enabled the deadline for the submission of the claim to be met in accordance with the conditions of the contract. This time pressure required the early identification of a legal strategy and recruitment of people to carry out the work. There were two main milestones to be met, the first being the preparation of the claim to comply with the contract conditions. The second was the data assembly, the taking of statements and preparation of material by expert witnesses for formal proceedings. How would you set up the dispute activities as a project management exercise?
2. The lessons learnt by the subcontractor from this experience are given below, and the subcontractor believes they are applicable to all contracts administration. Give your views on the lessons learnt.
   - Be familiar with the contract documents under which the work is being executed, having a clear understanding of rights and obligations.
   - Observe the conditions and time bars for notice provisions, such as notifications for extension of time, claims for variations or other payments.
   - Advise of possible items of dispute early and begin discussions or negotiation early rather than wait for a build up of a folio of claims or disputes, making formal proceedings the only viable dispute resolution device.
   - There is no bargaining position when the work is approaching completion; there is however when the work is afoot.

- Keep accurate records of all site or related events, such as labour records, issues of drawings or site instructions or periods of delay.
- Correspondence must be framed so that it is unambiguous, neutral in tone and capable of being clearly understood by a third party at some future time.
- Maintain a dialogue with the other side without resorting to pettiness or personal insults.
- The cost of pursuing legal remedies is high both in terms of legal costs and also the cost of company employees' time. The process is time consuming. Legal processes are a method of last resort.

11.2.26 CASE STUDY — DESIGN STANDARDS

Introduction

The case study looks at a contractual dispute on a design-and-construct (D&C) onshore pipeline project. The owner (funders) of this project included international and local oil companies. The pipeline was to pump oil via a number of pumping stations.

Contractual arrangement and project organisation

The contractual and management relationships of this project are shown in Figure 11.1.

A principal consultant (PCon), based overseas, was responsible for process design, pipe network design and project management of the civil and tanker contract work. The civil and tanker work, which was delivered on a design-and-construct (D&C) basis, was awarded to two contractors (HC1 and HC2) after competitive tendering. The principal

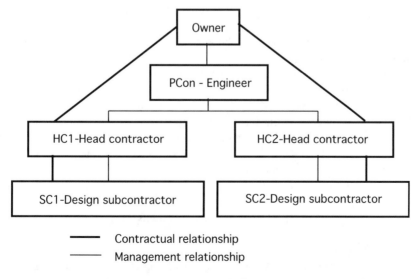

Figure 11.1. Civil and tanker D&C delivery.

consultant (PCon) had prepared the tender documents and specification for the civil and tanker contract.

In the absence of their own design facilities, the two main contractors subcontracted the design part of the work to two local consultancy firms (SC1 and SC2), following negotiations. The subcontractors had to complete their work within a set period, otherwise they had to pay an agreed sum of liquidated damages to the head contractors.

The scope of work for the civil and tanker contract included the design and construction associated with buildings, building services, tanks, pipes, roads, bridges and drainage.

The main reasons for the selection of the particular delivery method using multiple contractors were as follows:

- The complexity of the project required at least two head contractors to work simultaneously. Since the work was centred at different development sites, there would not be any interface problems between the contractors.
- It would be difficult for an engineer (principal consultant), who is based overseas, to supervise construction activities on the other side of the world. Therefore this gave more autonomy to the contractor; consultation was not required during the construction phase, since the design prepared by the contractor was approved by the consultant at an early stage.
- By adopting such a delivery method, the owner was effectively passing the potential risk to two main D&C contractors.

The disputes

Acceptable design standard

The project specification prepared by PCon stated that international standards be used for the civil and tanker work. Apart from a few tank design exemptions, structural, civil, mechanical, electrical, and HVAC should comply to international standards. The design engineers in SC1 and SC2 were not happy with the project specification, since they were not familiar with such standards. The local engineers were not exposed to foreign standards and were only familiar with local standards. Much design office software was based on local standards.

Both the main contractors and the subcontractors had overlooked this problem at the contract negotiation phase.

The two design subcontractors (SC1 and SC2) wrote to HC1 and HC2 explaining the matter and requesting acceptance of local standards for the project. In the meantime, the design subcontractors commenced the work, to avoid liquidated damages, anticipating an approval of their request. In the absence of a reply from the head contractors, SC1 and SC2 went ahead and completed the preliminary design based on local standards. They submitted the necessary drawings and documentation to HC1 and HC2 for approval. The main contractors passed those documents to PCon for final approval.

PCon refused to accept the submission and returned all documents to HC1 and HC2 with many comments marked in red. Some of the comments were related to the design standards and others were related to minor disagreements.

Claim for extra work

Both design subcontractors were under pressure to complete the design within the stipulated time. Although both design subcontractors had many years experience with the

traditional delivery method, this was the first time they had worked under a main contractor in a complex contractual arrangement.

They submitted and re-submitted the same design several times before getting final
approval from PCon.

SC1 and SC2 submitted several claims for extra work to HC1 and HC2. The reasons
cited by them were as follows:

- The re-design and submission of the same work several times, due to comments by
  PCon. According to their understanding this was a deviation from the original scope
  of work.
- The failure of HC1 and HC2 to resolve contractual disputes, for example on standards, at an early stage of the project.

The subcontractors successfully claimed a small amount as extra work. The consultant's
and contractors' faults were transferred to the owner, although not at fault itself.

Dispute resolution

It took many months to resolve the project standards dispute and commence the construction. After consultation with the two main contractors, the engineer (principal
consultant) appointed a locally based person to supervise the design activities, and
gave this person wide powers to resolve problems arising from the contractual arrangements.

This action eased many of the initial problems encountered by the contractors. After
negotiation with SC1 and SC2, the engineer accepted local standards as an alternative
to international standards for the civil and tanker work. This created a new working
relationship between the engineer and the two design subcontractors.

Exercise
1. It would appear sensible for any party, which is involved in an overseas contract,
   to be familiar with local conditions before preparing a specification for work in a
   foreign land. In this case there was a preference to use local standards, although such
   standards may not be accepted overseas. Under what circumstances should a foreign-
   based consultant enforce international standards on a project in another country?
2. Since there is no single international standard accepted by all countries in the world,
   care has to be taken when preparing a specification for work in a foreign land. Would
   adding a simple statement such as 'International Standard or accepted equivalent
   standard' have averted the whole dispute?
3. On the other hand, should contractors also learn the practices of other countries, and
   prepare to undertake work in an international environment? Contractors are increasingly looking for overseas work. Therefore, should they not fully equip themselves
   with the necessary technical and management tools to handle any situation?
4. The design subcontractors list the following as lessons learnt. Give your view.
   - In an international environment, it is very important to have intimate knowledge
     of work practices and cultures of other countries. It is not good enough to say,
     'That's how we do it in our country.'
   - Consultants and contractors should have a knowledge of several internationally
     accepted design standards.

- Educational institutions should take a pro-active role in promoting internationally accepted standards among their graduates. In this way, disputes arising from specifications, conditions of contract etc. would be minimised.
- When involved with international contracts, the parties should treat those on the other side of the table with due respect. Comments should be chosen to establish a good rapport between working parties.
- In a complex contract environment, all the parties involved should face each other and resolve their disputes. In an international contract, it is important to have a local nominee or surrogate. Although today the world has advanced communication methods, none of the methods available surpass direct communication between two people. Teleconference technology could overcome many problems in communication. However these technologies are still expensive to adopt in developing countries.
- In an international contract, the specification should cover several equivalent standards.
- It is very important to resolve any contractual problem at an early stage in a project.
- Lack of foresight of so called 'project managers' was one of the reasons for this particular dispute. The intrusion of non-technical people as project managers, especially to contracting organisations, and lack of technical background of those personnel is believed to be a key factor in causing this dispute.

11.2.27 CASE STUDY — CONTRACT DOCUMENTATION

As part of the civil work for a condominium development, reinforced earth retaining walls were proposed by the designer to support newly constructed access roads. The developer preselected a subcontractor for the design and certification of the walls, with the subcontractor working under the contractor responsible for the infrastructure.

The contract stated that the contractor was responsible for the design, construction and certification of each of the retaining walls. It was understood that the subcontractor would both design and certify the walls, while the contractor built the walls.

Documentation for the part of the contract dealing with the walls included the following ambiguities, discrepancies and omissions:

- The master drawings showed the retaining walls to have an incline similar to the road, however the civil documents showed two of the walls, and in particular Wall-H2, with horizontal wall tops.
- The specification indicated the civil drawings to be the drawings that had precedence, even though these were in fact wrong. It was established later that the site master drawings were updated by the owner's architect but not by the owner's civil design consultant.
- The contract stated that the contractor had design responsibility for the walls which it constructed. However a clause in the specification stated that the contractor should submit drawings for approval by the owner's consultant before proceeding.
- To complicate matters, the subcontractor responsible for the design of the walls was given superseded drawings by the contractor for the design of the wall. This aspect was corrected, however the owner's consultant twice approved the two walls with horizontal top profiles.

- Foundation conditions for the wall varied significantly for Wall-H2, making it higher overall than initially designed. This was incorporated in the redesign.

The outcome was that Wall-H2 was constructed to almost full design height before the mistake was noticed. From an analysis of the additional height of the wall (on the high end), approximately half of the wall was under-reinforced for the working loads.

The various parties involved had the following reactions:

- The wall subcontractor claimed that the fault was not its, as it had merely designed for the wall geometry provided by the contractor.
- The contractor claimed that the drawings given to it were not correct at the time of the tender (and this was true). The contractor also noted that the contract required approval of the wall by the owner's consultant, and therefore the responsibility for checking rested with the consultant; the consultant was also responsible for the original omissions on the master drawings.
- The consultant indicated that the full responsibility for the walls rested with the contractor. It conceded that the drawings were incorrect at issue, however amended drawings were provided to the contractor and appropriate variation work should have been carried out. They noted that sufficient time was available for these changes to have been carried out before construction.
- The owner was concerned with the delays in the work, as all of the above parties refused to take responsibility for the wall, each believing it was the others' responsibility.

The wall was eventually strengthened by reinforcing the already constructed lower sections of the wall with soil anchors, together with a reinforced concrete face over the anchors to strengthen the face of the wall. This was the cheapest solution and removed the need to rebuild the wall.

The design of the reinforcing requirements was carried out by the subcontractor as designer. The construction was carried out by the contractor.

The owner accepted that it had to pay the cost of installing additional reinforcement for the wall and this money was allowed the contractor as an extra. The error in the wall design was agreed to be the contractor's fault, and therefore the additional work was solely at its cost.

It was agreed that the civil contractor was at fault by not recognising the error in the wall geometry on two separate occasions. However, as it was understood that it had no design responsibility, it was not required to pay for the reinforcement of the wall.

Exercise

1. While the result from the above all seems reasonably straightforward in hindsight, the disagreement on this issue was protracted over a period of many months. As all parties claimed the mistake was not due to them, none were willing to take the lead in resolving the issue. All believed they had strong cases, and argued these over many meetings on this item. Resolution was only obtained by the owner taking a strong lead and forcing a resolution. Ambiguities, discrepancies and omissions blur the lines of responsibility, and so how, in general terms, do you untangle the mess after it has happened?

2. In hindsight, the major problem in the contract documents was the blurring of responsibility for the approval of the walls. By including the owner's consultant in the proc-

ess, this placed doubts on who was ultimately responsible for the walls. (The walls were the only item not subject to the approval of the resident engineer.) The contractor used these doubts to prolong the issue for an extended period. Can you have the design approval to the consultant, and the contractor/subcontractor responsibility at the same time?

3. It is considered that, while the errors and omissions were a problem, these would have been resolved more readily had there not been the above confusion over responsibility in the documentation. On seeing how the process evolved, it is considered that the mistake would have still been made, however the responsibility for the mistake could have been squarely placed on the contractor in the first instance. This would have resolved the issue in a significantly shorter time period. What steps are necessary in order that these errors and omissions should not occur?

11.2.28 CASE STUDY — MECHANICAL AND ELECTRICAL

Government charges

A foreign mechanical and electrical (M&E) contractor won an M&E installation project. At the tender interview stage, the owner had stated (and the tender documents stated) that the total tender should include all the related local government charges. The contractor assumed a certain percentage of the contract sum to cover such expenses, because a proper list of all these charges was not available. In order to secure the contract, the contractor had accepted the related risk.

During the project execution stage, the contractor imported equipment from its home country to carry out the installation work. In the mid-project period, the owner deducted part of the payments owing to the contractor, claiming that the deducted amount was for payment of a government inspection fee, for some specific imported equipment. The contractor rejected the deduction and a dispute arose. As the sum was quite substantial, the contractor threatened to withdraw from the site.

The owner stated that the contractor had already accepted paying all government charges at the tender interview stage and such acceptance was stated clearly in the contract documents. The contractor, however, claimed that, as the inspection fee was only announced by the government after the contract signing date, it should not be the responsibility of the contractor. Finally, after mediation by the project architect, quantity surveyor and engineer, the owner accepted that such charges should not be borne by the contractor.

Exercise

1. As the date of the latest changes to government legislation and rules was beyond the contract signing date, some people would say that the contractor could be expected to not be responsible for the extra charges. But then the owner is faced with additional charges. How do you resolve this commercial dilemma?

Fire services

The specification stated that the final fire services installation should comply with local regulations. However, the design drawings were prepared according to international

rules. There were discrepancies between the international rules and the local requirements for sprinkler head layout. For some areas, a double layer of sprinkler heads was
required by local requirements, because of the clearance between the floor slab and the
ceiling. However, the design drawings only showed a single layer of sprinkler heads.
The M&E contractor was not granted a variation, even though there were discrepancies
between the design drawings and the specification.

Exercise
2. The M&E contractor was not aware of the local requirements during tender preparation. However, the contract documents were inconsistent, and hence you would
   expect some responsibility to lay with the owner. How do you deal with such cases
   where different outcomes may occur in different countries because of different regulations?

11.2.29 CASE STUDY — SOME CONTRACTUAL DIFFICULTIES

Introduction

The infrastructure project included a large amount of temporary work associated with
underground construction, complicated utility relocation measures and traffic management in congested urban areas, as well as the construction of main civil structures
and all finishing, and interior electrical and mechanical work. A joint venture (JV)
consisting of several international contractors and one local company, together with
an international design consultant tendered for the D&C work. After the tender was
won, a range of tasks including staffing of the new JV, and negotiations with authorities, prospective suppliers, sub-contractors etc., in parallel with initial planning and
design, started. One important component was a contract between the JV and the
design consultant.

Design

The main contract listed three design stages for the project:
* Stage 1: Preliminary design containing the structural, electrical and mechanical
  design basics for the project as well as the main ideas for traffic management, utility
  diversions, preliminary soil and survey data.
* Stage 2: Definitive design including preparation of all surveys, investigations and
  testing of materials and equipment, calculations and analysis, drawings based on all
  previous items etc.
* Stage 3: Working drawings to include all drawings from Stage 2, plus site sketches,
  reinforcing steel (rebar) reference drawings, rebar bending schedules, fabrication and
  shop drawings, construction erection sequences etc.
From this followed as-constructed drawings.
   A very important aspect of the contract between the JV and the owner was the time
given to the owner's design consultant for design checks. During negotiation with the
owner, the JV pointed out that it would be very favourable for the project planning and

progress to have a fixed maximum duration for the owner's design approval and check. The owner's consultant, however, insisted that no exact limits on the design check and approval be included in the contract.

A program of design work, including assumptions for design approval periods, was created based on a construction program and sent to the owner for approval.

Contract between the JV and its design consultant

Negotiation between the JV and its design consultant concerned the type of contract and payments. The following possibilities, amongst others, were considered:
- A lump sum contract for all design work.
- Separate contracts for each design stage.
- A contract which would include only selected items.

After some consideration, a contract was signed with the design consultant, and this split the responsibilities between the JV and the design consultant.

The contract between the JV and its design consultant was a lump-sum contract.

Some problems which arose in connection with the contract between the JV and its design consultant

The contractual responsibilities of the different parties gave a complicated interconnected network regarding design and approvals of different parts of the project.

In the beginning of the project it became clear that:
- There would be a lot of design changes from the owner's side caused mostly by the difficulties in land acquisition.
- The owner's consultant didn't really intend to limit its check and approval time according to the site progress rate.

These features determined the way design progressed for the first several months of the project, and led to the situation where changes were issued with such frequency that there was difficulty incorporating them all on time. It was also a reason for the first dispute between the JV contractor and its design consultant.

The first problem

Pointing to the slow progress in design compared to the original schedule, the contractor asked the design consultant to increase its staff and to accelerate the most urgent (for the site) design packages. The design consultant considered such staff increases totally unnecessary, because unless the frequency of the owner's changes as well as its check and approval time were not sufficiently reduced, there was no chance catching up. As well, the design consultant informed the contractor that a number of claims for additional design work were under consideration. It pointed out that there was an ambiguity in the definition of the design stages, such that the design consultant believed it was producing more than it was contracted to do.

The second problem

A problem arose concerning the interpretation of the part of contract which stated that the design consultant should review the working drawings. The JV insisted that the review meant a complete check and endorsement of the drawings, but the design consultant did not agree with such an interpretation. Eventually it was agreed that the design consultant should check whether the working drawings were in full agreement with the definitive design. That did not mean, however, that it had to check whether these drawings were correct.

The third problem

This problem was related to the previous problem, and its essence depended on the answer to the question — should the changes in the definitive design, caused by the changes in the working drawings, be paid extra? The answers given by the JV and its design consultant were opposite. As a result, it was agreed that small changes would not be mentioned, but for any significant change, additional payment should be considered.

The fourth problem

Long, and sometimes not very pleasant, discussion took place regarding the question of how detailed the M&E preliminary design should be. The JV insisted on more detailed drawings and calculations, which had to be used for the M&E subcontract tender. The design consultant argued that there was nothing in its contract with the JV about this, and therefore all such details should be a part of the detailed design. Should the JV want such details to be produced, the design consultant should be paid additional.

There were a number of other problems, concerning the design, between the JV and the design consultant, but the points mentioned above were typical.

Exercise

1. An unclear design brief, together with some land acquisition problems, were the main items from the owner's side which created some difficulties for design progress and design-construction coordination. Under such conditions, is a design-and-construct approach applicable?
2. The case is an illustration that fine tuning between contractor(s) and design consultant(s) is critical for the success of design-and-construct delivery. What is your view?
3. The main problems in this case were associated with:
   - The owner's approach to the project.
   - The contract between the JV and its design consultant.
   In particular:
   - Both contracts (between the owner and the JV, and between the JV and the design consultant) had a lot of ambiguities regarding design.
   - The number of changes were extremely high.
   - The time given for design was not enough. (The milestones for the different design stages to be completed were part of the contract between the owner and the JV.)
   - The contract between the JV and the design consultant was too complicated with too many unclear points concerning the exact separation of responsibilities between the two parties, penalties for delays, changes and variations etc. The design consultants were perhaps also too inexperienced.
   - Design and planning for some parts of the project were completed in parallel with site work, and this resulted in misunderstandings and mistakes, and led to a time extension and cost increase on the project.
   Under such conditions, is a design-and-construct approach applicable?

11.2.30 CASE STUDY — ERRORS IN CONTRACT DOCUMENTS

The case study refers to a building construction project where the contract conditions used were a standard local form. The first matter at issue was the relatively simple one of a discrepancy between the drawings and the bill of quantities (BOQ). The former indicated the strength of certain reinforced concrete floors and beams as 25 MPa but the bill gave the mix as 20 MPa. The matter was raised first at a site meeting when the contractor's agent asked which strength of concrete it was to use, to which it got the reply, 'As shown on the drawings.'

The agent was correct in posing the question, because of the wording of the conditions which required the contractor to '... carry out the work ... described by or referred to in the contract bill.' Normally work would be carried out in accordance with the drawings and specification but, in the absence of the latter, the bill acted as the specification.

The matter of an increased rate was raised by the contractor's quantity surveyor (QS) with the consultant QS, who pointed out that, before he could deal with the matter, the contractor needed to give notice to the owner's representative (architect), under the relevant contract clause. Admittedly the matter had been raised (and possibly minuted) at a site meeting, but this was not a notice.

After writing to the owner's representative and receiving a reply, the matter was then taken up between the consultant QS and the contractor QS. The contractor was claiming for the additional cost of the stronger mix. The consultant QS asked the contractor QS to substantiate the claim for the higher unit rate of the stronger mix. At first the contractor was disinclined to produce this. However, because early settlement was desired, the consultant QS was shown the build-up of the rate. The rate was then agreed.

Exercise

1. Errors such as a simple discrepancy between the BOQ and specification/drawings are unlikely to cause problems, if the errors are noticed. Standard conditions of contract make provision for correction. Problems can arise where the price put against the (incorrect) BOQ item is either in error or is 'loaded'. The method of correction illustrated in the claim above would leave the original rate (with its error or loading etc.) intact, and might be preferred to constructing a new rate. Assuming the original rate tendered by the contractor was 'loaded', what strategy should the contractor now adopt when a new rate is requested?

2. It is important to emphasise the need for care and coordination in the planning and preparation of all contract documents to avoid the risk associated with discrepancies. Each document has a special purpose, special content and intended use. A complete and properly prepared package is required to serve as a proper basis for tendering and construction. The coordination of information in the contract documents is a critical responsibility of the design professional in order to minimise the exposure to claims from the contractor due to ambiguities, discrepancies and inconsistencies. Why then do people not devote the time and attention to document preparation that it deserves?

11.2.31 Case study — Road contract documentation

The case study concerns a dispute over contract documents on a project for upgrading and sealing of a road in a mine lease. The material supply was experiencing delays and impacting on other contractors. The particular contractor was responsible for the supply of sub-base, base and aggregate material for the road construction work. The owner provided crushing equipment and access to a pit in the mine lease. The contractor claimed that the contract documents were misleading in their description of the site conditions, the suitability of the crushing equipment, and the suitability of material available from the pit. The contractor also claimed that it was hampered in its ability to do the work because of this.

Both the contractor and the owner contributed to the problems which hampered the work. On the owner's side:

- The owner did not provide a specification that was clear in its intent. The selection of wording was often ambiguous. The owner had not really thought through what it was trying to achieve.
- The owner did not fairly distribute the risk. The owner, realising that there would be problems associated with winning the material from an operating mine lease, attempted to place all the risk on to the contractor. The contractor was directed to make itself familiar with the site, but it did not have the experience to appreciate the difficulties posed by the special site conditions.
- There was no detail within the contract which described how the testing of material quality and delivery of material would be undertaken. The contractor wanted to keep its stockpiling area to a minimum, whereas the owner wanted to avoid double handling of the material.
- At the time of contract award, the design was incomplete. The final quantities required were in the order of half as much again as that indicated in the tender documents.
- The specification tolerances could have been considered unreasonable to achieve in an area which was fairly remote and had high rainfall.
- The contract called up standard specifications for which there were no copies on site.
- The owner's project manager failed poorly in the area of contracts administration. Minutes of meetings were poorly written and recorded, and there was continual failure to follow up verbal site instructions with written directions.
- The project manager was also responsible for the design work and the specification. He was not receptive to challenges or discussions on the design or project, leading to a very adversarial scene. He did not have the necessary communication skills to deal with the contractor.
- The owner failed to provide sufficient staff with the necessary experience to manage the work. The project manager and site staff also had other commitments which led to slow responses to queries and inaction towards outstanding disputes.

On the contractor's side:

- The contractor never properly understood the implications of the narrow tolerances that were demanded by the specification.
- The contractor used staff who did not have the necessary technical qualifications and had no understanding as to the requirements of the work. There was considerable staff turnover during the work.

- The contractor undertook a strategy of achieving profit by challenging all aspects of the contract documents. At the height of the dispute, more effort was being put into claims than ensuring production targets were reached.
- The contractor's project manager contributed to the confrontational and abusive style of communication.

Exercise

1. Of the above views given in hindsight, which ones can be most readily addressed? Which ones would more than likely recur on another similar project?
2. A lot of trouble occurs because insufficient time is spent on the document preparation, or it is given as a job for a junior to do. Too many people are in a rush to start work. Is this false economy? It is believed that time spent early on is never wasted, though many people think it is. What is your view?

11.2.32 Case study — Contracting practices

Introduction

This case study is about polarised practices and interests of the various parties.

Project background

The project

The project was a development of a holiday unit resort for private investors. It involved the construction of apartment blocks, a car park and a shopping complex. The site was located in very steep terrain, and required the excavation of significant material to form a bench for construction. The final earthworks required the forming of a high cut batter, with some of the foundations for the structure located on steep slopes.

The developer

Units were sold off-the-drawings to finance the project. This enabled the developer to carry out the project with limited borrowing.

The project was the second attempted in the area by the developer. The first project, located adjacent to the current project site could not be considered a success, as it took many years to complete, and was dogged by disputes throughout its construction. The building had numerous maintenance and other problems, which continued to require attention and resources.

Recognising prospective investor's doubts about the project, the developer included a damage clause in all investor-contracts for the new building — a continual reduction in the purchase price with delays to possession of the units would occur if the project was not completed on time. These costs were significant, and could not be transferred to the contract with the contractor as damages for delay. The risks were therefore with the developer on delays in the project.

The consultants

A consultant was engaged on a fee based on a percentage of the total project cost. The percentage reflected the project size and the type of service provided by the consultant.

It was therefore in the consultant's interests, once design had been completed, not to change it as this reduced its margins. As the project was fast-tracked, the consultant preferred to wait till the last minute to provide drawings, thus minimising the reworking of the design.

The contractors

Eager to obtain the lowest price for the project, the developer obtained tenders from a large number of contractors. Preselection of contractors was not carried out, in spite of the difficult terrain and possible latent site conditions.

Two main contracts were let for the project — infrastructure work, and building work.

The following discussion focuses on these contracts, and the outcomes, in terms of disputes and other matters.

Infrastructure contract

The infrastructure contract comprised the construction of roads, retaining walls, earthworks and all foundations. The location of the project in steep hillside terrain required a contractor with previous experience working in such conditions. This advice from the consultant was not heeded by the developer.

The successful tenderer for the infrastructure work was an international construction company. Its strategy was to 'buy' the infrastructure contract, so as to secure the more lucrative main building contract. The company had no experience in earthworks in steep terrain, and did not bring in specialist personnel for the project, even though this had been promised to the developer on securing of the contract.

By accepting, the developer demonstrated its aim of reducing costs in the project as its prime concern. While it did retain specialist consultants to provide advice, this could not make up for the lack of expertise of the contractor.

As the project proceeded, the difficult terrain and poor weather conditions underlined the contractor's limited experience. Work was carried out poorly, or out of sequence, resulting in significant time delays on the project. The equipment required for the work was not on site, or was in poor repair and continually broke down. Poor site control, or short cuts, resulted in significant quantities of excavated materials being dumped on others' land, with significant erosion problems experienced by neighbours. The clean up of this item was the largest earthworks exercise in the entire project.

The relationship between the developer and the contractor became increasingly hostile as the delays continued, and work was slowed or carried out poorly. The developer now faced significant damages to people who had bought property off the drawings, as the proposed completion date could not be met.

Realising that the relationship with the developer was lost, the contractor set about minimising its loss. The result was a significant reduction in plant and equipment at the site, progress was slowed further. The delay in the infrastructure work was significant. Initially scheduled as six months, it eventually took more than eighteen months.

Poor site control and coordination with the consultant compounded delays. Meetings were scheduled at fortnightly intervals, however, frequently items were not resolved for periods of months. The poor management, compounded by increased interference by the developer in contractual matters, clouded lines of responsibility and allowed the contractor to drag out the project duration. Reluctance on the part of the consultant to amend drawings to suit site conditions also provided the contractor with legitimate reasons for delay.

The relationship between the developer and the contractor by the end of the project was almost openly hostile, the developer refusing to give any additions to the work scope, regardless of practicalities, in order to remove the contractor from the site. When finally the infrastructure contractor was removed from the site, the main building contractor was already on site.

Main building contract

Given the poor relationship, the infrastructure contractor did not tender for the main building contract, accepting the earlier contract work as a loss. Tenders were received by the developer for the main building work also from a large number of contractors. Again the developer selected the lowest bid.

The successful tenderer for the main building contract was a local company, well experienced in local contract dealings. A rumour at the start of the project was that it had bid low, its aim, to get the difference in contract variations by the end of the project. It is likely that this practice was decided after viewing the progress of the infrastructure project, and the end relationships. The poor control demonstrated in the infrastructure project may have been in its mind in adopting this approach.

The practice paid off for the contractor. The contractor used the contract as a lever to extract extra money from the developer at all opportunities. This included actively exploiting inadequacies of other contractors, including the infrastructure contractor, to claim delays. Some examples were:

- Deliberately building above an incomplete retaining wall, so that work stopped in that area, allowing the contractor to claim delays.
- After being advised, by the consultant, that the survey was not accurate for the foundations for the building, refusing to take possession of the site until the pile cap levels were amended on the drawings.

The above practice for the site possession eventually backfired, as the developer forced possession of the site, stating that the contractor had tendered to build the pile caps at the levels given on the drawings. This required the contractor to excavate for each pile cap. The reinstatement of excavated materials on a steep slope would have cost more than that gained by extras from the delay in taking possession of the site. This appeared to be the only time that the contractor lost out on its practices.

Overall the practice was rumoured to be financially successful for the contractor because of the extras won.

The different practices adopted by the contractors (infrastructure and main building) gave varying degrees of success. The main building contractor was on the whole more professional in its approach to the project, even though its approach was one of an adversarial nature.

Exercise

1. The developer for this project was clearly the low-bid owner type. The fundamental aim of the owner was clearly to save money, and to increase the profitability of the project. By singularly focusing on cost reduction, it resulted in a lower quality end-product at ultimately a higher cost.

    While it is not considered unreasonable to maximise profits, the developer's approach to the project was considered shortsighted in a number of areas. These include areas in which it would have been considered in its interests to manage, such as quality, time and overall cost, and long-term maintenance issues.

    Why cannot people see beyond short-term cost issues to the bigger picture?

2. Without exception, all decisions relating to the project were made on the basis of economy. This included the selection of a consultant, contractors and materials. Decisions that would have a long-term impact, such as those on the long-term maintenance of slopes, were made on price and not value. These decisions will eventually cost the developer more in the longer term.

    Examples included the selection of the consultant on the basis of price only. Given a lowest bidder consultant, there was limited incentive for the consultant to spend additional resources to reduce the cost to the developer, as this simply reduced the fees generated. When a value consultant was asked to review the design for this project, a significant reduction was achieved for all floor slabs and walls. Unfortunately, the reductions were not translated to foundations, as they had already been constructed by that time.

    The cost of a consultant is a small portion of the total project cost. Why then skimp on a consultant's fees? Suggest a better method for engaging consultants.

3. The case study demonstrates that the method of engaging services by the developer was in conflict with its aims and requirements. The engagement of all services for the project should have been carried out with more regard for the developer's aims.

    The contracts for both the infrastructure and main building should have been formed such that they worked with the interests of the developer in mind. This could have included:

    - Penalties for delays and bonuses for early completion.
    - Provision of a schedule of progress and resources to be used on the project for reference in tender evaluation.
    - A defects liability period or defects bond.

    Suggest other measures that could have been adopted.

4. Given the significant investor claims that the developer was liable for, if delays occurred, the adoption of the lowest bidder policy wouldn't assist avoiding these claims. Typically contractors are reluctant to increase the speed of work on a project because of the cost penalties arising out of acceleration of the schedule using, for example, overtime. This is especially true of tenderers with a low bid, and where there is no incentive for early completion.

    Suggest an incentive scheme for early completion.

5. While these methods may assist in bringing the goals of the contractor in line with that of the developer, these will have limited impact unless the aims of the developer change to view the project from other than a singular cost perspective. It is considered that if issues of quality, experience and overall value had been used to assess the tenders for the project, it is likely that the developer would have obtained a project

with less disputes, with better overall quality, and at a lower overall cost. What is your view?

6. The practice adopted by the main contractors was, it is suggested, as a direct result of developer's fixation with cost reduction. Recognising that price was the determining factor in winning the tender, the approach was to submit the lowest price, and recoup as much money as possible through claims, to increase the margin to an acceptable one. An outcome of this practice was that while the building contractor won significant additional money out of the contract, this money essentially went towards the contractor's bottom line. The quality of the structures constructed was considered poor, based on observations, and was a direct result of selecting the contractor on the basis of cost only.

   While it may be argued that a tenderer with a higher price may have also adopted the same strategy of pursuing extras to the contract work, resulting in an even higher total project cost, it is unlikely that such an aggressive policy would have been adopted. It is considered that while some extras were inevitable in the project, the total cost of the project using a higher bidder (selected on quality and experience) would have resulted in the project completed at the same cost as the current one, with less disputes and improved quality. What is your view?

7. Owners have significant influence on the direction contract work takes, and should use this influence to encourage the realisation of their goals by the contractors. It is generally in the owner's interest to form the contract such that the contractor's approach to completing the project is similar to its own. It is considered that similar strategies, while not eliminating contract abuses, would foster teamwork between all parties in achieving a successful outcome of a contract. What is your view?

11.2.33 CASE STUDY — CONTRACTOR'S VIEWPOINT

Situation

In a power plant project, being a long-term and complicated project, there were many problems that had to be dealt with. These problems related to:
* Specification details.
* Unit prices (particularly in respect of work extra to the contract).
* Schedule performance.
* Contract conditions.
* Dealing with other contractors.

Design and specification details
Problems were mainly caused by inconsistencies, discrepancies and ambiguities within the contract documents. For example, a drawing detail was different to that stated in the specification, or the site results were not as expected according to the drawings or design.

Unit prices
The volume of the actual work was greater than the estimated volume stated in the contract.

Schedule performance
The project ran behind schedule.

Contract conditions
Problems stemmed from the contract conditions. For example, the contract mentioned that materials on site would not be entitled to payment before being erected. These material costs were large, and so the contractor asked the owner to consider payment. The owner agreed that the material on site would have a initial value of a quarter of the price tendered, but the material would be owned by the owner. These conditions were added to the contract.

Design and specification details

To resolve problems relating to the design or specification, the owner provided an on-site consultant for this purpose. The contractor had to deal with the consultant if there were any technical problems. If the problem was small, the consultant and the contractor's personnel could solve the problem quickly. However even though the problem may be small, both parties had to still complete the necessary paperwork; for example, changes to the drawings might be made in order to record the problem. Usually the action did not wait for the paperwork to be completed, but the paperwork followed.

For bigger problems, such as extensive modifications, the action had to wait until the modification was approved by the owner, because the modification could have had a significant cost impact on the project.

Unit prices

Management level personnel usually dealt with the negotiation of unit prices.

Usually the contractor proposed very high unit prices, as an initial bargaining strategy. There was always a small possibility that the owner may accept the proposed price in the first offer. The owner usually would ask the contractor to reduce the proposed price for differing reasons. If the contractor failed to defend its proposed price, a lower price was agreed.

The unit price determination could be a long process, with several meetings needed before reaching final agreement.

Schedule performance

When the scheduled period for the subproject work was exceeded, the owner threatened the contractor with damages. The contractor's project manager then approached the owner's project manager, who had the right to grant time extensions, with completed documents supporting an extension of time.

Contract conditions

The contractor had to prepare its argument as to why changes in the contract conditions were needed, and had to convince the owner that these changes would not have a negative effect on the project cost and project progress

Exercise

1. All dispute resolution was handled through negotiation:
    - The first offer made by either party was never accepted.
    - Concessions were made to improve overall positions.
    - Concessions were only given in return for concessions.
    - A possible future working relationship was acknowledged.

    How much of this is culture dependent?

2. For large issues, negotiation teams were used. The teams were made up of management level people. The approach in handling disputes was not only through formal meetings, but also through informal meetings, for example by playing golf. These informal meetings also assisted in increasing the knowledge of the other party's position in relation to the disputed matters.

    Is golf a culturally driven form of socialisation used to assist working business relationships, or is it individual-specific?

3. The contractor tried to anticipate the reaction of the owner, for example by initially offering high prices. The contractor knew that the owner would not accept the first offer.

    How much of this approach is culture dependent?

4. Negotiation was a very good dispute resolution tool on this project, because agreement could be reached without sacrificing the relationship between the parties. The cost was much less than, for example, if formal resolution through the courts had been used, and it had a less negative effect in the project progress. However both parties had to be able to be trusted in the implementation of any agreement. What happens if trust is not there between the parties? How do you establish trust?

References and Bibliography

Adams, S. (1996). *Dogbert's Management Handbook*, Harper Business, New York.

Adams, S. (1997). *The Dilbert Future*, Harper Business, New York.

American Society of Civil Engineers (1991). *Avoiding and Resolving Disputes During Construction*, ASCE, New York.

Antill, J.M. (1970). *Civil Engineering Management*, Angus and Robertson, Sydney.

Antill, J.M. (1975). Some problems of construction contracting. *Contracting and Construction Engineer, 29* (11), 11-13.

Antill, J.M. (1979). Construction contracts administration. *Contracting and Construction Engineer, 33* (11), 10-16.

Antill, J.M. (1978). *The Arbitration of Commercial Disputes*, The Institute of Arbitrators Australia, June, Sydney.

Antill, J.M. (1983). Litigate or arbitrate. *Local Government Engineers Association of NSW Journal*, May, pp. 25-33.

Australian Constructors Association (1999). *Relationship Contracting*, ACA, North Sydney.

Carmichael, D.G. (1996). Flat organisational structures. *Journal of Project and Construction Management, 2*, 61–68.

Carmichael, D.G. (1996). Management fads. *Journal of Project and Construction Management, 3* (1), 115-125.

Carmichael, D.G. (1998), Gurus of faddish management. *Journal of Project and Construction Management, 4* (2), 77-84. Also appeared in *Proceedings of the International Conference on Construction Process Re-Engineering, Construction Process Re-engineering*, K. Karim, M. Marosszeky, S. Mohamed, S. Tucker, D.G. Carmichael, and K. Hampson (eds.), 1999, ACCI, UNSW, Sydney, pp. 365-371.

Carmichael, D. G. (2000). *Contracts and International Project Management*, A.A. Balkema, Rotterdam, 208 pp.

Collins, R. (1989). Alternative dispute resolution — Choosing the best settlement option. *Australian Construction Law Newsletter*, November, pp. 17-27.

Cooke, J.R. (1991). *Building and the Law*, University of New South Wales Press, Sydney.

Elashmawi, F. and Harris, P.R. (1994). *Multicultural Management*, S. Abdul Majeed & Co., Malaysia.

Fisher, R. and Ury, W. (1983). Getting to yes — *Negotiating agreement without giving in*, Arrow Books, London.

Fitch, R. (1989). *Commercial Arbitration in the Australian Construction Industry*, The Federation Press, Sydney.

Ford, B.J. (1982). *The Cult of the Expert*, Hamish Hamilton Ltd., London.

Fulton, M.J. (1989). *Commercial Alternative Dispute Resolution*, The Law Book Co., Sydney.

Gudykunst, W.B. and Ting-Toomy, S. (1988). *Culture and Interpersonal Communication*, Sage, Newbury Park, California.

Hall, E.T. (1976). *Beyond Culture*, Anchor Doubleday, New York.

Harrison, G. (1994). Culture and management. In: *Australian Accountant, 64*, 14-22.

Hawkins, L. and Hudson, M. (1986). *Effective Negotiation*, ENS Business Publications, Melbourne.

Hilmer, F.G. and Donaldson, L. (1996). *Management Redeemed*, The Free Press, New York.

Hofstede, G.H. (1980), *Culture's Consequences: International Differences in Work-Related Values*, Sage, Beverly Hills, California.

Hofstede, G.H. (1983). The cultural relativity of organisational practices and theories. In: *Journal of International Business Studies,* 75-89.

Hofstede, G.H. and Bond M.H. (1988). The Confucius connection: From cultural roots to economic growth. In: *Organisational Dynamics*, 5-21.

Jones, D. (1996). *Building and Construction Claims and Disputes,* Construction Publications, Sydney.

Mitchell (1995). Original source unknown.

National Public Works Conference / National Building and Construction Council (1990). *No Dispute*, Canberra.

New South Wales Government (1993). *Capital Project Procurement Manual*, Construction Policy Steering Committee, Public Works Department, Sydney.

Pears, G. (1989). *Beyond Dispute*, Cooperative Impacts Publications, Sydney.

Pendell, S.D. (1995). *Teaching Interculturally: Crossing Cultural Frontiers in Education*, 12th National Forum of the Open and Distance Learning Association of Australia, Port Vila, Vanuatu.

Richards, C. and Walsh, F. (1990). *Negotiating*, Australian Government Publishing Service, Canberra.

Rose, C. (1987). *Negotiate and Win*, Lothian, Melbourne.

Shapiro, E.C. (1995). *Fad Surfing in the Boardroom*, Harper Business, Sydney.

Sharkey, J.J.A. and Dorter, J.B. (1986). *Commercial Arbitration*, The Law Book Company, Sydney.

Tillett, G. (1991). *Resolving Conflict*, Sydney University Press, Sydney.

Trompenaars, F. (1997). *Riding the Waves of Culture*, Irwin/Nicholas Brealey, London.

Uher, T.E. and Davenport, P. (1998). *Fundamentals of Building Contract Management*, Ticaw, Sydney.

Subject index